CLASSROOM TRAINING HANDBOOK

NONDESTRUCTIVE TESTING

Ultrasonic

Published by **PH D**iversified, Inc.
5040-B Highway 49 South
Harrisburg, NC 28075
704-455-3717

Copyright © 1997 by **PH D**iversified, Inc.
Second Printing January 2001
Third Printing August 2003

All Rights Reserved

No part of this book may be reproduced in any form without written permission from the publisher.

Printed in the United States of America
ISBN 1-886630-16-X

PREFACE

Level II Classroom Training Handbook - Ultrasonic Testing, CT-4 is one in a series of training handbooks designed for basic Level II training when NDT is taught in a traditional classroom setting. The instructor would typically assign chapters and discuss the material in a classroom lecture format. To provide complete Level II classroom training the instructor should supplement this text with assignments that include industry specific applications and related practical examinations.

This Level II Classroom Training Handbook also serves as an excellent reference book during on-the-job training of nondestructive testing personnel.

This handbook is most effective when used by those persons who have successfully completed the **Level I Programmed Instruction Handbook, PI-4, Ultrasonic Testing (3 volumes)**. The Programmed Instruction Handbooks present entry level material in a **self-study format**. A cross-reference guide is printed in this Level II Classroom Training Handbook so that the student can read corresponding information in the self-study handbook to provide a more structured approach for individual learning.

Other **Classroom Training Handbooks** in the series include:

      CT-2 Liquid Penetrant Testing

      CT-3 Magnetic Particle Testing

      CT-5 Eddy Current Testing

      CT-6 Radiographic Testing

It is recommended that PI-1, Introduction to Nondestructive Testing, be completed before starting this book in order to have the benefit of certain basic metallurgy information that will make this book easier to understand.

ACKNOWLEDGMENTS

Publishing and Printing

    Revision Editor: . . . . Dr. George Pherigo, **PH D**iversified, Inc.

    Production Editor . . Ms. Mary Lou Hollifield, **PH D**iversified, Inc.

    Proofreading . . . . Ms. Jean Pherigo, **PH D**iversified, Inc.
    Proofreading . . . . Ms. Dana Smilie

Technical Content Revision

    Technical Editor . . . . Mr. Robert W. Smilie

This handbook was originally prepared by the Convair Division of General Dynamics Corporation under contract to NASA and was identified as N68-28790. This book is part of a series of books, commonly known as the General Dynamics Series, that has been the basis of many industrial NDT training programs for over 20 years.

Now, after several decades of widespread use, the entire series has undergone a major revision. The revised material no longer concentrates on applications in the aerospace industry, but instead, covers a wider range of industrial applications and discusses the newest techniques and applications.

Mr. Robert W. Smilie has been the principal technical editor of the revised material in this text. Using his nondestructive testing experiences in several industries, including work at the EPRI NDE Center, he has updated the text to better suit the entry-level technician/engineer.

CLASSROOM TRAINING MANUAL
ULTRASONIC TESTING

TABLE OF CONTENTS

CHAPTER 1	INTRODUCTION
CHAPTER 2	PRINCIPLES
CHAPTER 3	EQUIPMENT
CHAPTER 4	TECHNIQUE AND APPLICATIONS
CHAPTER 5	CALIBRATING TRANSDUCERS
APPENDIX A	COMPARISON AND SELECTION OF NDT PROCESSES
APPENDIX B	GLOSSARY
APPENDIX C	TRIGONOMETRY TABLES
APPENDIX D	ACOUSTIC PROPERTIES OF MATERIALS
APPENDIX E	SAMPLE UT CALIBRATION SHEET
APPENDIX F	SAMPLE UT DATA RECORD

CROSS REFERENCE TO PROGRAMMED INSTRUCTION HANDBOOKS

If this handbook is being used in a classrooom lecture format, the instructor may choose to make additional reading assignments from the Programmed Instruction series. The Programmed Instruction handbooks often provide a little more detail, especially in those areas that are most difficult to comprehend.

CT Text (Chapter & Page)	PI Text (Volume/Chapter/Page)
Chapter 1	N/A
Ch. 2: 2-1 to 2-4	Vol. I/Ch. 1; All: Ch. 3; 3-1 to 3-8
Ch. 2: 2-4 to 2-10	Vol. I/Ch. 3; 3-9 to 3-24, 3-43, 3-45
Ch. 2: 2-11 to 2-17	Vol. I/Ch. 2; All
Ch. 2: 2-18 to 2-27	Vol. I/Ch. 3; 3-26 to 3-49
Ch. 2: 2-27 to 2-32	Vol. I/Ch. 4; 4-1 to 4-9
Ch. 2: 2-33	Vol. I, Ch. 4, 4-10 to 4-27
	Vol. II, Ch. 5; All
Ch. 2: 2-34 to 2-39	Vol. I/Ch. 5; All
	Vol. III/Ch. 6; 6-20 to 6-35
Ch. 3: 3-1 to 3-14	Vol. II/Ch. 1
Ch. 3: 3-14 to 3-16	Vol. II/Ch. 2
Ch. 3: 3-17 to 3-23	Vol. II/Ch. 3
Ch. 3: 3-24 to 3-40	Vol. II/Ch. 4
Ch. 3: 3-41 to 3-43	Vol. II/Ch. 5
Ch. 4: 4-1 to 4-3	Vol. III/Ch. 1
Ch. 4: 4-3 to 4-8	Vol. III/Ch. 4
Ch. 4: 4-9 to 4-21	Vol. III/Ch. 1
Ch. 4: 4-22 to 4-30	Vol. III/Ch. 5 and Ch. 6
Ch. 4: 4-31 to 4-41	Vol. III/Ch. 2, Ch. 3; 3-1 to 3-70
Ch. 4: 4-42 to 4-48	Vol. III/Ch. 3; 3-71 to 3-92
Ch. 5	N/A

CHAPTER 1: INTRODUCTION

TABLE OF CONTENTS

                                                      Page

	Page
GENERAL	1-1
PURPOSE	1-1
DESCRIPTION OF CONTENTS	1-2
Arrangement	1-2
Locators	1-2
INDUSTRIAL APPLICATIONS OF ULTRASONIC TESTING	1-2
TESTING PHILOSOPHY	1-3
PERSONNEL	1-3
TESTING CRITERIA	1-3
TEST PROCEDURES	1-4
TEST OBJECTIVE	1-4

CHAPTER 1

INTRODUCTION

General

The complexity and expense of today's machines, equipment, and tools dictate the use of fabrication and testing procedures that will ensure maximum reliability. Nondestructive testing (testing without destroying) provides many of these procedures. Of the number of nondestructive testing procedures available, ultrasonic testing - of which this handbook is concerned - is one of the most widely used.

Purpose

The purpose of this handbook is to provide the fundamental knowledge of ultrasonic testing required by quality assurance and test personnel to enable them to:

- ascertain that the proper test technique, or combination of techniques, is used to assure the quality of the finished product.

- interpret, evaluate, and make a sound decision as to the results of the test.

- recognize those areas exhibiting doubtful test results that require either retest or assistance in interpretation and evaluation.

Description of Contents

- Arrangement

    The material contained in this handbook is presented in a logical sequence and consists of:

    - Chapter 1:   Introduction
    - Chapter 2:   Principles
    - Chapter 3:   Equipment
    - Chapter 4:   Technique and Applications
    - Chapter 5:   Calibrating Transducers
    - Appendix A: Comparison and Selection of NDT Processes
    - Appendix B: Glossary
    - Appendix C: Trigonometry Tables
    - Appendix D: Acoustic Properties of Materials

- Locators

    - At the front of each chapter is a table of contents referencing the major paragraphs in that chapter. Also included is a list of figures and tables, where applicable.

Industrial Applications of Ultrasonic Testing

Because of the basic characteristics of ultrasonic testing, it is used to test a variety of both metallic and nonmetallic products such as welds, forgings, castings, sheet, tubing, plastics, ceramics, etc. Since ultrasonic testing is capable of economically revealing subsurface discontinuities (variations in material composition) in a variety of dissimilar materials, it is one of the most effective tools available to quality assurance personnel.

Testing Philosophy

Nondestructive testing is used to assure maximum reliability of machines, equipment and tools. To accomplish such reliability, test standards have been set and test results must meet these standards.

Personnel

It is imperative that personnel responsible for ultrasonic testing be trained and highly qualified with a technical understanding of the test equipment and materials, the item under test (specimen), and the test procedures. Quality assurance personnel must be equally qualified. To make optimum use of ultrasonic testing, personnel conducting tests must continually keep abreast of new developments. There is no substitute for knowledge.

Testing Criteria

Modern manufacturing procedures dictate that faulty articles be discovered as early in the manufacturing process as possible. This means that each item must be tested individually before it is required to perform in a subassembly and that each subassembly be tested before it is required to perform in an assembly, etc. This building-block approach requires that test processes be selected and test procedures be generated at the lowest level in the manufacturing process in order that the highest reliability may be obtained with lowest cost.

Test Procedures

Approved procedures for ultrasonic testing are formulated from analysis of the test specimen or article, review of its past history, experience on like or similar specimens, and information available concerning discontinuities in similar specimens. It is the responsibility of personnel conducting or checking tests to ensure that test procedures are adequately performed, and that the test objective is accomplished. Procedures found to be incorrect or inadequate must be brought to the attention of responsible supervision for correction and incorporation into revised procedures.

Test Objective

The objective of ultrasonic testing is to ensure product reliability by providing a means of:

- obtaining a visual recorded image related to a discontinuity in the specimen under test.
- disclosing the nature of the discontinuity without impairing the material.
- separating acceptable and unacceptable material in accordance with predetermined standards.

No test is successfully completed until an evaluation of the test results is made. Evaluation of test procedures and results requires understanding of the test objective as well as a knowledge of the material from which the test article is made. It also requires a knowledge of the manufacturing processes that were involved.

CHAPTER 2: PRINCIPLES

TABLE OF CONTENTS

                              Page

EARLY SONIC TESTS	2-1
ULTRASONIC WAVE GENERATION	2-1
GENERAL	2-3
MODES OF VIBRATION	2-3
WAVE MODES	2-4
General	2-4
Comparison of Longitudinal and Shear Wave Modes	2-5
Shear and Surface Waves	2-6
Transducer Beam Angles	2-8
Rayleigh Waves	2-8
Lamb Waves	2-9
Lamb Wave Types	2-9
SOUND BEAM VELOCITIES	2-10
ACOUSTICAL IMPEDANCE	2-11
General	2-11
Acoustical Impedance	2-11
Impedance Ratio	2-11
ULTRASOUND REFLECTION	2-12
ULTRASOUND PATTERNS	2-13
General	2-13
Near Zone	2-14
Far Zone	2-15
REFLECTION AND MODE CONVERSION	2-18
General	2-18
Mixed Mode Conversion	2-19
Shear Wave Generation	2-20

TABLE OF CONTENTS (CONT'D.)

Surface Wave Generation	2-21
Summary	2-22
SNELL'S LAW	2-22
General	2-22
Snell's Law Calculations	2-23
Typical Problem Solving Method	2-23
CRITICAL ANGLES OF REFRACTION	2-25
General	2-25
First Critical Angle	2-25
Calculation of Critical Angles	2-26
CONTACT TESTING	2-27
General	2-27
Straight Beam Techniques	2-28
Angle Beam Techniques	2-31
Surface Wave Techniques	2-32
IMMERSION TESTING	2-32
General	2-32
ULTRASONIC DISPLAYS	2-33
General	2-33
INFLUENCE OF THE TEST SPECIMEN ON THE SOUND BEAM	2-34
General	2-34
REFERENCE TABLE	2-39

LIST OF FIGURES

Figure		Page
2-1	Sound Wave Generation	2-2
2-2	Ultrasonic Wave Generation	2-2

LIST OF FIGURES (CONT'D.)

2-3	Longitudinal Wave Mode	2-5
2-4	Longitudinal and Shear Wave Modes Compared	2-6
2-5	Shear and Surface Waves	2-7
2-6	Rayleigh or Surface Waves	2-9
2-7	Sound Beam Reflection	2-13
2-8	Beam Profile	2-14
2-9	Beam Spread in Steel	2-16
2-10	Sound Beam Radiation Patterns	2-17
2-11	Normal Incident Beam	2-18
2-12	5° Incident Beam	2-19
2-13	First Critical Angle	2-20
2-14	Second Critical Angle	2-21
2-15	Application of Snell's Law for the Calculation of Refracted Angles	2-24
2-16	Single Transducer Pulse-Echo Technique	2-29
2-17	Dual-Element Transducer Pulse-Echo Technique	2-29
2-18	Through-Transmission Technique	2-30
2-19	Angle Beam Technique	2-31
2-20	Surface Wave Technique	2-32
2-21	Irregular Back Surface Effect	2-35
2-22	Convex Surface Effect	2-36
2-23	Concave Surface Effect	2-37
2-24	Mode Conversion Caused by Beam Spread	2-38

LIST OF TABLES

Table		Page
2-1	Typical Ultrasonic Properties	2-10
2-2	Critical Angles, Immersion Testing	2-27

LIST OF TABLES (CONT'D.)

2-3   Critical Angles, Contact Testing 2-27

CHAPTER 2

PRINCIPLES

Early Sonic Tests

For centuries, men tested parts by hitting them with a mallet and listening for a tonal quality difference. Around the turn of this century, railroad men inspected parts by applying kerosene to the part and covering it with a coat of whiting (chalk). In areas where the whiting looked wet, the part was assumed to be cracked. In the early 1940's, Dr. F. A. Firestone developed the first pulse-echo instrument for detecting deep-seated flaws. The establishment of basic standards and the development of the first practical immersion testing system are credited to W. C. Hitt and D. C. Erdman.

Ultrasonic Wave Generation

Ultrasound (or any frequency sound) is the mechanical vibration of particles in a medium (material). When a sound wave travels in a material, the particles in the material vibrate about a fixed point at the same frequency as the sound wave. The particles do not travel with the wave but only react to the energy of the wave. It is the energy of the wave that moves through the material. When a tuning fork is struck, it vibrates and produces sound waves by compressing the air. These waves travel through air to the ear of the listener as shown in Figure 2-1. The tuning fork vibrations soon die out and no longer produce waves. Similarly, in ultrasonic testing, a transducer (crystal) is electrically excited which then vibrates. The ultrasound from the transducer then travels through a couplant, which may be water, oil, etc., to the front surface of the test piece. Figure 2-2 shows the transducer in contact with the test piece and the ultrasound pulses traveling through the piece.

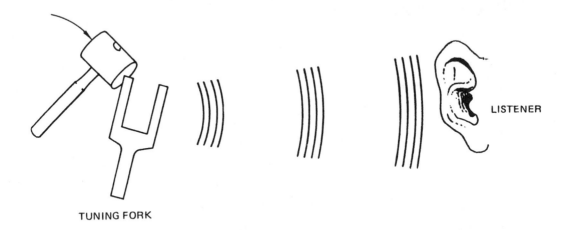

Figure 2-1. Sound Wave Generation

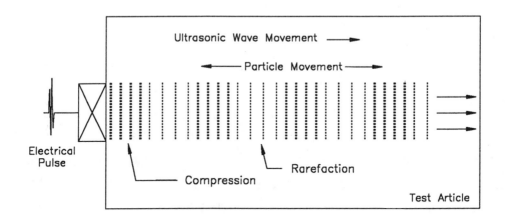

Figure 2-2. Ultrasonic Wave Generation

General

Ultrasonics is the name given to the study and application of sound waves having frequencies higher than those which the human ear can hear. Adults with normal hearing can hear frequencies in a range from 20 Hz to 20,000 kHz (thousand cycles per second). Ultrasonic nondestructive testing is the use of ultrasonics to examine or test material without destroying the material. An ultrasonic test may be used to measure the thickness of a material or to examine the internal structure of a material for possible discontinuities such as voids and/or cracks. Testing frequencies commonly range from 0.5 MHz (0.5 million hertz) to 25,000,000 Hz (25 MHz).

Modes of Vibration

If the length of a particular sound wave is measured from trough to trough, or from crest to crest, the distance is always the same. This distance is known as the wavelength (λ) and is defined in the following equation. The time it takes for the wave to travel a distance of one complete wavelength, λ, is the same amount of time it takes for the source to execute one complete vibration.

$$\lambda = \frac{V}{f}$$

Where:

λ = the wavelength of the wave
V = velocity
f = the frequency of the wave

Several types of sound waves travel through solid matter. There are longitudinal or compression waves where the particles vibrate back and forth in the same direction as the motion of the sound (as in Figure 2-2).

There are also shear or transverse waves where the particles vibrate back and forth in a direction that is at right angles to the motion of the sound. It is also possible, within certain limits, to produce waves that travel along the free boundary or surface of a solid.

These surface, or Rayleigh (pronounced "ray-lee"), waves penetrate the material to a depth of about one wavelength. Ultrasonic vibrations in liquids or gases are only propagated as longitudinal waves because of the absence of rigidity in liquids and gases. Longitudinal, shear, and surface waves can be propagated in solids.

The shortest ultrasonic wavelengths are of the order of magnitude of visible light. For this reason, ultrasonic wave vibrations possess properties very similar to those of light waves; i.e., they may be reflected, focused, and refracted.

The high frequency particle vibrations of ultrasound waves are propagated in homogeneous solid objects in the same manner as directed light beams. Ultrasound is reflected (partially or totally) at any surface acting as a boundary between the test object and a gas, liquid, or another type of solid. As with echo-sounding in sonar applications, the ultrasonic pulses reflect from discontinuities, thereby enabling detection of the presence and location of the discontinuity.

Wave Modes

- General

    All materials are made up of atoms (or tiny particles) lined up in straight lines to form lattices, as shown in Figure 2-3. If we strike the side of this lattice, we find that the first column of atoms strikes the second column rebounding and reverberating back and forth and striking the third column and so on in sequence. This particle motion produces a wave movement in the direction shown. In this case, the particle movement direction is the same as the wave movement direction. This type of ultrasound wave motion (or mode) is called the longitudinal or compression wave mode. This wave mode travels the fastest.

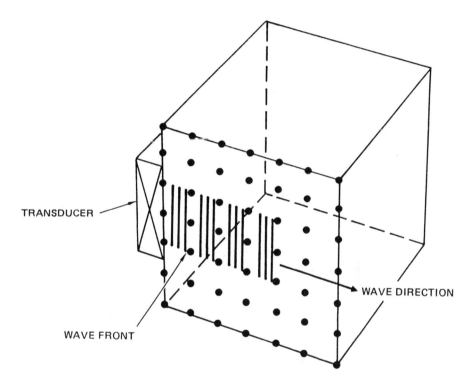

Figure 2-3. Longitudinal Wave Mode

- Comparison of Longitudinal and Shear Wave Modes

Figure 2-4 shows two transducers generating ultrasonic waves in the same piece. Note that the transducer on the left is producing longitudinal waves and that the transducer on the right is producing a different kind of wave. This different kind of wave is called a shear wave because the particle movement direction is at right angles to the wave movement direction. The velocity of shear waves through a material is approximately half that of the longitudinal waves. Note also that the transducer on the right is mounted on a plastic wedge so that the ultrasonic waves generated by the crystal enter the material at a specific angle. The specific angle required to produce only shear waves depends on the velocity of ultrasound within the material.

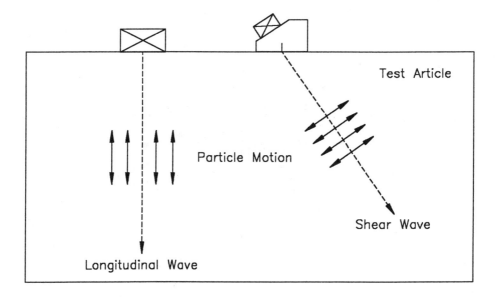

Figure 2-4. Longitudinal and Shear Wave Modes Compared

- Shear and Surface Waves

The particle displacements of shear waves are oriented in a plane normal to the direction of propagation. A special type of shear wave is confined to a thin layer of particles on the free boundary of a solid material. These waves are called surface or Rayleigh waves, and propagate with a velocity between 3 percent and 15 percent less than shear waves. As shown in Figure 2-5, when a transducer is mounted on a steeply-angled plastic wedge, the longitudinal beam in the wedge strikes the test surface at an angle resulting in a surface wave mode of sound travel in the test specimen. As shown, a surface wave can travel around a curve. Reflection of the surface wave occurs only at a sharp corner or at some discontinuity. The contact transducers that produce shear waves and surface waves are called angle beam transducers.

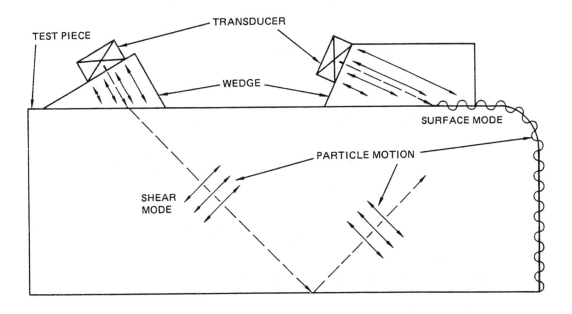

NOTE: The ultrasonic beams are longitudinal in each wedge. Mode conversion occurs when the ultrasound enters the test piece.

Figure 2-5. Shear and Surface Waves

- Transducer Beam Angles

Confusion may be encountered when angle beam transducers designed to produce a specific angle in one kind of material are applied to other materials with different acoustic velocities. A transducer designed to produce a shear wave beam at 45° in steel, for example, will produce a shear wave beam at 43° in aluminum or 30° in copper. Refer to "Refraction and Mode Conversion" later in this chapter.

- Rayleigh Waves

Rayleigh waves travel over the surface of a solid and bear a rough resemblance to waves on the surface of water; they were studied by Lord Rayleigh (c. 1875) because they are the principal component of disturbance from an earthquake at a distance from the epicenter. Reflections of Rayleigh waves from cracks in the surface or from discontinuities lying just beneath the surface may be seen on an oscilloscope. Rayleigh waves traveling on the top face of a block are reflected from a sharp edge corner, but if the edge is rounded off, the waves continue down the side face and are reflected at the lower edge and return to the origination point. These waves may travel the entire way around a cube if all of its edges are rounded off. They also travel around a cylinder. Rayleigh waves are almost completely absorbed by touching a finger to the surface, so the path of any reflection can be easily traced by observing the oscilloscope while moving the finger over the surface of the work. Rayleigh waves are also called surface waves, since their depth of penetration along the surface direction of travel is usually no more than one wavelength. The ultrasound travels along the surface with an elliptical particle motion as illustrated in Figure 2-6.

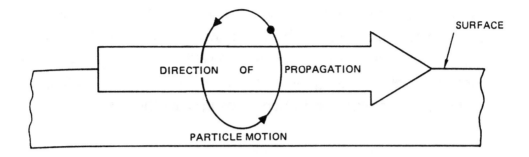

Figure 2-6. Rayleigh or Surface Waves

- Lamb Waves

If a surface wave is introduced into a material that has a thickness equal to three wavelengths or less of the beam, a different kind of wave results. The material begins to vibrate as a plate; i.e., the wave encompasses the entire thickness of the material. When this occurs, the normal rules for wave velocity along the plate break down. The velocity is no longer dependent upon the type of material and the type of wave. Instead, we get a wave velocity that is dependent on the frequency of the wave, the angle of incidence, and, of course, the type of material. The theory describing Lamb waves was developed by Horace Lamb (c. 1916), hence the name.

- Lamb Wave Types

There are two general types of Lamb (or plate) waves, depending on the way the particles in the material move as the wave moves along the plate. These are symmetrical and asymmetrical. Each type of Lamb wave has an infinite number of modes that the wave may attain. These modes are differentiated by the manner in which the particles in the material are moving. The ability of Lamb waves to propagate in thin plates makes them applicable to a wide variety of problems requiring the detection of subsurface discontinuities. Examples of practical problems for which plate waves are useful are: 1) immersion inspection of thin-walled tubing and plates for internal defects or grain size determinations; 2) testing for laminations in plate; and 3) changes in plate thickness.

Sound Beam Velocities

Ultrasonic waves travel through solids and liquids at relatively high speeds, but are more readily attenuated, or die out, in gases. The velocity of a specific mode of ultrasound is a constant through a given homogeneous material. The velocities of vibrational waves through various materials related to ultrasonic testing are listed by many authorities in centimeters per second x 100,000 (cm/s x 10^5) or inches per second x 100,000 (ips x 10^5). For convenience, velocities are given in this manual in centimeters per microsecond (cm/µs). Table 2-1 lists the velocities of a longitudinal wave through several different types of material to illustrate the wide range of velocities. These differences in velocity are due largely to differences in the density and elasticity of each material. Density alone cannot account for the extremely high velocity of sound in beryllium, which is less dense than aluminum. The acoustic velocity of water and mercury are almost identical, yet mercury is thirteen times as dense as water.

Table 2-1. Typical Ultrasonic Properties

MATERIAL	DENSITY (gm/cm³)	LONGITUDINAL WAVES	
		VELOCITY cm/µs	IMPEDANCE (gmx10⁵/cm²-s)
AIR	0.001	0.033	---
WATER	1.000	0.149	149
PLASTIC (ACRYLIC)	1.180	0.267	315
ALUMINUM (2117-T4)	2.800	0.625	1,750
BERYLLIUM	1.820	1.280	2,330
MERCURY	13.000	0.142	1,846

Acoustical Impedance

- General

    When a transducer is used to transmit an ultrasonic wave into a material, only part of the wave energy is transmitted; the rest is reflected at the interface between two different materials as ultrasound passes from one to the other. How much of the sound beam is reflected depends on a factor called the *acoustical impedance ratio*.

- Acoustical Impedance

    Acoustical impedance is a material property and can be generally referred to as the resistance of a material to the passage of sound waves. The specific acoustical impedance (Z) of any material may be computed by multiplying the density of the material (ρ) by the velocity of sound (V) through the material.

    $$Z = \rho V$$

    Air has a very low impedance while the impedance of water is relatively higher than the impedance of air. Aluminum and steel have still higher impedances.

- Impedance Ratio

    Impedance ratio between two materials is simply the acoustical impedance of one material divided by the acoustical impedance of the other material. When a sonic beam is passing from material one into material two, the impedance ratio is the impedance of the second material divided by the impedance of the first material. As the ratio increases, more of the original energy will be reflected.

Since air has a very small impedance, the impedance ratio between air and any liquid or solid material is very high. Therefore, most if not all of the ultrasound will be reflected at any interface between air and any other material.

An impedance ratio is often referred to as "an impedance mismatch." If the impedance ratio, for example, was 5/1, the impedance mismatch would be 5 to 1. The impedance ratio for a liquid-to-metal interface is on the order of 20 to 1 (approximately 80 percent reflection) while the impedance ratio for air-to-metal is about 100,000 to 1 (virtually 100 percent reflection). This results in only a small percentage of the ultrasonic energy transmitting into the test article. Ideally, a 1 to 1 impedance ratio would be desirable for the optimum transmission of ultrasound.

Ultrasound Reflection

In many ways, high-frequency vibrations act in the same way as light beams. For example, when they strike an interrupting object, most of the energy is reflected. These reflections may then be picked up by a second or, in most cases, by the same crystal or transducer. Within the crystal the mechanical energy is transformed into electrical energy. The electrical energy is sent to the test system where it is amplified and presented. Ultrasonic testing does not give direct information about the exact nature of the discontinuity. This is deduced from several factors, the most important being a knowledge of the test piece material and its construction. Ultrasonic waves are reflected from both the discontinuity and the back surface of the test piece as indications which are also referred to as signals or echoes. The indication from the discontinuity is received before the back surface reflection is received. Figure 2-7 shows a situation where the time required for the ultrasound to travel through the test piece to the discontinuity and back is only two-thirds of the time required for the sound beam to reach to the back surface and return. This time differential indicates that the discontinuity is located two-thirds of the distance to the back surface.

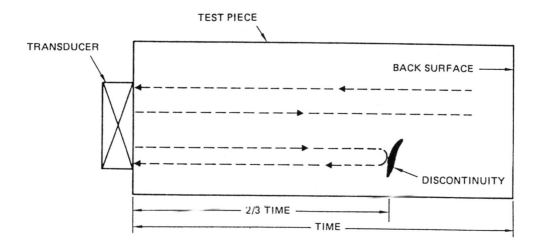

Figure 2-7. Sound Beam Reflection

Ultrasound Patterns

- General

    In ultrasonic testing, the sound beam is generally considered to be a straight-sided projection of the face of the transducer. In reality, the beam is not all that consistent. If the beam intensity is measured at various distances from the transducer, two distinct zones are found as shown in Figure 2-8. These zones are known as the *near zone* (or Fresnel Zone) and the *far zone* or (Fraunhofer Zone).

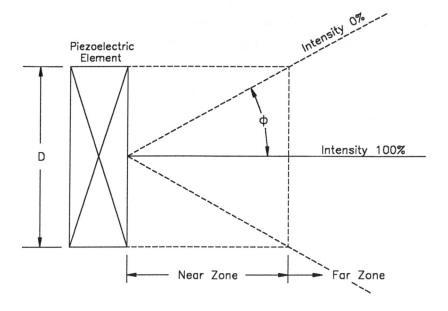

Figure 2-8. Beam Profile

- Near Zone

In the zone closest to the transducer, the measurement of the ultrasonic intensities reveals an irregular pattern of localized high and low intensities. This irregular pattern results from the interference between sound waves that are emitted from the face of the transducer.

The transducer may be considered to be a mosaic of crystals, or a plane source, each vibrating at the same frequency but slightly out of phase with each other. Near the face of the crystal, the composite sound beam propagates chiefly as a plane-front wave; but spherical-front waves, which emanate from the periphery of the crystal face, produce side-lobe waves that interfere with the plane-front waves to cause patterns of acoustical maximums and minimums where they cross.

The effect that the presence of these acoustical patterns in the near zone will have on the ultrasonic test varies, but if the operator has proper knowledge of the presence of the near field, the proper test block can be scanned and those indications correlated with the indications obtained from the test.

The length of the near zone is dependent on the diameter of the transducer and the wavelength of the ultrasonic beam and may be computed from the equation:

$$N = \frac{D^2}{4\lambda}$$

where:
N = length of the near zone
D = transducer diameter
λ = wavelength of the ultrasound

Since the wavelength of an ultrasonic beam in a particular material is inversely proportional to the frequency, the length of the near zone in a particular material can be shortened by lowering the frequency.

- Far Zone

In the region furthest from the transducer, the only effect of consequence is the spreading of the ultrasonic beam. Fraunhofer diffraction causes the beam to spread starting at the end of the near zone. At this distance, the beam appears to have originated at the center of the radiating face of the transducer and spread outward.

The degree of spread may be computed from the equation

$$\sin \phi = 1.22 \frac{\lambda}{D}$$

where:

ϕ = half-angle beam spread
λ = wavelength of the ultrasound
D = transducer diameter

Beam spread in steel, at various frequencies, is given in Figure 2-9. At any frequency, the larger the crystal, the narrower the beam; the smaller the crystal, the greater the beam spread. At any diameter, higher frequencies result in narrower beam spread than lower frequencies. The diameter of the transducer is often limited by the size of the available contact surface. Transducers smaller than 1/8-inch (3.2 mm) diameter have been used. For shallow depth testing, 3/8 inch diameter (9.5 mm) and 1/2 inch diameter (12.7 mm) transducers are used at the higher frequencies such as 5.0 to 25.0 MHz. The large-diameter transducer is usually selected for testing through greater depths of material due to its increased power.

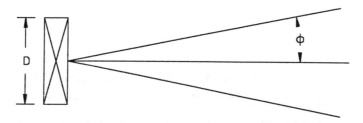

FREQUENCY MHz	λ in.	TRANSDUCER DIAMETER (D) INCHES			
		3/8	1/2	3/4	1.0
1.0	0.230	48°	34°	22°	16°
2.25	0.102	19°	14°	10°	7°
5.0	0.046	9°	6°	4°	3°

Figure 2-9. Beam Spread in Steel

Figure 2-10 shows the reduction in beam spread in steel for a 1/2-inch-diameter (12.7 mm) transducer when the frequency is raised from 1.0 MHz to 2.25 MHz. The secondary or side lobes shown in the figure are edge effects caused by the manner of crystal mounting. In practical work, the primary beam is the only one of consequence. Secondary beams are considered when the geometry of the test specimen is such that the secondary beams are reflected back to the transducer, creating spurious effects. The strongest intensity of the sound beam is along its central axis with a gradual reduction in amplitude away from the axis.

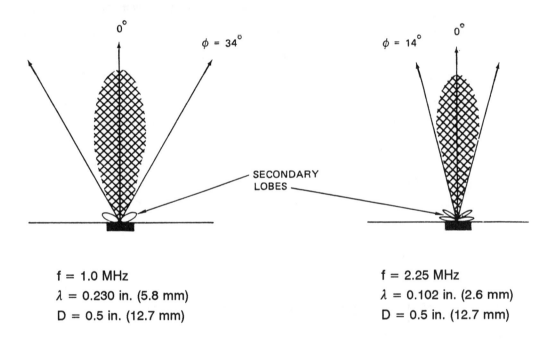

f = 1.0 MHz
λ = 0.230 in. (5.8 mm)
D = 0.5 in. (12.7 mm)

f = 2.25 MHz
λ = 0.102 in. (2.6 mm)
D = 0.5 in. (12.7 mm)

Figure 2-10. Sound Beam Radiation Patterns

Refraction and Mode Conversion

- General

    Refraction and mode conversion of the ultrasonic beam as it passes at an angle from one material to another are comparable to the refraction of light beams when passing from one medium to another. The angles given are based on typical ultrasonic velocities for these mediums and are approximate.

    Figure 2-11 shows a transducer inducing a longitudinal sound beam into water. The water transmits the beam to the test piece. When the longitudinal wave (L-wave) sound beam is incident to the surface of the test specimen in the normal (perpendicular) direction, the beam is transmitted through the first and second medium as a 100 percent longitudinal beam and no refraction occurs.

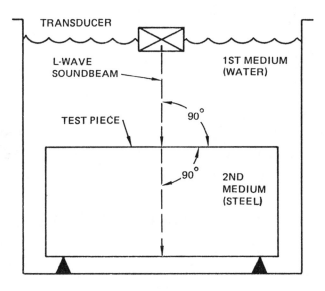

Figure 2-11. Normal Incident Beam

- Mixed Mode Conversion

As shown in Figure 2-12, as the incident angle is changed from the initial 90° position, refraction and mode conversion occur and the original longitudinal beam is transmitted into the second medium as varying percentages of both longitudinal and shear wave beams. As illustrated for water and steel, the refracted angle for the L-wave beam is four times the incident angle, and the S-wave beam angle is a little more than twice the incident angle. If the incident angle is rotated further, the refraction angles of the L-wave and the S-wave increase.

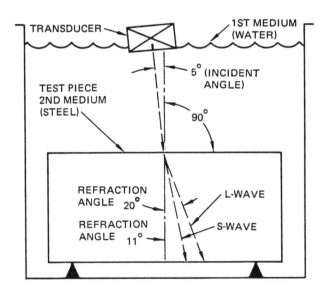

Figure 2-12. 5° Incident Beam

Refraction and mode conversion occur because the L-wave velocity changes when the beam enters the second medium. The velocity of the shear wave is approximately half that of the longitudinal wave. As the incident angle is rotated further, both refracted angles increase. The first beam to reach a refraction angle of 90° will be the L-wave.

2-19

- Shear Wave Generation

When the transducer is rotated to an incident angle of approximately 15°, the refracted angle of the L-wave is increased to 90° and the L-wave is totally reflected from the test surface as shown in Figure 2-13. This incident angle is called the First Critical Angle (the angle where the L-wave beam is totally reflected and only S-wave beams are transmitted through the second medium).

Further rotation of the transducer increases the angle of the refracted shear wave beam. When the S-wave beam reaches 90°, the resultant incident angle is termed the Second Critical Angle. In the entire region between the First and Second Critical Angle, only S-wave beams are produced.

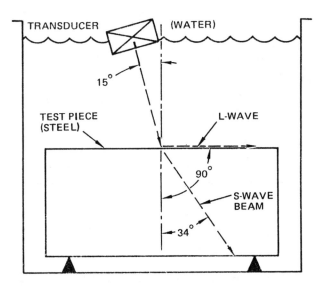

Figure 2-13. First Critical Angle

- Surface Wave Generation

When the transducer is rotated to an incident angle of about 27°, the S-wave refraction angle is increased to 90°. Figure 2-14 shows that the only reflected waves are L-waves. At slightly (approximately 5°) increased angles of incidence a surface wave is generated. The surface wave has some particle disturbance in the test surface; however, these waves are rapidly attenuated by the water medium. The shear waves are not reflected because they do not propagate in a liquid or gaseous medium.

In contact testing, where the transducer is placed directly on the test piece, surface waves are produced at angles just beyond the Second Critical Angle. Surface waves serve a function in ultrasonic testing and their presence can be readily detected.

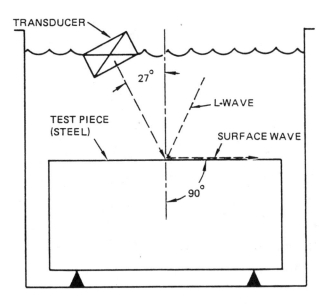

Figure 2-14. Second Critical Angle

- Summary

    To summarize, the critical angles are the incident angles at which either the longitudinal or shear wave travels along the surface of the second medium. For those incident angles beyond the Second Critical Angle there is total reflection of the beam, and no ultrasound energy is transmitted into the second medium.

    In contact testing, the incident angle slightly beyond the Second Critical Angle produces surface waves in the test specimen.

    Both critical angles may be calculated by an equation derived from Snell's Law if the velocities of the sound beam in the first and second medium are known or can be established.

Snell's Law

- General

    When the sonic velocities in the couplant used in immersion testing or the wedge material used in contact angle beam testing are different than the sonic velocity in the test specimen, the longitudinal waves passing through the wedge or couplant are refracted when the sound beam enters the test material. Refracted angles may be computed by an equation developed from Snell's Law. Reflected waves are governed by the law of reflection which states that the reflected angle equals the incident angle.

- Snell's Law Calculations

Snell's Law states that the ratio of the sine of the angle of incidence to the sine of the angle of refraction equals the ratio of the corresponding wave velocities.

Snell's Law can be expressed mathematically as follows:

$$\frac{\sin \phi_I}{\sin \phi_R} = \frac{V_I}{V_R}$$

Where:

ϕ_I = incident angle from normal in the liquid or wedge
ϕ_R = angle of the refracted beam in the test material
V_I = velocity of incident beam in the liquid or wedge
V_R = velocity of refracted beam in the test material

NOTE: The calculations for determining angles of incidence or refraction require the use of trigonometric tables (Appendix D). The sine (abbr: sin) ratios are given in decimal fractions. Velocities are given in centimeters per microsecond (cm/µs) in Appendix E. To convert cm/µs to cm/s x 10^5, move the decimal one place to the right. Divide cm/µs by 2.54 to obtain in/µs.

- Typical Problem Solving Method

Figure 2-15 shows a contact transducer mounted at an incident angle of 35° on a plastic wedge. The angle of the refracted beam may be calculated with Snell's Law since the incident angle and the velocity of the sound beam in the first and second medium are known. In this case, only shear waves are produced in the steel, as the incident angle is in the region between the First and Second Critical Angles.

Problem: Calculate the angle of refraction for longitudinal waves and shear waves as they enter a steel specimen from a plastic wedge at an incident angle of 35° as illustrated in the following test set-up.

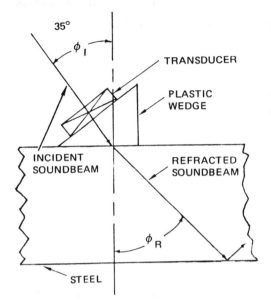

$$\frac{\sin \phi_I}{\sin \phi_R} = \frac{V_I}{V_R}$$

Given:

$V_I = 0.267$ cm/µs
$V_R = 0.585$ cm/µs (L-waves in steel)
$V_R = 0.323$ cm/µs (S-waves in steel)
$\sin \phi_I = \sin 35° = 0.574$

Solution for L-waves

By substitution into Snell's Law we have:

$$\frac{\sin 35°}{\sin \phi_R} = \frac{0.267}{0.585}$$

$$\sin \phi_R = \frac{0.585 \, (0.574)}{0.267}$$

$$\sin \phi_R = 1.258$$

Because the $\sin^{-1}$ of 1.258 is undefined, all L-waves are reflected.

Solution for S-waves

By substitution into Snell's Law we have:

$$\frac{\sin 35°}{\sin \phi_R} = \frac{0.267}{0.323}$$

$$\sin \phi_R = \frac{0.323 \, (0.574)}{0.267}$$

$$\sin \phi_R = 0.694$$

$$\phi_R = 44°$$

S-waves are refracted at an angle of 44°.

Figure 2-15. Application of Snell's Law for the Calculation of Refracted Angles

Critical Angles of Refraction

- General

    Sound beams passing through a medium such as water or plastic are refracted when entering a second medium at an incident angle. For small incident beam angles, sound beams are refracted and subjected to mode conversion resulting in a combination of shear and longitudinal waves. The region between normal incidence and the First Critical Angle is not used as commonly for ultrasonic testing as is the region beyond the First Critical Angle where only shear waves are produced. The presence of both the longitudinal and shear waves creates additional display interpretation difficulties for the ultrasonic operator.

- First Critical Angle

    As the angle of incidence is increased, the First Critical Angle is reached when the refracted longitudinal beam angle reaches 90°. At this point, only shear waves exist in the second medium. When selecting a contact shear wave angle beam transducer, or when adjusting an immersed transducer at an incident angle to produce shear waves, two conditions are considered. First, and of prime importance, is that the refracted longitudinal wave must be totally reflected (its angle of refraction must be 90° or greater) so that the penetrating beam is limited to shear waves only. Second, within the limits of the first condition, the refracted shear wave must enter the test piece in accordance with the requirements of the test procedure. In immersion testing, the First Critical Angle is calculated to make certain that the probe is positioned so that the sound beam enters the test material at the desired angle in the area of interest.

- Calculation of Critical Angles

    If the sound beam velocities for the incident wave and for the refracted wave are known (V_I and V_R), either critical angle may be calculated with the formula for Snell's Law using the sine of 90° (which is 1) as the sine of the refracted angle in the second medium. Thus, to compute the First Critical Angle in the case of the contact transducer mounted on a plastic wedge for testing steel:

    $$\text{Snell's Law:} \quad \frac{\sin \phi_I}{\sin \phi_R} = \frac{V_I}{V_R} \text{ (longitudinal wave)}$$

    $$\frac{\sin \phi_I}{\sin \phi_R (1.00)} = \frac{0.267 \text{ cm}/\mu s}{0.585 \text{ cm}/\mu s}$$

    Divide V_R into V_I = 0.456 = 27° for First Critical Angle. If the Second Critical Angle is desired, V_R is the sound beam velocity for a shear wave in steel: 0.323 cm/μs. V_R is again divided into V_I = 0.827 = 56° for the Second Critical Angle.

    - Table 2-2 lists approximate critical angles for various test materials when water is used as the first medium (V_I = 0.149 cm/μs).

    - Table 2-3 lists approximate critical angles for the same test materials when a plastic wedge is used as the first medium (V_I = 0.267 cm/μs). Note that uranium does not have a second critical angle in this case. This is because the shear wave velocity in uranium is less than the longitudinal wave velocity in plastic. Essentially this means that the incident angle would have to be greater than 90° to obtain a 90° refraction of the beam in uranium.

Table 2-2. Critical Angles, Immersion Testing

First Medium is H₂O (V = 0.149 cm/µs)

TEST MATERIAL	1ST CRITICAL ANGLE	2ND CRITICAL ANGLE	VELOCITY (cm/µs)	
			LONGITUDINAL	SHEAR
BERYLLIUM	7°	10°	1.280	0.871
ALUMINUM, 2117-T4	14°	29°	0.625	0.310
STEEL	15°	27°	0.585	0.323
STAINLESS, 302	15°	29°	0.566	0.312
TUNGSTEN	17°	31°	0.518	0.287
URANIUM	26°	51°	0.338	0.193

Table 2-3. Critical Angles, Contact Testing

First Medium is Plastic (V = 0.267 cm/µs)

TEST MATERIAL	1ST CRITICAL ANGLE	2ND CRITICAL ANGLE	VELOCITY (cm/µs)	
			LONGITUDINAL	SHEAR
BERYLLIUM	12°	18°	1.280	0.871
ALUMINUM, 2117-T4	25°	59°	0.625	0.310
STEEL	27°	56°	0.585	0.323
STAINLESS, 302	28°	59°	0.566	0.312
TUNGSTEN	31°	68°	0.518	0.287
URANIUM	52°	-	0.338	0.193

CONTACT TESTING

- General

    Contact testing is divided into three techniques which are determined by the sound beam wave mode desired: the straight beam technique for transmitting longitudinal waves in the test specimen, the angle beam technique for generating shear waves, and the angle beam surface wave technique for producing Rayleigh or Lamb waves. Transducers used in these techniques are held in direct contact with the material using a thin liquid film for a couplant. The couplant selected should be high enough in viscosity to remain on the test surface during the test. For most contact testing, the couplant should be relatively thin and should be selected to provide the proper impedance match.

- **Straight Beam Techniques**

The straight beam technique is accomplished by projecting a sound beam perpendicularly to the test surface on the test specimen to obtain pulse-echo reflections from the back surface or from discontinuities which lie between the two surfaces. To avoid confusion from dead zone and near zone effects encountered with straight beam transducers, parts with a thickness less than 5/8 inch (15.9 mm) are tested with straight beam probes that utilize a delay line or stand-off to absorb these effects. This technique is also used in the through-transmission technique using two transducers where the internal discontinuities interrupt the sound beam causing a reduction in the received signal.

— Pulse-Echo Techniques

Pulse-echo techniques may use either single or double straight beam transducers. Figure 2-16 shows the single-unit straight beam transducer in use. With the single unit, the transducer acts as both transmitter and receiver projecting a pulsed beam of longitudinal waves into the specimen and receiving reflections from the back surface and from any discontinuity lying in the beam path. The double transducer or dual-element transducer unit is useful when the test surface is rough or when the specimen shape is irregular and the back surface is not parallel with the front surface. One transducer transmits and the other receives as shown in

Figure 2-17. In this case, the receiver will receive discontinuity reflections and may receive back surface reflections.

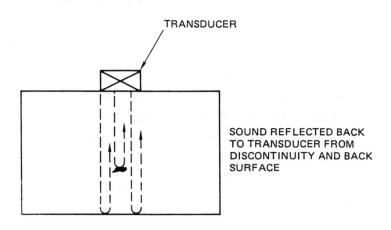

Figure 2-16. Single Transducer Pulse-Echo Technique

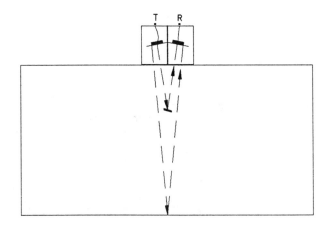

Figure 2-17. Dual-Element Transducer Pulse-Echo Technique

- Through-Transmission Techniques

Two transducers are used in the through-transmission technique—one on each side of the test specimen as shown in Figure 2-18. One unit acts as a transmitter and the other as a receiver. The transmitter unit projects a sound beam into the material. The beam travels through the material to the opposite surface, and the sound is picked up at the opposite surface by the receiving unit. Any discontinuities in the path of the sound beam cause a reduction in the amount of sound energy reaching the receiving unit. For best results in this technique, the transmitter utilizes a crystal that is the best available generator of acoustic energy, and the receiver utilizes a crystal that is the best available receiver of acoustic energy.

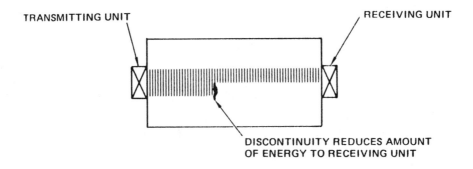

Figure 2-18. Through-Transmission Technique

- Angle Beam Techniques

The angle beam technique is used to transmit sound waves into the test material at a predetermined angle to the test surface. According to the angle selected, the wave modes produced in the test material may be mixed longitudinal and shear (bimodal), shear only, or surface modes. Usually shear wave transducers are used in angle beam testing. Figure 2-19 shows an angle beam unit scanning plate and pipe material. In the angle beam technique, the sound beam enters the test material at an angle and proceeds by successive zig-zag reflections from the specimen boundaries until it is interrupted by a discontinuity or an acoustical interface oriented perpendicular to it where it reverses direction and is reflected back to the transducer. Angle beam techniques are used for testing welds, pipe or tubing, sheet and plate material, and for specimens of irregular shape where straight beam units are unable to contact all of the surface. Angle beam transducers are identified by case markings that show the refracted angle (usually in steel) though no standard exists for marking probes.

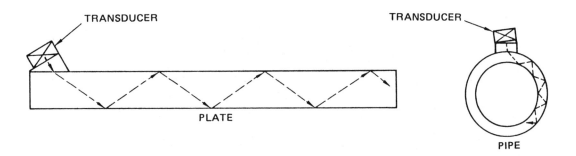

Figure 2-19. Angle Beam Technique

- Surface Wave Techniques

    The surface wave technique requires special angle beam transducers that project the sound beam into the test specimen at an angle slightly beyond the second critical angle. For test specimens where near-surface discontinuities are encountered, surface wave transducers are used to generate Rayleigh surface waves in the test material. The surface wave technique is shown in Figure 2-20.

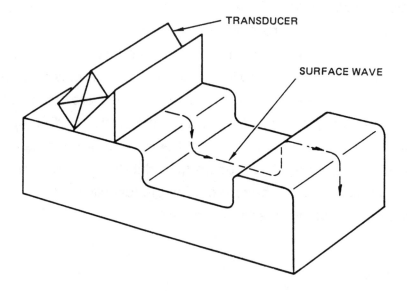

Figure 2-20. Surface Wave Technique

Immersion Testing

- General

    Any one of the following three techniques may be used in the immersion method.

- The immersion technique where both the transducer and the test specimen are immersed in water

- The bubbler or squirter technique where the sound beam is transmitted through a column of flowing water

- The wheel transducer technique, where the transducer is mounted in the axle of a liquid-filled tire that rolls on the test surface

The sound beam is directed through the water into the material using either a straight beam technique for generating longitudinal waves or one of the many angle beam techniques for generating shear waves.

Ultrasonic Displays

- General

    There are three common types of displays utilized with ultrasonic tests. These are the A-scan, B-scan and C-scan displays. The A-scan display presents time or distance horizontally and reflected sound (or amplitude) vertically. The B-scan represents an end or cross-sectional view of the specimen and the C-scan represents a plan (or top) view of the specimen. Other data display methods are available such as D-scan and P-scan which are somewhat specialized and will not be further discussed. Computerized equipment makes possible the simultaneous data collection and color-coded display of all three representations.

Influence of the Test Specimen on the Sound Beam

- General

    The highest degree of reliability in ultrasonic testing is obtained when the influence of test specimen variables and their effects are understood and considered. A shortcut for evaluating the effects of test specimen geometry and material properties is to drill flat-bottomed holes, or other suitable reflectors, in one of the test parts and then to use that part as a reference standard. With or without such a standard, the operator must be familiar with the influence of geometric and material variables. In one form or another, the operator will receive spurious or confusing indications from any of the following test specimen variables.

    - Surface Roughness

        Rough surfaces distort ultrasonic indications as follows:

        - Loss of echo amplitude from discontinuities within the part. This loss may be due to scatter at the surface of the part or to roughness of the surface on the discontinuity.

        - Loss of resolving power (the ability to distinguish between two closely spaced reflectors) which is caused by a widening of the initial pulse (the initial electrical pulse from the instruments pulser circuit) caused by reflection of transducer side- or secondary-lobe energy. Side-lobe energy is normally not reflected back into the transducer from smooth surfaces. This condition may mask the presence of a discontinuity just below the surface.

- Widening of the ultrasonic beam due to scatter caused by the rough surface or due to a requirement for a lower frequency to reduce scatter.

— Shape or Contour of Specimen

Angular (nonparallel) boundaries or contoured surfaces of the test specimen cause partial or total loss of reflection. Figure 2-21 shows a test specimen with an irregular back surface. In the area where the back surface is parallel to the front surface, the sound waves are returned to the transducer. On the left side, in the area where the back surface is sloped at an angle from the front surface, the sound waves are reflected from one boundary to another until they die out from attenuation. In actual practice, portions of the sound beam are spread from each reflection point so that a few weak signals may be received by the transducer. These signals create confusing indications.

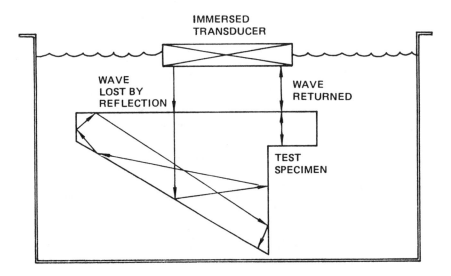

Figure 2-21. Irregular Back Surface Effect

- A convex surface on the test specimen is shown in Figure 2-22. The sound beam is widened by refraction after passing through the convex boundary. Considerable ultrasonic energy is lost by reflection at the test specimen surface as shown, and by beam spread. Signals reflected from the discontinuity have less amplitude than signals received from a discontinuity of the same size in a flat test specimen.

- Figure 2-23 shows a test specimen with a concave surface. After passing through the concave boundary, the sound beam is narrowed or focused. The discontinuity signals are relatively high in amplitude but may be difficult to identify because of unwanted reflections from the test surface.

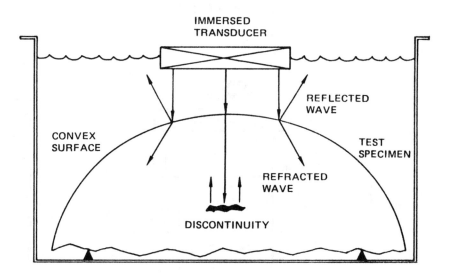

Figure 2-22. Convex Surface Effect

2-36

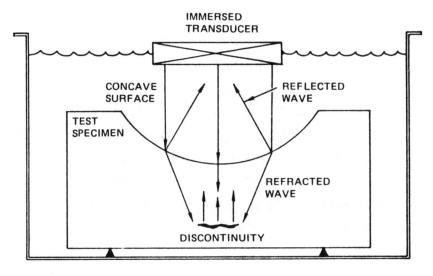

Figure 2-23. Concave Surface Effect

- Mode Conversion Within Test Specimen

When the shape or contour of the test specimen is such that the sound beam, or a portion of it as in the case of beam spread, is not reflected directly back to the transducer, mode conversion occurs at the boundary points contacted by the beam. If a direct reflection is obtained, mode conversion indications may be identified as they will appear behind the first reflection. These echoes are slow to appear because they are slowed by velocity changes when they are changed from longitudinal waves to shear waves and then back to longitudinal waves during mode conversion. Sound beams are reflected at angles which are governed by the law of reflection. Snell's Law can be used to calculate the refracted and reflected angles.

As the incident angle of the longitudinal beam is known or can be easily determined, the angle of the reflected longitudinal beam is known, since the angle of incidence is equal to the angle of reflection. The reflected shear beam angle will be about half the longitudinal beam angle, since the velocity of the shear beam is about half the velocity of the longitudinal beam. Figure 2-24 shows sound beam reflections within a long, solid test part. The spreading beam contacts the sides of the part with a specific angle of incidence. Depending on the material, the resulting mode conversion consists of mixed modes of longitudinal, shear, and possibly surface waves.

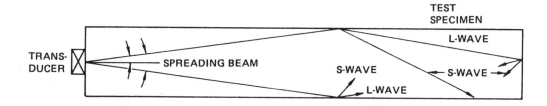

Figure 2-24. Mode Conversion Caused by Beam Spread

- Coarse Grains Within Test Specimen

    Coarse or large grains within the test specimen can cause scatter and loss of reflections or multiple nonrelevant indications, particularly when the size of the grain and the wavelength are comparable. If the frequency is lowered to the point where the wavelength is greater than the grain size, scattering losses are reduced but sensitivity is lowered.

- Orientation and Depth of Discontinuity

    The orientation and depth of the discontinuity may cause confusing indications, or may result in the loss of the discontinuity echo. In the case of orientation, the discontinuity may lie with its long axis parallel to the sound beam causing a small indication in proportion to the size of the discontinuity. If the discontinuity is angled from the sound beam, its reflections are directed away from the transducer. A sudden loss of a back surface reflection while scanning can indicate the presence of such a discontinuity. If the decrease in amplitude of the back surface deflection is proportional to the indication caused by the reflection from the discontinuity, the discontinuity is flat and parallel to the test surface. If the discontinuity indication is small compared to the loss of back surface reflection, the discontinuity is probably at an angle other than parallel to the test surface.

The extent of the near zone is determined by the formula presented earlier (page 2-15). Sound beam intensity is irregular in the near zone, causing a condition where varying indications may be obtained from the same discontinuity as the transducer is moved across it.

Beyond the near zone in the far zone, the amplitude of the indication from the discontinuity diminishes exponentially as the distance increases.

Reference Table

Appendix E lists the acoustic properties of several materials. Included in the table are the density of each material and the velocity and impedance of longitudinal, shear, and surface waves in each material. Density is given in grams per cubic centimeter (gm/cm^3) while the wave velocity is given in centimeters per microsecond ($cm/\mu s$) and the impedance is given in grams times 10^3 per square centimeter-second ($gm \times 10^3/cm^2$-s).

CHAPTER 3: EQUIPMENT

TABLE OF CONTENTS

Page

PIEZOELECTRICITY 3-1
SOUND BEAM FREQUENCIES 3-1
SOUND BEAM ATTENUATION 3-2
TIME/DISTANCE RELATIONSHIP 3-2
TRANSDUCERS 3-4
    General 3-4
    Sensitivity 3-4
    Resolution 3-4
    Materials 3-5
    Transducer Types 3-6
    Frequency Selection 3-13
COUPLANTS 3-14
    General 3-14
    Immersion Couplant Selection 3-16
    Contact Couplant Selection 3-16
STANDARD REFERENCE BLOCKS 3-17
    General 3-17
    Area Amplitude Block Set 3-18
    Distance Amplitude Block Set 3-19
    Basic Block Set 3-20
    Other Blocks 3-21
    Mockups 3-23
DISPLAY 3-24
BASIC INSTRUMENT OPERATION 3-25
    General 3-25
    Sweep Delay 3-27
    Sweep Length (Range) 3-28

i

TABLE OF CONTENTS (CONT'D.)

    Summary 3-30
PULSE-ECHO INSTRUMENTS 3-32
    General 3-32
    Controls 3-34
    A-Scan Equipment 3-37
    B-Scan Equipment 3-38
    C-Scan Equipment 3-39
    Digital Thickness Gauges 3-40
COMPUTER BASED ULTRASONICS 3-41
    General 3-41

LIST OF FIGURES

Figure		Page
3-1	Time/Distance Measuring	3-3
3-2	Straight Beam and Angle Beam Transducers	3-8
3-3	Typical Dual-Element Transducers	3-9
3-4	Typical Paintbrush Transducer	3-10
3-5	Flat and Contour-Corrected Transducers	3-11
3-6	A Focused Beam in Metal	3-12
3-7	Linear Array	3-13
3-8	Area Amplitude Reference Blocks	3-19
3-9	Other Reference Blocks	3-21
3-10	ASME Piping Calibration Block	3-22
3-11	Typical Mockup of Fatigue Crack Implant in Thick Vessel Material	3-23
3-12	Typical Mockup of Fatigue Cracks in Dissimilar Metal Piping Weld	3-23
3-13	Typical Contact Ultrasonic Test Display	3-24

LIST OF FIGURES (CONT'D.)

3-14   Typical Cathode-Ray Tube (CRT) 3-26
3-15   Sweep Delay Adjustment 3-28
3-16   Sweep Length Adjustment 3-29
3-17   Pulse-Echo Unit, Block Diagram 3-31
3-18   Typical Pulse-Echo Unit 3-33
3-19   A-Scan Presentation 3-38
3-20   B-Scan Presentation 3-39
3-21   C-Scan Presentation 3-40
3-22   Digital Thickness Gauges 3-41
3-23   Computer-Based System 3-43

LIST OF TABLES

Table   Page

3-1   Characteristics of Common Piezoelectric
       Materials 3-6
3-2   Metric Equivalents 3-18

CHAPTER 3

EQUIPMENT

Piezoelectricity

To produce an ultrasonic beam in a test piece, a transmitter applies high-frequency electrical pulses to a "piezoelectric" crystal. The prefix "piezo" is derived from a Greek word meaning "to press." The first two syllables should be pronounced like the words "pie" and "ease." Piezoelectricity refers to a reversible phenomenon whereby a crystal, when vibrated, produces an electric current or, conversely, when an electric current is applied to the crystal it vibrates. When energized with electrical pulses the crystal transforms the electrical energy into mechanical vibrations and transmits the vibrations through a coupling medium, such as water or oil, into the test material. These pulsed vibrations propagate through the object with a velocity that depends on the density and elasticity of the test material.

Sound Beam Frequencies

Most ultrasonic search units have frequencies available in a range from 0.5 MHz to 25 MHz. These vibrations, which are far beyond the audible range for humans, propagate in the test material as waves of particle vibrations. Ultrasound of all frequencies can penetrate fine-grained material without difficulty. However, when using high frequencies in coarse-grained material, interference in the form of scattering may be expected. Greater depth of penetration may be achieved by using lower frequencies. Selection of the test frequency is governed by the nature of the particular problem. Ultrasound with low frequencies (up to about 1 MHz) readily penetrate through most materials because of the small amount of attenuation of low frequencies. They are also scattered less by a coarse grain structure and can be used when the test surface is rough. On the debit side, the angle of divergence of low-frequency beams is large, making it difficult to resolve small flaws. High-frequency transducers emit a more concentrated beam with better resolving power. Limitations in the use of high frequencies are the extended near zone and the increased scattering by coarse-grained metals. All available frequencies may be used in immersion testing. The extended near zone is the main reason higher frequencies (above 10 MHz) are not generally used in contact testing. As the frequency of ultrasonic vibrations increases, the wavelength correspondingly decreases and approaches the dimensions of the grains in a metal.

Sound Beam Attenuation

High-frequency ultrasound passing through a material is reduced in power or attenuated by reflection and scattering of the beams at the grain boundaries. This loss is proportional to the grain size in the material and the wavelength of the beam. Scattering losses are greatest where the wavelength is less than one-third the grain size. As the frequency is lowered and the wavelength becomes greater than the grain size, attenuation is due primarily to absorption of the wave. Wave energy is lost through heat due to friction of the vibrating particles.

Time/Distance Relationship

The round trip distance that the sound beam travels to a reflecting surface can be measured on the cathode-ray tube (CRT) as illustrated in Figure 3-1. The initial pulse, or main bang, and the echo (sound beam traveling through water in this illustration) from the reflecting surface produce two sharp indications on the baseline. The indication at the left (A) results from the initial pulse while the indication at the right (B) is the indication from the front surface of the plate under test. The distance between the two indications is proportional to the distance between the transducer and the front surface of the specimen.

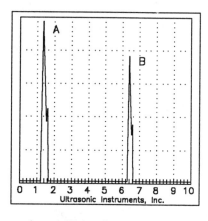
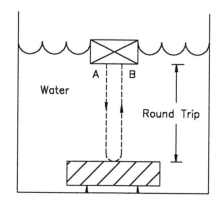

A — Initial pulse
B — Front surface reflection

Figure 3-1. Time/Distance Measuring

As mentioned earlier, time and distance measurements are related. In later discussion, it will be seen that the baseline may be adjusted to match the number of screen divisions involved in the round trip distance (as illustrated).

Transducers

- General

    In ultrasonic testing, the "eye" and "ear" of the system is the transducer. After transmitting the ultrasound energy, the transducer receives reflections that result from the condition of the material. It then relays the information back to the instrument where it is displayed. As is often the case, detailed evaluation of the characteristics of each transducer may be necessary. Chapter 5 describes that process. However, the capabilities of the transducer and the testing system are for the most part described by two terms: sensitivity and resolution.

- Sensitivity

    The sensitivity of a transducer is its ability to detect reflections from small discontinuities at given distances. Precise transducer sensitivity is unique to a specific transducer. Even transducers manufactured by the same company of the same size, frequency, and material do not always produce identical indications. Transducer sensitivity is the ability to detect and process the reflected energy from a given size discontinuity at a specific sound-path distance in a reference block.

- Resolution

    The resolution, or resolving power, of a transducer refers to its ability to separate indications from two targets close together in position or depth (for example, the front surface indication and the indication from a small discontinuity just beneath the surface). The time required for the transducer to stop "ringing" or vibrating after having been supplied with a large voltage pulse is a measure of its resolving power. Long "tails" or transmissions of sound energy from a ringing transducer cause a wide, high-amplitude front surface indication. In this case, a small discontinuity just beneath the surface is masked by the ringing signal.

- Materials

Developments in man-made polarized ceramics have produced the most efficient crystals for applying ultrasonic energy. They operate on low voltage, are practically unaffected by moisture, and are usable at high temperatures up to 1112°F (600°C). They are limited, however, by their relatively low mechanical strength, some mode conversion interference, and a tendency to age (i.e., to lose their polarization and become very brittle).

The most commonly used piezoceramics include sodium bismuth titanate, lead metaniobate, lead titanate and several variations within the lead zirconate titanate (PZT) family. These include PZT4, PZT5H, PZT7A and PZT8. These PZT materials are selected for applications dependent upon their specific characteristics. Additionally, PZT comes in the form of a composite.

Table 3-1 describes a few characteristics of these crystals. It is not important that you remember the details but rather that you understand that piezoceramics offer a tremendous array of possibilities.

Table 3-1. Characteristics of Common Piezoelectric Materials

Crystal Material	Characteristics/Applications
Lead Metaniobate #1	Good power; great receiver; broad band; high temperature applications
Lead Metaniobate #2	Good transmitter and receiver; narrow band; contact transducers ≥2 MHz
PZT4	Good transmitter and receiver; best at lower frequencies; good penetrator of coarse-grained materials; high power applications
PZT5H	Great penetrator; good receiver; multifrequency; accelerometers
Lead Titanate	Great receiver; good penetrator; good at higher frequencies; medical diagnostics
Sodium Bismuth Titanate	Extremely stable; ultrahigh temperature applications

- Transducer Types

Transducers are made in a limitless number of sizes and shapes from extremely small (pinpoint) to 6-inch-wide (15.2 cm) "paintbrush" transducers. The many shapes are the result of much experience and the requirement for many special applications. Size of a transducer is a contributing factor to its performance. For instance, the larger the transducer, the narrower the sound beam (less beam spread) for a given frequency. The narrower beams of the larger (beyond 1 inch or 25.4 mm) high-frequency transducers have greater ability for detecting very small discontinuities. The larger transducers also transmit more energy into the test part, so they can be used to gain deep penetration. The large, single-crystal transducers are generally limited to lower frequencies (≤5 MHz) because the very thin high-frequency crystals are fragile.

- Contact Transducers

    Contact transducers are made for both straight and angle beam testing. When the straight beam unit is faced with a wear plate, an electrode on the front face of the crystal provides for an internal ground. All angle beam and immersion-type transducers are internally grounded. In addition, the immersion-type transducers, including the coaxial cable connection, are waterproof since, in use, they are completely submerged.

- Angle Beam Transducers

    Transducers are also classified as either straight beam transducers or angle beam transducers. The term "straight beam" means that the energy from the transducer is transmitted into the test specimen in a direction normal (perpendicular) to the test surface. Angle beam transducers direct the sound beam into the test specimen surface at an angle other than 90°. Angle beam transducers are used to locate discontinuities oriented at angles to the surface and to determine the size of discontinuities oriented at angles other than parallel to the surface. Angled transducers are also used to propagate shear, surface, and plate waves into the test specimen by mode conversion. In contact testing, angle beam transducers use a wedge, usually of plastic (Lucite), between the transducer face and the surface of the test specimen to direct the ultrasound into the test surface at the desired angle. In immersion testing, angulation of the beam is accomplished by varying the angle of a straight beam transducer to direct the beam into the test part at the desired angle. Both straight and angled transducers are shown in Figure 3-2.

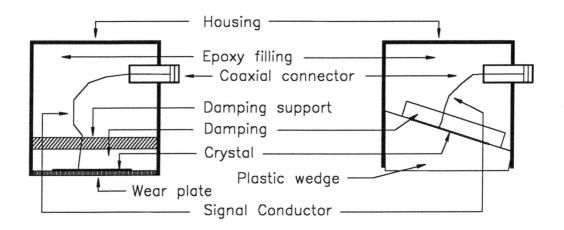

Figure 3-2. Straight Beam and Angle Beam Transducers

- Dual-Element (Double) Transducers

The dual-element transducer differs from the single transducer in that, while the single transducer may be a transmitter only, a receiver only, or both transmitter and receiver, the dual-element probe is in essence two single transducers mounted in the same housing for pitch-and-catch testing. In the dual-element (dual) search unit, one transducer is the transmitter and the other is the receiver. They may be mounted side by side for straight beam testing and mounted stacked or tandem for angle beam testing. In all cases, the crystals are separated by a sound barrier to block cross-talk. Figure 3-3 shows both types of dual-element transducers.

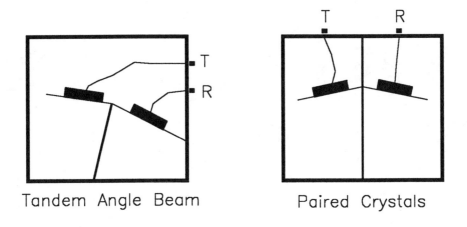

Tandem Angle Beam   Paired Crystals

Figure 3-3. Typical Dual-Element Transducers

- Paintbrush Transducers

The wide (6 inch or 15.2 cm) paintbrush transducers are made up of a mosaic pattern of smaller crystals carefully matched so that the intensity of the beam pattern varies very little over the entire length of the transducer. This is necessary to maintain uniform sensitivity. Paintbrush transducers provide a long, narrow, rectangular beam (in cross section) for scanning large surfaces. Their purpose is to quickly discover discontinuities in the test specimen.

Smaller, more sensitive transducers are then used to define the size, shape, orientation, and exact location of the discontinuities. Figure 3-4 shows a typical paintbrush transducer.

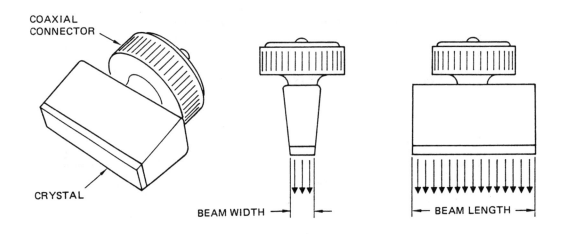

Figure 3-4. Typical Paintbrush Transducer

- Faced Unit or Contour Focused Transducers

In addition to wedges, other frontal members are added to the transducer for various reasons. On contact transducers, wear plates are often added to protect the fragile crystal from wear, breakage, or the harmful effects of foreign substances and to protect the front electrode. Frontal units shaped to direct the ultrasound perpendicular to the surface at all points on curved surfaces and radii are known as contour-correction lenses. These lenses sharpen the front surface indication by equalizing the sound-path distance between the transducer and the test surface. A comparison of flat and contoured transducers is shown in Figure 3-5.

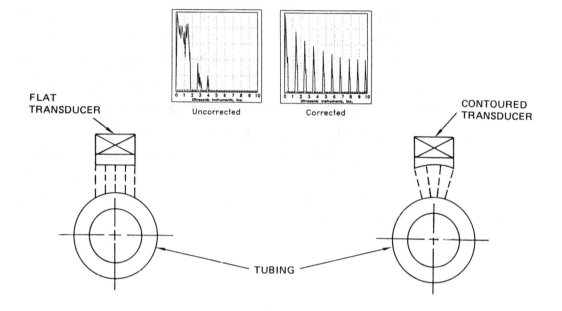

Figure 3-5. Flat and Contour-Corrected Transducers

Other acoustic lenses focus the sound beam from the transducer much as light beams are focused. Cylindrically focused transducers concentrate the ultrasound into a long, narrow, blunt-pointed beam of increased intensity which is capable of detecting very small discontinuities in a relatively small area. Spherically focused transducers focus the sound beam to a point within a test article. Focusing the sound beam moves its point of maximum intensity towards the transducer, but shortens its usable range. The test specimen has the effect of a second lens in this case.

When the beam enters the test surface it is refracted, as shown in Figure 3-6. The increased intensity produces increased sensitivity at and near the focal point. Moving the point of maximum intensity closer to the transducer (which is also closer to the test surface) improves the near surface resolution. The disturbing effects of rough surface and material noise are also reduced. This is true simply because a smaller cross-sectional area is being tested. In a smaller cross-sectional area, the true discontinuity indications will be relatively large compared to the combined noise of other nonrelevant indications. The useful thickness range of focused transducers is approximately 0.010 to 2.0 inches (0.25 to 50.8 mm).

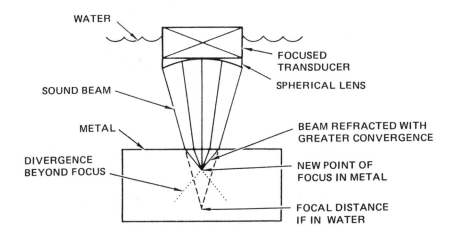

Figure 3-6. A Focused Beam in Metal

- Electronically-Focused Transducers

    Special transducers, known as "linear arrays," can be focused using electronic phase control. A typical linear array is shown in Figure 3-7. The linear array is a collection of very small transducers with each transducer able to act as a transmitter or receiver. Transducer width may be as small as 0.020 inches (0.5 mm) and an array may consist of several hundred small transducers. By selectively pulsing each transducer or group of transducers in a given order, the ultrasonic beam may be focused in sound path or may cause the focused beam to sweep across the test specimen without moving the array.

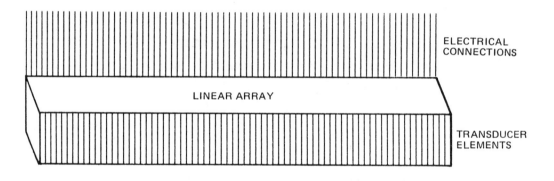

Figure 3-7. Linear Array

- Frequency Selection

    The frequency of a transducer is a determining factor in its use. Basic characteristics are affected by the need for sensitivity. Sensitivity is related to wavelength: the higher the frequency, the shorter the wavelength; the shorter the wavelength, the higher the sensitivity. Transducer frequency and crystal thickness are also related. The higher the frequency, the thinner the crystal. For example, a 15-MHz transducer has a crystal only about 0.007 inch (0.2 mm), which is fragile for contact testing. If such a crystal were used in the contact testing technique, the near zone length would have to be evaluated. It is possible to place a "delay line" or "standoff" on the end of a contact probe to compensate for the near zone. In this way, the sound beam entering the test object would be limited to the far zone and would therefore be useful. Immersion testing makes this compensation with the water path. For this reason, the immersion testing technique would be the better choice when testing conditions require the use of a 15-MHz or higher-frequency transducer. However, contact transducers of 25 MHz and higher are commercially available. Other considerations are:

- The higher the frequency of a transducer, the narrower (less beam spread) the sound beam and the greater the sensitivity and resolution, but the attenuation is also greatest and the penetration is poor.

- The lower the frequency of a transducer, the deeper the penetration and the less the attenuation; but the greater the beam spread, the less the sensitivity and resolution.

- At any given frequency, the larger the transducer, the narrower the sound beam and the greater the sensitivity.

Couplants

- General

    One of the practical problems in ultrasonic testing is the

transmission of the ultrasonic energy from the source into the test specimen. If a transducer is placed in contact with the surface of a dry part, very little energy is transmitted through the interface into the material because of the presence of air between the transducer and the test material. The air causes a great difference in acoustic impedance (impedance mismatch) at the interface.

A couplant is used between the transducer face and the test surface to ensure efficient sound transmission from transducer to test article. The couplant, as the name implies, couples the transducer ultrasonically to the test specimen by filling the minor irregularities of the test surface and by excluding all air from between the transducer and the test surface. The couplant can be any of a vast variety of liquids, semiliquids, pastes, and even some solids, that will satisfy the following requirements.

- A couplant wets (fully contacts) both the surface of the test specimen and the face of the transducer and excludes all air from between them.

- A couplant is easy to apply.

- A couplant is homogeneous and free of air bubbles, or solid particles in the case of a nonsolid.

- A couplant is harmless to the test specimen and transducer.

- A couplant has a tendency to stay on the test surface, but is easy to remove.

- A couplant has an acoustic impedance value between the impedance value of the transducer face and the impedance value of the test specimen, preferably approaching that of the test surface.

- Immersion Couplant Selection

   In immersion testing, nothing more than clean, deaerated tap water with a corrosion inhibitor and a wetting agent is used for a couplant. For operator comfort, the water temperature is usually maintained at 70°F (21°C) by automatic controls. Wetting agents are added to the water to ensure that the surface is thoroughly wet, thereby eliminating air bubbles.

- Contact Couplant Selection

   In contact testing, the choice of couplant depends primarily on the test conditions; i.e., the condition of the test surface (rough or smooth), the temperature of the test surface, and the position of the test surface (horizontal, slanted, or vertical).

   One part gelatin (cellulose) with two parts water and a wetting agent is often used on relatively smooth, horizontal surfaces. For slightly rough surfaces, light oils (such as engine oil) may also be used. Rough surfaces, hot surfaces, and vertical surfaces require the use of a heavier or special-purpose couplants. In all cases, the couplant selected must be as thin as possible and allow for consistent, effective results. Couplant manufacturers offer a wide variety of special couplants.

Standard Reference Blocks

- General

    In ultrasonic testing, all discontinuity indications are compared to indications received from testing a reference standard. The reference standard may be any one of many reference blocks or sets of blocks specified for a given test. Ultrasonic standard reference blocks, often called test blocks, are used in ultrasonic testing to calibrate the ultrasonic equipment and to evaluate the discontinuity indications received from the part under test. Calibrating does two things: it verifies that the instrument/transducer combination is performing as required, and it establishes a sensitivity or gain setting at which all discontinuities of the size specified, or larger, should be detected. Evaluation of discontinuities within the test specimen is accomplished by comparing their indications with the indication received from artificial discontinuities of known size and at the same sound path in a standard reference block of the same material. Remember, other factors (type of reflector, orientation of the reflector, etc.) affect this relative "size" estimate.

    Standard test blocks are made from stock which has been carefully selected and ultrasonically inspected and which meets predetermined standards of sound attenuation, grain size, and heat-treat. Discontinuities can be represented by flat-bottomed holes (FBHs) which are carefully drilled, side-drilled holes (SDH's) or notches. Test blocks are made and tested with painstaking care so that the only discontinuity present is the one that was added intentionally. Three such sets of standard reference blocks are: (1) the area amplitude blocks; (2) the distance amplitude blocks and; (3) the ASTM basic set of blocks that combine area amplitude and distance amplitude blocks in one set. To streamline the text, refer to Table 3-2 to convert to metric units as we continue our discussion of reference blocks.

Table 3-2. Metric Equivalents

Fraction (inch)	Metric Equivalent (mm)
1/64	0.4
2/64; 1/32	0.8
3/64	1.2
4/64; 1/16	1.6
5/64	2.0
6/64; 3/32	2.4
7/64	2.8
8/64; 1/8	3.2

- Area Amplitude Block Set

This set consists of 8 blocks, each 3-3/4 inches (95.3 mm) long and 1-15/16 inches (49.2 mm) square. A 3/4 inch deep (19 mm), flat-bottomed hole is drilled in the bottom center of each block. The hole diameters are 1/64 inch in the No. 1 block through 8/64 inch in the No. 8 block, as illustrated in Figure 3-8. As implied, the block numbers refer to the FBH diameter; e.g., a No. 3 block has a 3/64 inch diameter flat-bottomed hole. Similar area amplitude reference blocks are made from 2-inch-diameter (50.8 mm) round stock.

Area amplitude blocks provide a means of checking the linearity of the test system. They confirm that the amplitude (height) of the indication in the display increases in proportion to the increase in size of the discontinuity.

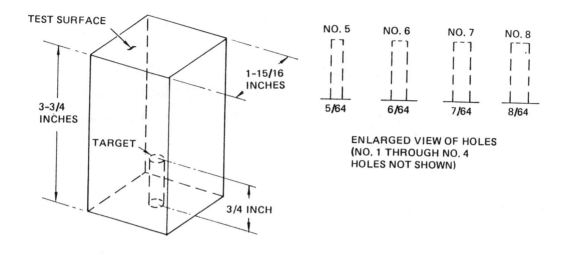

Figure 3-8. Area Amplitude Reference Blocks

- Distance Amplitude Block Set

This set of blocks consists of nineteen 2-inch-diameter (50.8 mm) cylindrical blocks, all with 3/4 inch deep (19 mm) flat-bottomed holes of the same diameter drilled in the center at one end. These blocks are of different lengths to provide sound-path distances of 1/16 inch to 5-3/4 inches (146.1 mm) from the test surface to the flat-bottomed hole. Sets with 3/64, 5/64, or 8/64 inch-diameter holes are available. The sound-path distances in each set are: 1/16 inch, 1/8 inch through 1 inch in 1/8-inch increments, and 1-1/4 inches (131.8 mm) through 5-3/4 inches (146.1 mm) in 1/2-inch (12.7 mm) increments.

Distance amplitude blocks serve as a reference by which the size of discontinuities at varying sound-path distances within the test material may be evaluated. They also serve as a reference for establishing the sensitivity or gain of the test system so that the

3-19

system will display readable indications for all discontinuities of a given size, but will not flood the screen with indications that are of no interest. On instruments so equipped, these blocks are used to set the STC (sensitivity time control) or DAC (distance amplitude correction) so that a discontinuity of a given size will produce an indication of the same amplitude regardless of its sound-path distance from the examination surface.

- Basic Block Set

The ASTM basic set consists of 10 blocks which are 2 inches in diameter (50.8 mm) that have 3/4 inch deep (19 mm) flat-bottomed holes drilled in the center at one end. One block has a 3/64 inch-diameter FBH and a sound-path distance of 3 inches (76.2 mm) from the test surface to the flat-bottomed hole. The next 7 blocks each have a 5/64-inch FBH but sound-path distances are 1/8, 1/4, 1/2, 3/4, 1-1/2, 3, and 6 inches (3.2, 6.3, 12.7, 19, 38.1, 76.2, and 152.4 mm) from the test surface to the FBH. The 2 remaining blocks each have an 8/64 inch diameter FBH and sound-path distances of 3 inches (76.2 mm) and 6 inches (152.4 mm). In this basic set the 3 (No. 3, 5, and 8) blocks with the 3-inch (76.2 mm) sound-path distance provide the area amplitude relationship, and the 7 blocks with the 5/64 inch diameter FBH (No. 5) and varying sound-path distances provide the distance amplitude relationship.

It is important that the test block material be the same or similar to that of the test specimen. Alloy content, heat-treatment, degree of hot or cold working from forging, rolling, etc., all affect the acoustical properties of the material. If test blocks of identical material are not available, they must be similar in ultrasound attenuation, surface condition, velocity, and acoustic impedance.

- Other Blocks

  The IIW (International Institute of Welding) reference block and the miniature angle beam field calibration block (Rompas block) shown in Figure 3-9 are examples of other reference standards in common use.

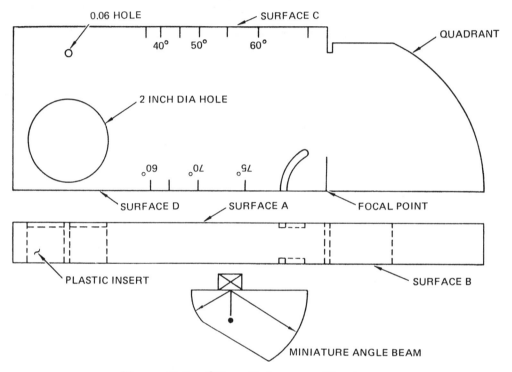

Figure 3-9. Other Reference Blocks

For irregularly-shaped articles, it is often necessary to make one of the test articles into a reference standard by adding artificial discontinuities in the form of flat-bottomed holes, saw cuts, notches, cracks, etc. In some cases, these artificial discontinuities can be placed so that they will be removed by subsequent machining of the article. In other cases, a special individual calibration technique is developed by carefully studying an article ultrasonically and then verifying the detection of discontinuities in the article by destructive investigation. The results of the study then become the basis for the testing standard.

Another block that is used for calibration is the ASME piping calibration block as illustrated in Figure 3-10. This block is specified in the ASME Boiler and Pressure Vessel Code and is used by various industries. Many specialty calibration blocks exist for various material types, discontinuity orientations and applications.

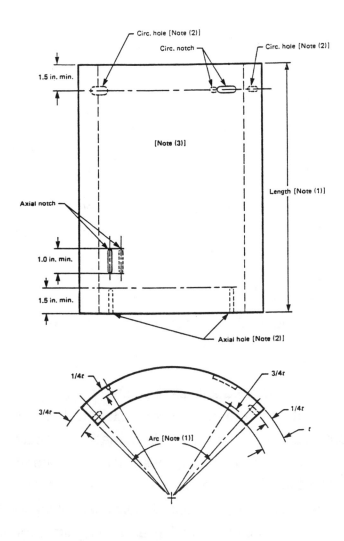

Figure 3-10. ASME Piping Calibration Block

- Mockups

  In some cases real flaws are implanted at critical locations in a mockup to simulate the ultrasonic examination that will later be conducted on actual components. Unlike reference and calibration blocks, these mockups are often used as a capability demonstration to qualify UT procedures and personnel. As shown in Figure 3-11 and 3-12, mockups and other intentionally flawed specimens are available from several commercial sources including the publisher of this text.

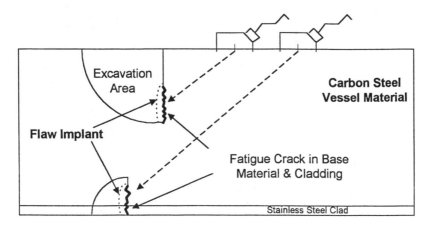

Figure 3-11. Typical Mockup of Fatigue Crack Implant in Thick Vessel Material

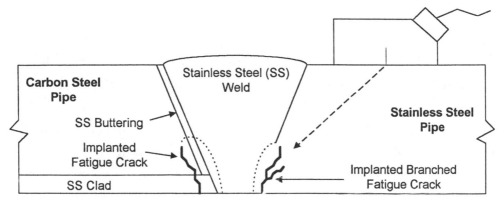

Figure 3-12. Typical Mockup of Fatigue Cracks in Dissimilar Metal Piping Weld

Display

Figure 3-13 shows a typical ultrasonic contact test setup and the resulting CRT display. Notice the position of the displayed indications on the screen in relation to the actual positions of the test piece front surface, discontinuity, and back surface.

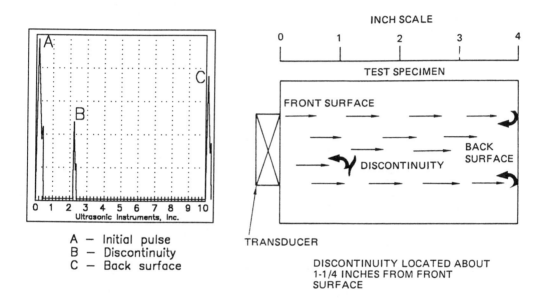

A — Initial pulse
B — Discontinuity
C — Back surface

DISCONTINUITY LOCATED ABOUT 1-1/4 INCHES FROM FRONT SURFACE

Figure 3-13. Typical Contact Ultrasonic Test Display

The CRT is divided into *horizontal screen divisions* located below the horizontal sweep. There are normally 10 *major* screen divisions, each broken down into 5 *minor* divisions. The horizontal screen divisions are used as relative units of time or distance. We can adjust the ultrasonic instrument so that these screen divisions represent a specific distance (or time) between the front surface reflection and the back surface reflection, in inches, cm., etc. There may be 5 or 10 horizontal screen divisions, depending on the instrument manufacturer. These are used to compare indication amplitude.

In Figure 3-13, the indications were adjusted to position the initial pulse, or front surface indication, on the 0 major screen division. The back surface indication is on the 10th major screen division. The discontinuity indication appears just to the right of the 2nd major screen division (at 2.2 screen divisions). The indication positions were accomplished by varying two controls on the instrument; the DELAY and the SWEEP LENGTH, MATERIAL CALIBRATION, or RANGE controls.

Basic Instrument Operation

- General

    A cathode-ray tube (CRT) is often used to display ultrasonic indications in a manner similar to a television picture tube. Figure 3-14 shows a typical cathode-ray tube and its electron gun. This tube comes in many sizes and shapes. It is made of specially-tested glass that is constructed with a screen at one end for the picture display. The screen is coated with material called a phosphor compound which varies in composition to produce various brightnesses, colors, and time persistence. The phosphor compound glows and produces light when bombarded by high-speed electrons directed at the screen from the electron gun in the base of the tube.

    At the opposite end of the tube, electrons are produced in the electron gun. The electrons are emitted from a hot filament similar to the filament in an ordinary light bulb. By electromagnetic means, these electrons are accelerated and focused to form a beam the size of a pinhead when it strikes the phosphor screen. The position of the spot on the screen is altered by changing the direction of the electron beam. Changing the direction of the electron beam is accomplished by changing the electrical charge on the horizontal and vertical deflection plates.

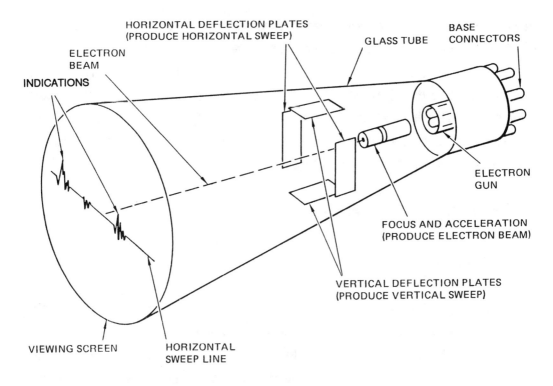

Figure 3-14. Typical Cathode-Ray Tube (CRT)

In ultrasonic testing, the CRT usually shows a bright horizontal line when there is no signal received. This horizontal line is called the sweep or baseline. An electronic circuit energizes the horizontal deflection plates to cause the electron beam to sweep from the left edge of the screen to the right edge at a certain fixed speed. As soon as the beam reaches the right edge, it is caused to return to the left edge at a very high speed, too fast to be seen on the screen. In operation, the electron beam draws a line of light across the screen. The line length is a measure of the time required for the spot on the screen to move from left to right.

Distance may be determined when time and speed are known. The distance along the line represents the lapse of time since the initial pulse (sound entry surface) and this lapse of time multiplied by speed equals distance from the sound entry surface. By adjusting the sweep speed of the electron beam, the baseline may be adjusted to represent a particular distance.

When a signal is relayed to the CRT from the transducer, a voltage is applied to the vertical deflection plates, causing a vertical indication to appear from the line. When the transducer receives signals reflected from the test piece front and back surface, voltages are again applied to the vertical deflection plates and the front surface indication appears first and the back surface indication appears some time later on the baseline. The spacing between these indications is a measure of the sound-path distance between the surfaces.

Digitized displays are also utilized and offer great advantages in size and weight over the typical CRT described above. Additionally, digitization of the data permits data transfer to computer-based systems for further interpretation, evaluation, multiscan display, and data archiving.

- Sweep Delay

The DELAY control of the instrument permits the baseline and the indications on it to be shifted either to the right or to the left while the spacing between the indications remains constant. Figure 3-15 shows the result of adjusting the DELAY control to shift the baseline to the left in order to see the indications related to the material under test (see Figure 3-13 for test setup).

In Figure 3-15, the operator first picked up the front surface indication (which is also the initial pulse in this example) and the discontinuity indication. By adjusting the DELAY, the front surface indication is moved to the far left bringing the discontinuity and back surface indications into view. Notice that the distance between the first two indications has not changed.

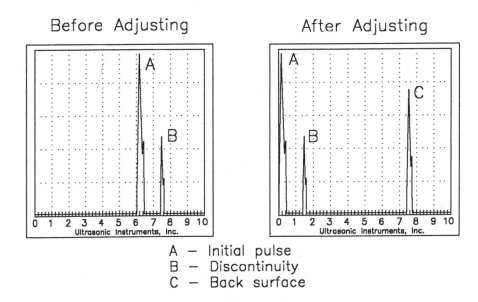

A — Initial pulse
B — Discontinuity
C — Back surface

Figure 3-15. Sweep Delay Adjustment

- Sweep Length (Range)

Now that the delay has been adjusted the RANGE adjustment must be considered. The operator may wish to display the front and back surface indications so that the distance relationship and horizontal positions on the screen is directly related to the actual dimension of the test piece. To do this, the horizontal trace or baseline is expanded or contracted to change the distance between the indications displayed. The discontinuity indication is always located in the same relative position with respect to the front and back surface indications.

3-28

- The expansion or contraction of the baseline is away from or toward the left side of the screen. That is, if the DELAY is set so that the start of the desired presentation is at the left side of the screen, adjustment of the RANGE moves the right-hand indications away from or toward the left-hand indication. The DELAY control also makes it possible to view the responses from any desired segment of the sweep. In effect, the DELAY control allows the viewing screen to be moved along the sound path of the part. In conjunction with the RANGE control, the DELAY makes it possible to examine a magnified segment of the part by spreading the segment across the entire width of the display.

- Figure 3-16 shows the CRT presentation before and after the RANGE has been adjusted to expand the view of the entire part across the screen with the front surface and back surface indications aligned with the 0 and 10th screen divisions respectively. Assume that the CRT represents 10 inches (254 mm) full screen width (FSW). The discontinuity is located at 2.2 screen division which represents 2.2 inches (55.9 mm) from the sound entry surface.

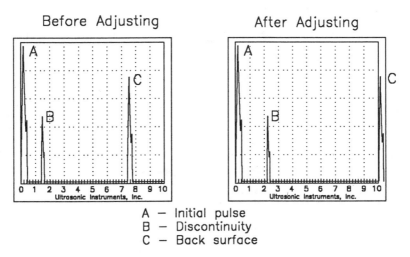

A — Initial pulse
B — Discontinuity
C — Back surface

Figure 3-16. Sweep Length Adjustment

- Two controls, the DELAY and the RANGE, regulate how much of the test part is presented at one time on the screen and what portion, if not the whole, of the part is presented.

- Summary

After the instrument is turned on, allowed to warm up, and/or has completed the self-test, the first adjustments made on the instrument concern scale illumination, baseline intensity focus, horizontal positioning, and vertical positioning. The power ON switch may contain a control for the brightness of the scale scribed on the CRT screen overlay. This brightness is considered a matter of personal choice. The intensity control determines the brightness of the spot moving across the screen to form the sweep line. Sweep line intensity is kept at a minimum with no bright spot at either end. The astigmatism and focus controls adjust the sharpness of the screen presentation. The horizontal positioning control determines the starting point of the sweep line on the CRT, which is usually at the left edge of the screen. The vertical positioning control raises and lowers the sweep line or baseline to coincide with the screen graticules. The exact procedures to operate and adjust the many controls of various ultrasonic instruments are described in the operation manual for the individual instrument. The precise capabilities of each instrument is contained in the same source.

A simplified block diagram of a typical A-scan pulse-echo ultrasonic testing instrument is shown in Figure 3-17. The timer or rate generator is the heart of the system. In the contact testing setup illustrated, the energizing pulse from the pulse unit is routed to the receiver amplifier unit at the same time that it is sent to the transducer so that the initial pulse and the front surface indication occur at nearly the same time. In an immersion testing setup, the

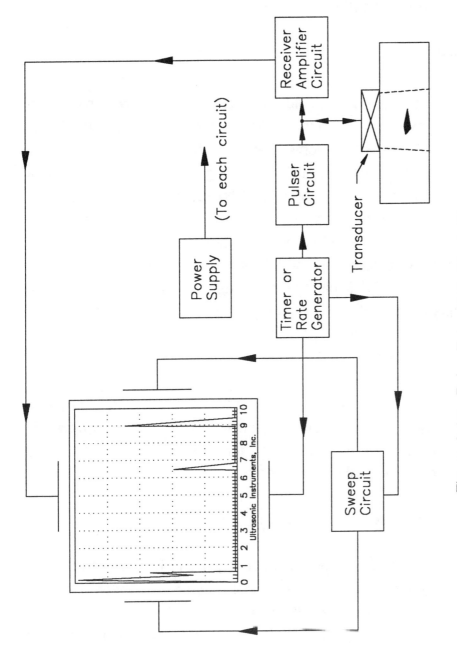

Figure 3-17. Pulse-Echo Unit, Block Diagram

3-31

initial pulse and the front surface indication are separated by the water travel distance to the test piece.

Pulse-Echo Instruments

- General

    All makes of pulse-echo equipment have similar electronics circuitry and provide basic common functions. A typical pulse-echo unit is shown in Figure 3-18. Nomenclature of the given functions varies from one instrument to another according to the manufacturer. The manufacturer's manual provides operation and maintenance instructions for the unit, a review of theory, and other more specific information. Manufacturers' recommendations concerning instrument operation supersede this text in the event of conflicting information. Each ultrasonic system provides the following essentials:

    - Power Supply

        Circuits for supplying current for all basic functions of the instrument constitute the power supply. Electrical power is served from line supply or, for some units, from a battery contained in the instrument.

    - Transducer

        The transducer consists of the crystal, its housing, and cable. The crystal converts electrical energy to ultrasonic energy and introduces vibrations into the test specimen. It also receives reflected vibrations from within the test specimen and converts them into electrical signals for amplification and display.

- Pulser/Receiver

    The pulser, or pulse generator, is the source of short high-energy bursts of electrical energy (triggered by the timer) which are applied to the transducer. Return pulses from the test specimen are received, amplified, and routed to the display unit.

- Display/Timer

    The display is usually a CRT or a digitized display with a sweep generator and the controls required to provide a visual image of the signals received from the test specimen. The timer is the source of all timing signals to the pulser and is sometimes referred to as the rate generator or clock.

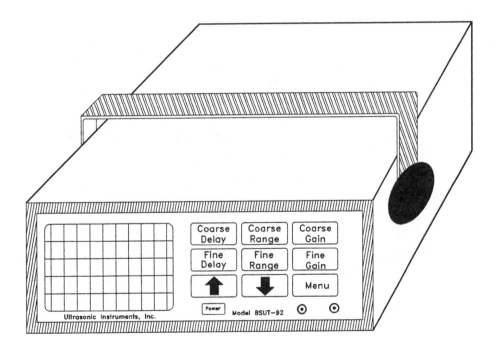

Figure 3-18. Typical Pulse-Echo Unit

- Controls

    Controls are provided for the various systems of the instrument, such as power supply, pulser, receiver, timer, and display. The nomenclature used in the following description of the controls may vary from one type of unit to another.

    – Power Supply

        The power supply is usually controlled by an ON-OFF switch and a fuse. After turning power on, there are certain time-delay devices which protect circuit elements during instrument warm-up.

    – Pulser/Receiver

        The pulse of ultrasonic energy transmitted into the test specimen is adjusted by the PULSE LENGTH control. For single-transducer testing, the transmit and receive circuits are connected to one jack on the same transducer. For double-transducer testing, called through-transmission or pitch-and-catch testing, a T (transmit) jack is provided to permit connecting one transducer for use as a transmitter and an R (receive) jack is provided for connecting another transducer for receiving only. A MODE switch for THRU or PULSE-ECHO transmission is provided for control of the T and R jacks. A selector for a range of operating frequencies is usually marked FREQUENCY with the available frequencies given in megahertz. Gain controls usually consist of FINE and COARSE sensitivity selectors or one control marked SENSITIVITY. For a cleaner video display with low-level noise minimized, a REJECT control is provided.

- Display/Timer

    The display controls and their functions for the display unit are as follows:

    - Vertical - Controls the vertical position of the indications displayed.

    - Horizontal - Controls the horizontal position of the indications displayed.

    - Intensity - Varies the brightness of the display as desired.

    - Focus - Adjusts the focus of the trace.

    - Astigmatism - Corrects for distortion or astigmatism introduced by changing transmit time of the electron beam across the oscilloscope screen.

    - Power - Turns power on and off for entire unit.

    - Scale Illumination - Adjusts illumination of the grid lines when provided.

    Timer circuits usually consist of a pulse repetition rate device which controls the rate at which pulses are generated to other circuits. Pulse repetition rate is varied to suit the material and thickness of the test specimen. The DELAY control is also used to position the initial pulse on the left side of the display screen with a back surface reflection, or multiples of back surface reflections, visible on the right side of the screen.

- Other Controls

    Other controls, which are refinements not always provided, include the following:

    - DAC or STC

        DAC (Distance Amplitude Correction), STC (Sensitivity Time Control), and other like units called TCG (Time Corrected Gain), or TVG (Time Varied Gain) are used to compensate for a drop in amplitude of signals from reflectors of long sound paths within the test specimen.

    - Damping

        The pulse duration is shortened by the DAMPING control which adjusts the length of the wave train applied to the transducer. Resolution is improved by increasing the damping.

    - Display Selector

        The DISPLAY SELECTOR switch is used to select the desired type of display, RADIO FREQUENCY or VIDEO.

    - Gated Alarm

        Gated alarm units enable the use of automatic alarms when discontinuities are detected. This is accomplished by setting up specific gated or zoned areas within the test specimen. Signals appearing within these gates may be monitored automatically to operate visual or audible alarms. These signals are also passed on to process feedback control devices. Gated alarm units usually have three controls as follows:

-- Start or Delay

The gate START or DELAY control is used for adjustment of the location of the leading edge of the gate.

-- Length or Width

The gate LENGTH or WIDTH control is used for adjustment of the width of the gate or the location of the gate trailing edge.

-- Alarm Level or Threshold

The alarm LEVEL or THRESHOLD control is used for adjustment of the gate's vertical threshold (either + or -) to turn on signal lights or to activate an alarm.

- A-scan Equipment

The A-scan system is a data presentation technique that displays the returned signal amplitude from the material under test as illustrated in Figure 3-19. The horizontal baseline indicates elapsed time (from left to right), and the vertical deflection shows signal amplitudes. For a given ultrasonic velocity in the specimen, the sweep can be calibrated directly across the screen in terms of sound-path distance or depth of penetration into the sample. Conversely, when the dimensions of the sample are known, the sweep time may be used to determine ultrasonic velocities. The height of the indications represent the intensities of the reflected sound beams. These may be used to determine the relative size of the discontinuity, the depth or distance to the discontinuity from the front or back surface, the sound beam spread, and other factors. The chief advantage of this equipment is that it provides amplitude information needed to evaluate the relative size and position of the discontinuity.

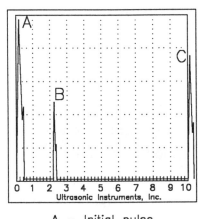

A — Initial pulse
B — Discontinuity
C — Back surface

Figure 3-19. A-scan Presentation

- B-scan Equipment

The B-scan system shown in Figure 3-20 is particularly useful where the distribution, shape and location of large discontinuities within a sample cross section is of interest. A chief advantage of the B-scan equipment is that a cross-sectional view of the sample and the discontinuities within it are displayed. In computerized high-speed scanning, the cross-section image is used to evaluate several cross sections of the sample in the area(s) containing the discontinuity(s).

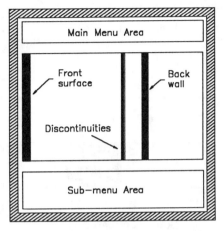

Typical Computer Monitor B-scan Display Illustrates the depth (sound path) and length of the discontinuities.

Figure 3-20. B-scan Presentation

- C-scan Equipment

C-scan equipment displays the test article and associated discontinuities in a plan (or top) view. The C-scan recording indicates the projected length and width of the discontinuity and the outline of the test specimen as if viewed from directly above the specimen. The C-scan recording indicates the depth of the discontinuity in the test specimen. The same signals that generate the indications on the A-scan produce a change on the C-scan recording. The front and back surface signals from the specimen are eliminated from the recording by the instrument gating circuits, and the alarm sensitivity control setting determines the amplitude of the signal (indication) required to produce a change on the recording. Commonly used displays are high-resolution color computer monitors and liquid crystal displays (LCD's) The data can be readily printed on a multicolor printer or stored on magnetic or optical media. Figure 3-21 shows a C-scan presentation.

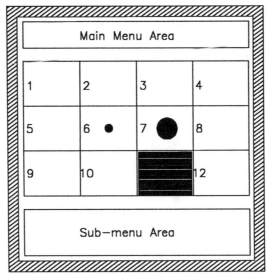

Typical Computer Monitor C-scan Display
Illustrates the use of 12 gates simultaneously.

Figure 3-21. C-scan Presentation

- Digital Thickness Gauges

The digital thickness gauge is a specialized A-scan instrument that utilizes the DAC to determine when the first indication of interest returns to the search unit. Time is kept by an electronic counter which displays the sound-path distance digitally on the LCD. The operation is straightforward and is based on the velocity of the material under test (range control) and the thickness of a standard (delay setting). Surface condition problems can cause these types of instruments to yield misinformation. Often times a digitized typical A-scan presentation is presented with the LCD digital display to better inform the operator of test conditions. Two typical digital thickness gauges are illustrated in Figure 3-22.

3-40

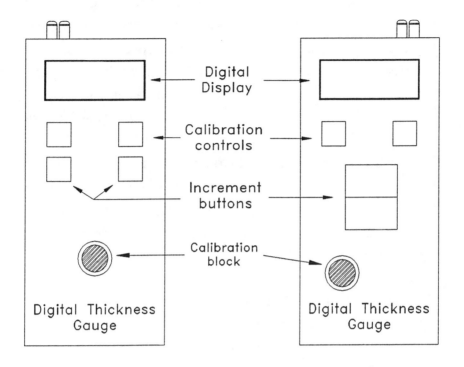

Figure 3-22. Digital Thickness Gauges

Computer-Based Ultrasonics

- General

    There are many types of components that can be utilized to design and assemble a system. As shown in Figure 3-23, many ultrasonic systems are composed of a personal computer, a high-resolution color monitor, modular ultrasonic circuit components, a scanner and scanner controller, an immersion tank outfitted with a carriage and bridge, and a search unit manipulator, and often, an in-tank turntable.

Generally probes mounted in scanners similar to the one on the following page are capable of three axes of movement. They can move in the X, Y, and Z (up and down) directions individually or simultaneously. Specialized systems can provide movement in five axes and utilize robotics to load and unload test articles. These systems can be "trained" to move at a constant water-path distance from the surface of a curved object such as a large-diameter vessel.

Most of these scanners are driven by stepper motors, but some can be operated manually if necessary. The stepper motors can provide small enough increments of change (0.01" which is 0.25 mm or less) to essentially guarantee complete coverage of the test object, regardless of the probe chosen.

One can quickly see that the possibilities for systems is limited only by the imagination and the availability of resources. Manufacturers will gladly supply the potential client with catalogs. They would surely entertain customizing a system for your application should you prefer.

Many systems can display the A-scan, B-scan, and C-scan information simultaneously. Such a feature is a useful aid to the data analysts.

The ultrasonic systems described so far have been associated with immersion systems. It must be said that many of the same data presentations and displays are an integral part of systems utilizing portable scanners such as those used for the testing of pipe, vessels, storage tanks, and similar components. Much NDT done in the field is on as-found components, so it is important that we adapt our test apparatus to the actual component configurations.

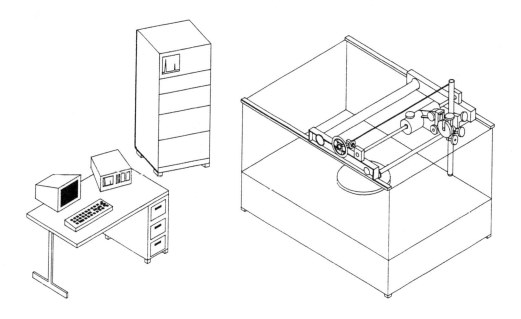

Figure 3-23. Computer-Based System

Ultrasonic systems and scanners, their design, attachments and drive mechanisms range from simple to complex. Consequently, so do their costs. It is beyond the scope of this text to describe the various types of equipment available. There are organizations that will design a specialized system to your specifications or perform application analysis.

CHAPTER 4: TECHNIQUE AND APPLICATIONS

TABLE OF CONTENTS

Page

GENERAL 4-1
TESTING PROCEDURES 4-1
    General 4-1
    Calibration of the Test System 4-2
    Performance of the Test 4-2
    Interpretation of Results 4-3
IMMERSION TESTING 4-3
    General 4-3
    Immersion Techniques 4-4
    Bubbler Techniques 4-5
    Wheel Transducer Techniques 4-6
ULTRASONIC TANK AND BRIDGE/MANIPULATOR 4-7
    General 4-7
    Ultrasonic Tank 4-7
    Bridge/Manipulator 4-8
TYPICAL CALIBRATION PROCEDURE 4-9
    General 4-9
    Area Amplitude Check 4-11
    Distance Amplitude Check 4-13
    Transducer Check 4-15
PREPARATION FOR TESTING 4-18
    General 4-18
    Frequency Selection 4-19
    Transducer Selection 4-20
    Reference Standards 4-20
INTERPRETATION OF TEST RESULTS 4-21
    General 4-21

TABLE OF CONTENTS (CONT'D.)

    Typical Immersion Test Indications 4-22
    Typical Contact Test Indications 4-31
TIP DIFFRACTION 4-42
    General 4-42
ANGLE BEAM FLAW LOCATOR 4-44

LIST OF FIGURES

Figure		Page
4-1	Bubbler and Wheel Transducer Techniques	4-4
4-2	Water-Path Distance Adjustment	4-5
4-3	Stationary and Moving-Wheel Transducer	4-6
4-4	Wheel Transducer Angular Capabilities	4-7
4-5	Bridge/Manipulator	4-8
4-6	Typical Area Amplitude Response Curve	4-13
4-7	Typical Distance Amplitude Response Curve	4-15
4-8	Transducer Axial Distance Amplitude Characteristics	4-17
4-9	Transducer Beam Pattern	4-18
4-10	Mockup with Nonrelevant Indications	4-22
4-11	Force-Oriented Discontinuity Indication	4-24
4-12	Amplitude Range of 1/64 to 8/64 Inch Flat-Bottomed Holes	4-25
4-13	Large Discontinuity Indication	4-26
4-14	Reduced Back Reflection Due to Porosity	4-27
4-15	Nonrelevant Indication from Contoured Surface	4-29
4-16	Grain Size Indications	4-31
4-17	Long and Short Pulse Effects on Display	4-33
4-18	Typical Contact Test Discontinuity Indications	4-33

LIST OF FIGURES (CONT'D.)

4-19	Effect of Lamination on Back Surface Reflection Multiples	4-34
4-20	Weld Indications Using Angle Beam Contact Techniques	4-35
4-21	Porosity and Slag Indications in Weld Groove	4-36
4-22	Surface Crack Indication Using Angle Beam Technique	4-37
4-23	Two-Transducer Indications	4-37
4-24	Indications of Near Surface Discontinuity	4-38
4-25	Coarse Grain Indications	4-38
4-26	Nonrelevant Indication from Cylindrical Specimen	4-39
4-27	Nonrelevant Indication from Long Bar Specimen	4-39
4-28	Nonrelevant Surface Wave Edge Reflection	4-40
4-29	Nonrelevant Indication from Plastic Wedge	4-41
4-30	Nonrelevant Indication from Defective Transducer Crystal	4-42
4-31	Tip Diffraction Concept	4-43
4-32	Corner Trap and Tip Diffracted Indications	4-44
4-33	Typical Angle Beam Inspection	4-45
4-34	Sound-Path Angles for 0.375 Inch (9.5 mm) Thick Plate	4-46
4-35	Transparency Sheet for 0.375 Inch (9.5 mm) Thick Plate	4-46
4-36	Sound-Path Angles for Use With 0.50 Inch (13 mm) Thick Plate	4-47
4-37	Transparency Sheet for 0.50 Inch (13 mm) Thick Plate	4-47
4-38	Sound-Path Angles for Use With 0.625 Inch (16 mm) Thick Plate	4-48
4-39	Transparency Sheet for 0.625-Inch (16 mm) Thick Plate	4-48

CHAPTER 4

TECHNIQUE AND APPLICATIONS

General

Ultrasonic testing, like other NDT testing techniques, follows a set pattern of events that are designed to give consistent results. Ultrasonic testing consists of the following basic steps.

- Calibration of the test system
- Performance of the test
- Interpretation of results

The paragraphs that follow outline these basic steps.

Testing Procedures

- General

    The type of test and technique required for a particular component is usually specified in a test procedure that tells the operator the type discontinuities one is to look for, the type of test required to locate the discontinuities, the definition of limits of acceptability and gives other basic facts pertinent to the test. Either the contact or immersion technique can be utilized in the test procedure. It is the responsibility of the operator to follow the procedure. The paragraphs that follow are intended to familiarize the operator with basic steps that are required to conduct satisfactory ultrasonic tests.

Prior to any ultrasonic test it is important to verify that the test instrument is internally (or electronically) calibrated. This ensures the proper performance of the test instrument and the linearity of the response of the instrument to discontinuities of different sizes and sound-path distances.

- Calibration of the Test System

    Calibration of the testing system is a vital step in the test procedure. Calibrating is the adjustment of the equipment controls so that the operator can be sure that the instrument will detect the discontinuities one is expected to find. Calibrating the system consists of setting up the instrument system as it is to be used in the test and adjusting the controls to give an adequate response to discontinuities of known size and sound-path distance in specific reference standards. The type, size, and position of the artificial discontinuities are specified in the test specification.

- Performance of the Test

    Once the ultrasonic system is calibrated, the actual testing can begin. Limited access to the instrument controls are permitted during actual testing since adjusting certain controls negate the calibration and may require recalibration.

    Ultrasonic tests are accomplished with one of two basic techniques: contact or immersion testing. In contact testing, the transducer is placed in direct contact with the test specimen with a thin liquid film used as a couplant. On some contact units, plastic wedges, wear plates, or flexible membranes are mounted over the face of the search unit. The display from a contact search unit usually shows the initial pulse and the front surface reflection very close together. In immersion testing, a waterproof transducer is used at some distance from the test specimen and the ultrasonic beam is transmitted into the material through a water path or column. Because of the reduced velocity of ultrasound in water, the water distance appears on the display with a fairly wide space between the initial pulse and the front surface reflection. The particular procedure to be used is specified in the test procedure.

- Interpretation of Results

    Once the ultrasonic test has been performed, the results must be interpreted. Many factors must be taken into consideration in interpreting the results of an ultrasonic test. These are outlined later in this chapter.

Immersion Testing

- General

    Any one of several techniques may be used in immersion testing: (a) the immersion technique where both the transducer and the test specimen are immersed in water; (b) the bubbler or squirter technique where the sound beam is transmitted through a column of flowing water; and (c) the wheel transducer technique where the transducer is mounted in the axle of a liquid-filled tire that rolls on the test surface. An adaptation of the wheel transducer technique is a unit with the transducer mounted in the top of a water-filled tube. A flexible membrane on the lower end of the tube couples the unit to the test surface. In all of these techniques, a further refinement is the use of focused transducers that concentrate the sound beam (much like light beams are concentrated when passed through a magnifying glass). The bubbler and wheel transducer techniques are illustrated in Figure 4-1.

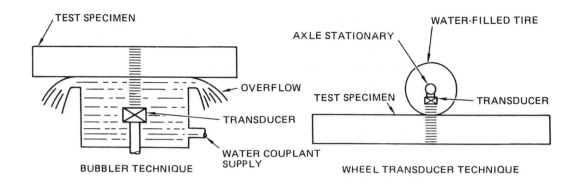

Figure 4-1. Bubbler and Wheel Transducer Techniques

- Immersion Techniques

    In the immersion technique, both the transducer and the test specimen are immersed in water. The sound beam is directed through the water into the material using either a straight beam technique for generating longitudinal waves or one of the many angle beam techniques for generating shear waves in the test article. In many automatic scanning operations, focused beams are used to detect near surface discontinuities or to define minute discontinuities with the concentrated sound beam.

    The transducers usually used in immersion testing are straight beam units that accomplish both straight and angle beam testing through manipulation and control of the sound beam direction. The water-path distance must be considered in immersion testing. This is the distance between the face of the transducer and the surface of the test specimen. This distance is usually adjusted so that the time required to send the sound beam through the water is greater than the time required for the sound to travel through the test specimen. When done properly, the second front surface reflection will not appear in the display between the first front and first back surface reflections. In water, sound velocity is about 1/4 that of aluminum or steel; therefore, 1 inch (25.4 mm) of water path will appear on the monitor as equal to 4 inches (101.6 mm) of sound path in steel. As a rule of thumb, position the transducer so that the water distance is equal to 1/4 the thickness of the part plus 1/4 inch (6.4 mm). The correct water-path distance is particularly important when the test area shown in the display is gated for automatic signaling and recording operations. The water-path distance is carefully set to clear the test area of unwanted signals that cause confusion and possible misinterpretation. Figure 4-2 illustrates the relationship between the actual water path and the display.

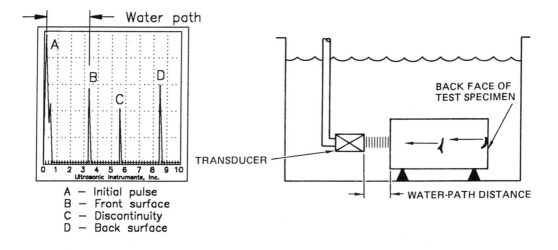

Figure 4-2. Water-Path Distance Adjustment

- Bubbler Techniques

The bubbler technique is a variation of the immersion method. In the bubbler technique the sound beam is projected through a water column into the test specimen. The bubbler is usually used with an

4-5

automated system for high-speed scanning of plate, sheet, strip, cylindrical forms, and other regularly-shaped parts. The ultrasound is projected into the material through a column of flowing water, and is directed in a normal direction (perpendicular) to the test surface to produce longitudinal waves or is adjusted at an angle to the surface to produce shear waves.

- Wheel Transducer Techniques

The wheel transducer technique is an aspect of the immersion method in that the ultrasound is projected through a water-filled tire into the test specimen. The transducer, mounted in the wheel axle, is held in a fixed position, while the wheel and tire rotate freely. The wheel may be mounted on a mobile apparatus that runs across the material, or it may be mounted on a stationary fixture where the material is moved past it. Figure 4-3 illustrates the stationary and the moving-wheel transducer. The position and angle of the transducer mounting on the wheel axle may be constructed to project straight beams as shown in Figure 4-3, or to project angled beams as shown in Figure 4-4.

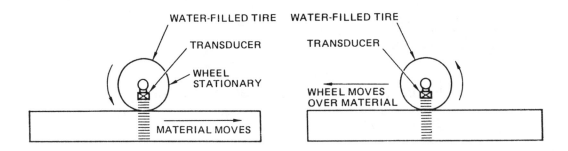

Figure 4-3. Stationary and Moving-Wheel Transducer

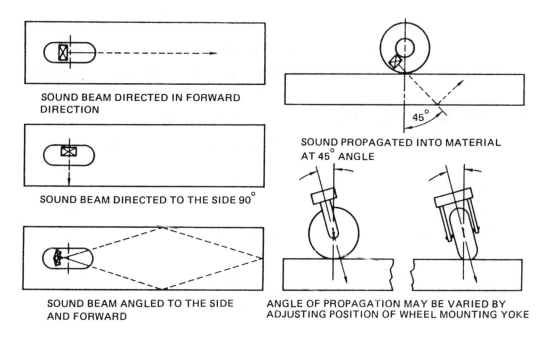

Figure 4-4. Wheel Transducer Angular Capabilities

Ultrasonic Tank and Bridge/Manipulator

- General

    Ultrasonic tanks and bridge/manipulators are necessary equipment for high-speed scanning of immersed test specimens. Modern units consist of a bridge and manipulator mounted over a fairly large water tank. Drive power units move the bridge along the tank side rails while traversing power units move the manipulator from side to side along the bridge. Most of these units are automated, although they can be manually operated.

- Ultrasonic Tank

    The ultrasonic tank may be of any size or shape required to accommodate the test specimen. Coverage of the specimen by

1 foot (30.5 cm) or more of water is usually sufficient. Adjustable brackets and rotational turntables are provided on the tank bottom for support and rotation of the test specimen. The water couplant in the tank is clean, deaerated water containing a wetting agent and corrosion inhibitor. For operator comfort, the water temperature is usually maintained at 70°F (21°C) by automatic controls.

- Bridge/Manipulator

The bridge/manipulator unit is primarily intended to provide a means of scanning the test specimen with an immersed transducer. The stripped-down version in Figure 4-5 has a bridge with a carriage unit at each end so the bridge may be easily moved along the tank side rails.

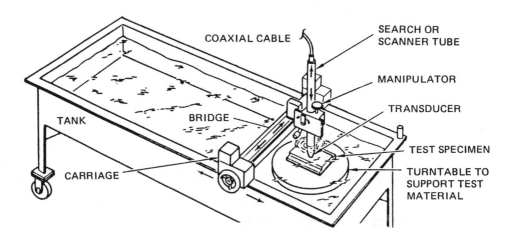

Figure 4-5. Bridge/Manipulator

The manipulator is mounted on a traversing mechanism which allows movement of the manipulator from side to side. The traversing mechanism is an integral component of the bridge assembly. The search tube is usually held rigid, as illustrated, at right angles to the surface of the test specimen. Locking knobs are provided on the manipulator to allow positioning of the search tube in two planes for angle beam testing. When the equipment is automated, electric motors are added to power the bridge carriage, the traversing mechanism, and the up/down movement of the search tube. The pulse-echo unit and the recording unit may also be mounted on the bridge with all power cords secured overhead to allow movement of the bridge along the full length of the tank.

Typical Calibration Procedure

Frequent periodic factory (electronic) calibration of ultrasonic testing units is required to establish the linearity of displayed indications and to ensure proper instrument performance. Once the equipment is electronically calibrated to known standards, the operator may confidently adjust or calibrate the unit to the values of the test material. Once the test unit is electronically calibrated and calibrated to known standards, the operator can expect an accurate display of discontinuities within the test sample. When acceptance of the test sample is based on a rigid test procedure, considerable attention is given to calibration of the instrument system. Calibrating the instrument is accomplished through the use of special test blocks sometimes referred to as standard reference blocks. These blocks are made of the same material as the test sample and match the acoustical properties and dimensions of the test sample as closely as possible. Additionally, real flaws are often implanted in a mockup of the component to replicate anticipated discontinuities.

- General

    A typical calibration procedure is outlined in the paragraphs that follow. The procedures assume conditions and equipment as follows:

- Test Instrument

  Any of several commercially available pulse-echo ultrasonic testing instruments.

- Transducer

  An immersion transducer of 3/8-inch (9.5 mm) diameter with an operational frequency of 15 MHz.

- Power Source

  AC line voltage with regulation ensured by a voltage-regulating transformer.

- Immersion Tank

  Any container that holds couplant and is large enough to allow accurate positioning of the transducer and the calibration block is satisfactory.

- Couplant

  Clean, deaerated water is used as a couplant. The same water, at the same temperature, is used when comparing the responses from differing reference blocks.

- Bridge and Manipulator

  The bridge is strong enough to support the manipulator and rigid enough to allow smooth, accurate positioning of the transducer. The manipulator adequately supports the transducer and provides fine angular adjustment in two vertical planes normal to each other and in the Z-axis (up and down) direction.

- Reference Blocks

  An area amplitude set and a distance amplitude set of reference blocks are required. A basic set which combines both area and distance responses may be used; for example, the ASTM basic set consisting of 10 reference blocks.

- Fundamental Reference Standard
  When calibrating area amplitude responses of the test set, an alternate to the reference blocks described in the preceding step is the ASTM set of 15 steel balls, free of corrosion and surface marks and of ball-bearing quality, ranging in size from 1/8-inch (3.2 mm) diameter to 1-inch (25.4 mm) diameter in 1/16-inch (1.6 mm) increments. A suitable device, such as a tee pin, is necessary to hold each ball.

- Area Amplitude Check

The linear range of the instrument is determined by obtaining the ultrasonic responses from each of the area amplitude type reference blocks (the steel balls may be used as an alternate for the reference blocks) as follows:

a. Place a No. 5 area amplitude reference block (a block containing a 5/64-inch-diameter hole) in the immersion tank with the drilled hole down. Position the transducer over the upper surface of the block, slightly off-center, at a water-path distance of 3±1/32 inches (76.2±0.8 mm) between the face of the transducer and the surface of the block. This accurate distance is obtained by using a gauge between the block and the transducer.

b. Adjust the transducer with the manipulator to obtain a maximum indication height from the front surface reflection of the block. This indication assures that the ultrasound is perpendicular to the top surface of the block. A maximum number of back surface reflection indications serves the same purpose.

c. Move the transducer laterally until the maximum response is received from the flat-bottomed hole.

d. Adjust the instrument gain control until the hole indication height is 31 percent of the maximum obtainable. Do not repeat this step for the remaining blocks in the set.

e. Replace that reference block with each of the other blocks in the set. Repeat steps "b" and "c" for each block and record the indications. Maintain a water-path distance of 3 inches (76.2 mm) for each block except for the No. 7 and No. 8 blocks which require a water distance of 6 inches (152.4 mm).

f. Plot a curve of the recorded indications as in Figure 4-6. In the example, the points where the "curve" of responses deviates from the ideal linearity line defines the limit of linearity in the instrument. Amplitudes plotted below the "limit of linear response" (in this example) are in the linear range of the instrument and no correction is required. Amplitudes of indications above the limit of linear response are in the non-linear range and are increased to the ideal linearity curve. This is done by projecting a vertical line upward from the actual height (AH) of indication until the ideal linearity curve is intersected. The point of intersection defines the correct height (CH) of indication in percent of maximum amplitude that the instrument can display. The difference between the corrected height (CH) and the actual height (AH) is the correction factor (CF). For each indication that appears in the non-linear range a different correction factor (CF) is plotted because the deviation is not constant.

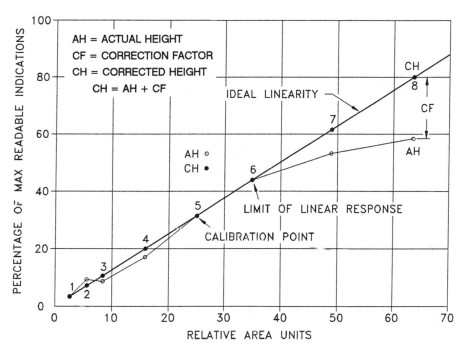

Figure 4-6. Typical Area Amplitude Response Curve

When the actual indication height is displayed, the corrected indication height is computed by adding the correction factor directly to the actual indication height as follows.

$$AH + CF = CH$$

- Distance Amplitude Check

The distance amplitude characteristics of the instrument are determined by obtaining the ultrasonic responses from each of the reference blocks in a set of blocks of varying sound-path distance with a 5/64 inch diameter hole in each block. The resultant

4-13

indications are recorded on a curve as outlined in the following procedure:

a. Select a reference block containing a 5/64 inch flat-bottomed hole with a sound-path distance of 3 inches (76.2 mm) from the top surface to the hole bottom and place it in the immersion tank. Position the transducer over the upper surface of the block, slightly off-center, at a water distance of 3 inches (76.2 mm) between the face of the transducer and the surface of the block. Adjust this distance accurately, within a tolerance of ±1/32 inch (±0.8 mm), by using a gauge between the block and the transducer.

b. Adjust the transducer with the manipulator to obtain a maximum indication height from the front surface reflection of the block. This indication assures that the sound beam is perpendicular to the top surface of the block. A maximum number of back surface reflections serves the same purpose.

c. Move the transducer laterally until the maximum response is received from the flat-bottomed hole. Adjust the instrument gain control until the indication height is 50 percent of the maximum obtainable.

d. Replace that reference block with each of the other blocks in the set. Repeat steps "b" and "c" for each block and record the indications. Maintain water-path distance of 3 inches (76.2 mm) for each block except if the basic set is being used. A water-path distance of 6 inches (152.4 mm) is required for the block containing an 8/64 inch diameter hole with a sound-path distance of 3 inches (76.2 mm).

e. Plot a "curve" of the recorded indications as in Figure 4-7. In the example shown, the near zone extends from the 1/2 inch (12.7 mm) sound-path distance indication to the 2 inch (50.8 mm) sound-path distance indication. As the sound-path distance increases beyond 2 inches (50.8 mm), the indications attenuate, or decrease, in height.

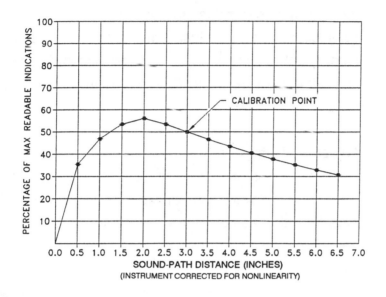

Figure 4-7. Typical Distance Amplitude Response Curve

- Transducer Check

To improve accuracy during test equipment calibration, the characteristics of the transducer, as modified or distorted by the test instrument, may be determined by recording a distance amplitude curve from a 1/2 inch diameter (12.7 mm) steel ball immersed in water. A beam pattern or plot can also be obtained from the same steel ball at a fixed water distance of 3 inches (76.2 mm). It is well to remember that the curve and beam plot recorded in this procedure are not valid if the transducer is subsequently used with any test instrument other than the one used in this procedure. A complete analysis of transducer characteristics cannot be accomplished with the commercial ultrasonic testing equipment used in this procedure. To ensure maximum accuracy, the transducer may be calibrated with special equipment. This information is outlined in Chapter 5: Calibrating Transducers. In the procedure that follows, the apparatus used for checking the transducer is the same as that prescribed in the previous paragraphs for calibrating the instrument with reference blocks. The manipulator is set to allow a range in water distance of 0 to at least 6 inches (152.4 mm) from the face of the transducer to the ball surface.

a. Adjust the instrument gain control until the indication height is 50 percent of the maximum obtainable with the transducer positioned at a water-path distance of 3±1/32 inch (76.2±0.8 mm) from the face of the transducer to the top surface of the ball. Exercise care in producing a true maximum indication by locating the transducer beam center on the center of the ball. Record the maximum indication. Do not readjust the instrument gain control in this or succeeding steps of the procedure.

b. Vary the water-path distance in 1/8 inch increments through a range of 1/4 inch (6.4 mm) to 6 inches (152.4 mm). Record the maximum indication for each increment of water distance, using care each time the transducer is moved back that the beam center remains centered on the ball.

c. As in Figure 4-8, plot the recorded indications (corrected for any non-linearity) on a graph to demonstrate the axial distance amplitude response of the transducer and particular test instrument used in the test. The curve for an acceptable transducer is similar to the curve illustrated in Figure 4-11. It is important that the peaks in the curve occur at water distances of 1-1/4, 1-3/4 and 3 inches (31.8, 44.5, 76.2 mm) as shown. The allowable deviation in water distance for the occurrence of these peaks is 1/16 inch (1.6 mm).

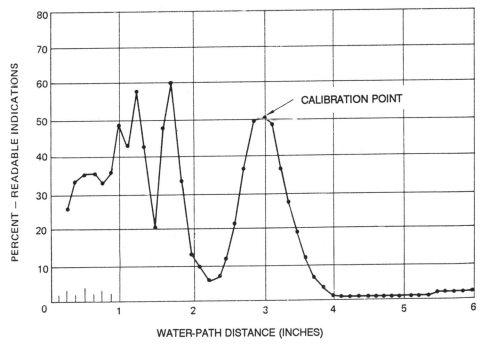

Figure 4-8. Transducer Axial Distance Amplitude Characteristics

d. Determine the transducer beam pattern by relocating the manipulator to obtain a 3±1/32 inch (76.2±0.8 mm) water-path distance from the 1/2 inch diameter (12.7 mm) steel ball to the face of the transducer. While scanning laterally, 3/8 inch (9.5 mm) total travel, the height of the indication from the ball is observed while the transducer passes over the ball. Three distinct lobes or maximums are observed. The symmetry of the beam is checked by making four scans and displacing each scan by rotating the transducer in its mounting 45°. The magnitude of the side lobes should not vary more than 10 percent about the entire perimeter of the sound beam. Generally, an acceptable transducer would produce a symmetrical beam profile which has side lobes with magnitudes no less than 20 percent nor more than 30 percent of the magnitude of the center lobe. The beam pattern or plot of an acceptable transducer is illustrated in Figure 4-9.

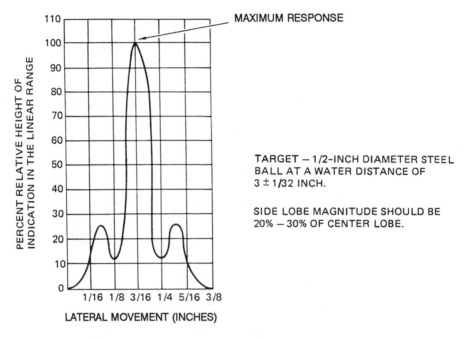

Figure 4-9. Transducer Beam Pattern

Preparation for Testing

- General

Ultrasonic test preparations begin with an examination of the test specimen to determine the appropriate technique. Components are then selected from available equipment to perform the test. Many variables affect the choice of technique. For example, the test specimen may be too large to fit in the immersion tank. In the case of large, fixed structures, the testing unit is moved to the test site. This may require portable testing equipment. Other factors are the number of parts to be tested, the nature of the test material, test surface roughness, methods of joining (welded, bonded, riveted, etc.), and the shape of the specimen. If the testing program covers a large number of identical parts and a permanent test record is desirable, an immersion technique with automatic scanning and recording may be suitable. One-of-a-kind or odd-lot jobs may be tested with portable contact testing units. Each case will require some study as to the most practical and efficient technique.

When setting up any test, an operating frequency is selected, a transducer is chosen, and a reference standard is established. The test specimen is carefully studied to determine its most common or probable discontinuities. For example, in forgings, laminar discontinuities are found parallel to the forging flow lines. Discontinuities (i.e., laminations) in plate are usually parallel to the plate surface and elongated in the rolling direction. Whenever possible, a mockup is manufactured to implant real flaws at critical locations in a specimen to replicate the intended inspection.

- Frequency Selection

High test frequencies are an advantage in immersion testing. In contact testing, 10 MHz is usually the maximum frequency. Low frequencies permit penetration of ultrasound to a greater depth in the material, but may cause a loss of near surface resolution and sensitivity. A sample test specimen is used to evaluate ultrasound penetration with a high-frequency transducer (10 to 25 MHz for immersion and 5 to 10 MHz for contact) and to observe the total number of back reflections. If there is no back surface indication, a lower frequency is required. Successively lower frequencies are applied until several back surface reflections are obtained. If near surface resolution is required, it may be necessary to turn the part over and retest from the opposite side, or to use a high-frequency search unit following the low-frequency scan.

- Transducer Selection

Transducer selection is largely governed by the optimum frequency, as determined in the previous paragraph.

In immersion testing, other considerations include the possibility of using a paintbrush transducer for high-speed scanning to detect gross discontinuities or the possibility of using a focused transducer for greater sensitivity in detecting small discontinuities in near surface areas (no deeper than 2 inches or 50.8 mm). Note that with a given transducer diameter, beam spread decreases as the frequency increases. For example, of two 3/8 inch diameter (9.5 mm) transducers, one 10 MHz and the other 15 MHz frequency, the 15 MHz unit has less beam spread. In contact testing, angle beam units are used for testing welds and relatively thin material.

- Reference Standards

Commercial ultrasonic reference standards have been previously described. These standards are adequate for many test situations, provided the acoustic properties are matched (or nearly matched) between the test specimen and the reference standard. In most cases, responses from discontinuities in the test specimen are likely to differ from the indications received from the reflector in the standard. For this reason, a sample test specimen is sectioned, subjected to metallurgical analysis, and studied to determine the nature of the material and its probable discontinuities. In some cases, artificial discontinuities in the form of holes or notches are introduced into the sample to serve as a basis for comparison with discontinuities found in specimens. For critical applications, a mockup can be manufactured to implant real flaws (cracks, slag inclusions, porosity, etc.) to replicate the component being examined. The implanted flaws usually have a known size (±0.040 inch or 1 mm) and are placed at strategic locations in the mockup. The UT operator is then required to distinguish the actual defects from geometric reflectors. From these studies, an acceptance level is determined that establishes the number and magnitude of discontinuities allowed in the component being examined. A sensible testing program is then established by an intelligent application of basic material failure theory.

Interpretation of Test Results

- General

    Ultrasonic test indications from subsurface discontinuities within the test specimen can be related or compared to indications from reference reflectors of varying depths or sizes in standard test blocks or mockups. These comparisons are a fairly accurate means of evaluating the size, shape, position, and orientation of discontinuities. These conditions, and the discontinuities themselves, are sometimes the cause of ultrasonic phenomena which are difficult to interpret. This type of difficulty can only be resolved by relating the ultrasonic indications to the probable type of discontinuity with reference to the test conditions. Impedance of the material, surface roughness, surface contour, attenuation, and angle of incidence are all to be considered when evaluating the size and location of an unknown discontinuity. The simplest method is to compare the indication of the discontinuity with indications from a reference standard or mockup similar to the test specimen. The experienced operator also learns to discriminate between the indications of actual defects and false or *nonrelevant* indications, as illustrated in Figure 4-10.

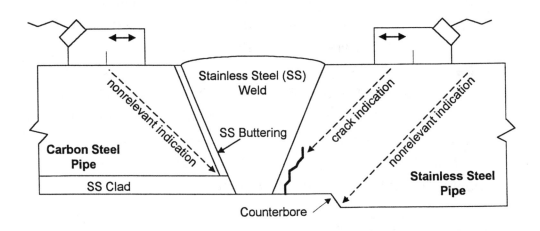

Figure 4-10. Mockup with Nonrelevant Indications

- Typical Immersion Test Indications

    Immersion test indications, generally displayed on A-scan pulse-echo units, are interpreted by analysis of three factors: the amplitude of the reflection from a discontinuity, the loss of back surface reflection, and the distance of a discontinuity from the sound entry surface of the article. Individual discontinuities that are small, compared with the transducer's beam spread usually profile the beam (as previously discussed in Transducer Check). Discontinuities larger than the beam spread are evaluated by noting the distance the probe is moved over the test specimen while an indication is maintained. In this case, the amplitude has no quantitative meaning. The length of time the amplitude is maintained does indicate the extent of the discontinuity in one plane. A loss or absence of back surface reflection is evidence that the transmitted sound has been absorbed, refracted, or reflected so that the energy has not returned. Evaluating this loss does not determine the extent of the discontinuity (except when using through-transmission).

When relatively large discontinuities are encountered, the discontinuity may eliminate the back surface reflection, since the sound beam is not transmitted through the discontinuity. Remember though, since the surface of the test specimen and the surface of a discontinuity within it are not as smooth as the surface of the test block and the flat-bottomed hole in the test block, the estimated size of the discontinuity is generally a bit smaller than the actual size.

- Small Discontinuity Indications

    A significant number of the discontinuities encountered in ultrasonic testing of wrought aluminum are relatively small. Foreign materials or porosity in the cast ingot are rolled, forged, or extruded into wafer-thin discontinuities during fabrication. The forces used in fabrication tend to orient the flat plane of the discontinuity parallel to the surface of the part. Such a discontinuity and its ultrasonic indication are illustrated in Figure 4-11.

    The relationship of the discontinuity indication and its amplitude is determined by comparison with a range of test block flat-bottomed hole reflections as illustrated in Figure 4-12.

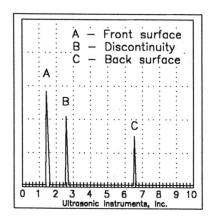

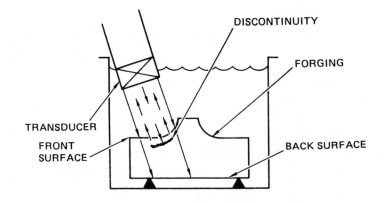

Figure 4-11. Force-Oriented Discontinuity Indication

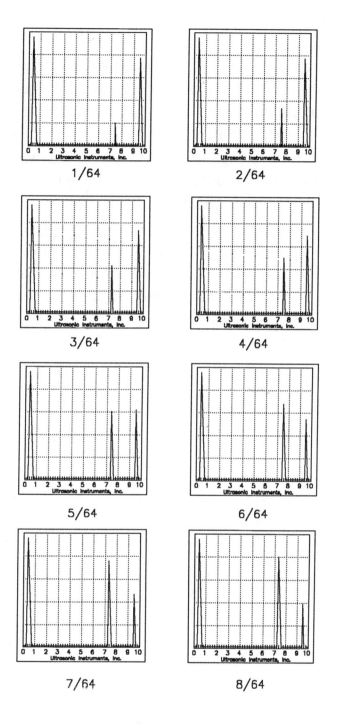

Figure 4-12. Amplitude Range of 1/64- to 8/64-Inch Flat-Bottomed Holes

4-25

- Large Discontinuity Indications

    Discontinuities that are large, when compared with the beam spread, usually produce a display as in Figure 4-13. Since the discontinuity reflects nearly all of the ultrasound energy, the partial or total loss of back surface reflection is typical. The dimensions of the discontinuity may be determined by measuring the distance that the transducer is moved while still receiving an indication. If the discontinuity is not flat but is three-dimensional, the extent of the third dimension may be determined by turning the article over and scanning from the back side. If the possibility of two discontinuities lying close together is suspected, the article may be tested from all four sides if possible.

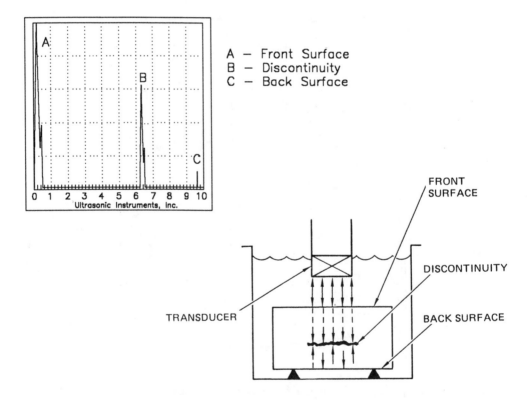

Figure 4-13. Large Discontinuity Indication

- Loss of Back Surface Reflection

Evaluating the loss of back surface reflection is most important when it occurs in the absence of significant individual discontinuities. In this case, among the causes of reduction or loss of back surface reflection are large grain size, porosity, and a dispersion of very fine precipitate particles. Figure 4-14 illustrates the indications received from a sound test specimen and the indications displayed from a porous specimen. Note that the back surface reflections obtained from the porous specimen are reduced considerably.

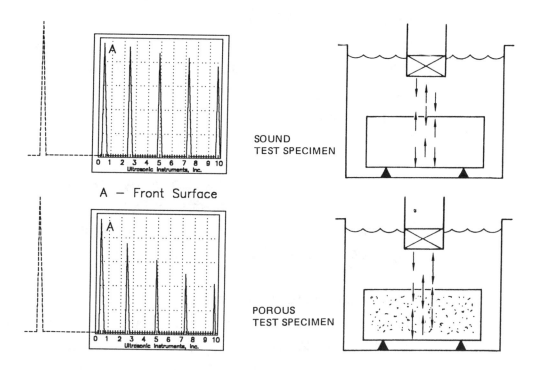

Figure 4-14. Reduced Back Reflections Due to Porosity

- Nonrelevant Indications

    When considering indications that may be nonrelevant, it is a good rule to be suspicious of all indications that are unusually consistent in amplitude and appearance while the transducer is passing over the test specimen. Reflections from fillets and concave surfaces may result in responses appearing between the front and back surfaces. These are sometimes mistaken for reflections from discontinuities. If a suspected indication results from a contoured surface (see Figure 4-15), the amplitude of the indication will diminish as the transducer is moved over the flat area of the front surface. At the same time, the amplitude of the indication from the flat area will increase. Moving the transducer back over the contoured surface will cause the flat area indication to decrease as the amplitude of the suspected signal increases. Where a reflection from an actual discontinuity is strong in localized areas, a nonrelevant indication will tend to be consistent as the transducer is moved along the contoured surface. Reflections which can follow around a contoured surface may be shielded off by interrupting the sound beam with a foreign object as in Figure 4-15. Broad-based indications, as contrasted to a sharp indication, are likely to be reflections from a contoured surface.

    Near the edges of rectangular shapes, edge reflections are sometimes observed with no loss of back surface reflection. This type of indication usually occurs when the transducer is within 1/2 inch (12.7 mm) of the edge of the part.

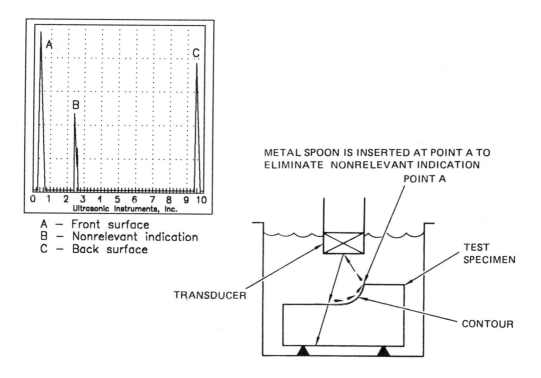

Figure 4-15. Nonrelevant Indication from Contoured Surface

Articles with smooth, shiny surfaces will sometimes give rise to false indications. For example, with a thick aluminum plate machined to a smooth finish, spurious indications which appeared to be reflections from a discontinuity located at about one-third of the article depth were received. As the transducer was moved over the surface of the plate, the indication remained relatively uniform in shape and magnitude. Apparently this type of indication results from surface waves generated on the extremely smooth surface and possibly reflected from a nearby edge. They can be eliminated or minimized by coating the surface with an ultrasonic couplant.

- Angled-Plane (Planar) Discontinuity Indications

    Discontinuities oriented with their principal plane at an angle to the examination surface are sometimes difficult to detect and evaluate. Usually it is best to scan initially at a comparatively high gain setting (high sensitivity) to detect planar discontinuities. Later the transducer is manipulated around the area of the discontinuity to evaluate its magnitude. In this case, the manipulation is intended to cause the sound beam to strike the discontinuity at right angles to its principal plane. With large discontinuities that have a relatively flat, smooth surface but lie at an angle to the surface, the indication moves along the baseline of the display as the transducer is moved. This happens because of the change in sound-path distance. Bursts in large forgings fit this category as they tend to lie at an angle to the surface.

- Grain Size Indications

    Unusually large grain size in the test specimen may produce "hash" or noise indications as illustrated in Figure 4-16. In the same illustration, note the clearer indications received from the same type of material with fine grain. In some cases, abnormally large grain size results in a total loss of back surface reflection. These conditions are usually brought about by prolonged or improper forging temperatures or high temperature during hot working and subsequent improper annealing of the test specimen. Such multisized grain structures are common in austenitic stainless steel materials and often reduce the signal-to-noise ratio significantly.

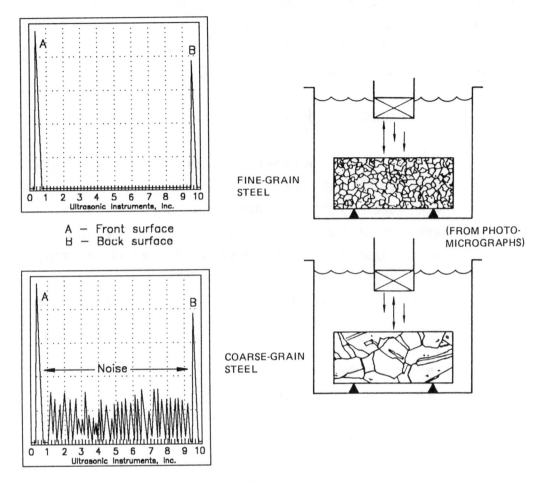

Figure 4-16. Grain Size Indications

- Typical Contact Test Indications

    Contact test indications, in many instances, are similar or identical to those discussed in the previous paragraphs on immersion test indications. Little additional discussion will be given when contact indications are similar to immersion indications. Interference from the initial pulse at the front surface of the test specimen, variations in efficiency of coupling, and poor wedge design in angle beam testing produce nonrelevant effects that are sometimes difficult to recognize in contact testing. As in immersion testing, signal amplitude, loss of back reflection, and distance of the discontinuity from the surfaces of the article are all major factors used in evaluation of the display.

- Dead Zone Indications

    The dead zone is an area directly beneath the front surface from which no reflections are displayed because of obstruction by the initial pulse. In most contact testing, the initial pulse obscures the front surface indication. Near surface discontinuities may be difficult to detect with straight beam transducers because of the initial pulse interference. Shortening the pulse may be effective when near surface discontinuities are obscured by the ringing "tail" of the initial pulse. Figure 4-17 shows a comparison of long and short pulses applied to the test specimen where the discontinuity is near the surface. Only by inserting a standoff, such as a plastic block, or utilizing a dual-element probe can separation of these responses be achieved in contact testing. In immersion testing, the initial pulse is separated from the front surface indication by the water path.

- Typical Discontinuity Indications

    Typical indications encountered in ultrasonic testing include those from discontinuities such as nonmetallic inclusions, seams, forging bursts, cracks, and flaking found in forgings as in Figure 4-18.

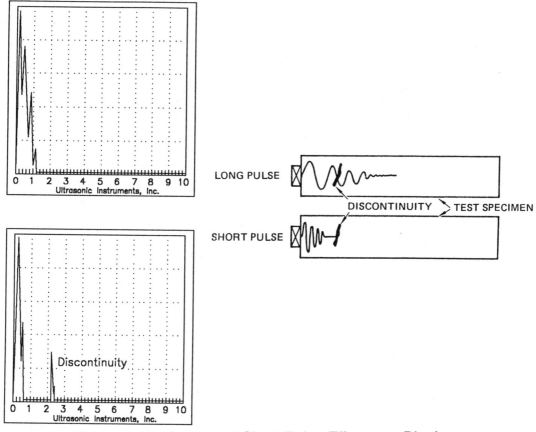

Figure 4-17. Long and Short Pulse Effects on Display

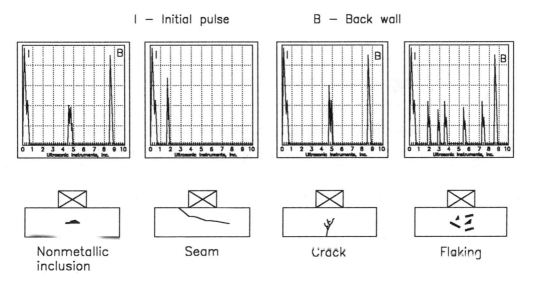

Figure 4-18. Typical Contact Test Discontinuity Indications

Laminations in rolled sheet and plate are defined by a reduction in the distance between back surface reflection multiples as illustrated in Figure 4-19. View A illustrates the display received from a normal plate and View B shows the back surface reflections received when the transducer is moved over the lamination.

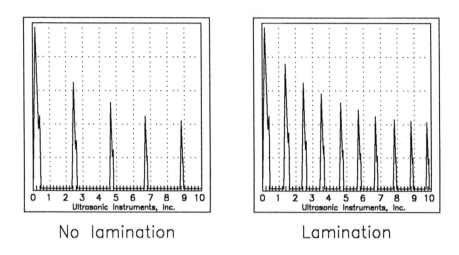

Figure 4-19. Effect of Lamination on Back Surface Reflection Multiples

In angle beam testing of welds, an acceptable weld area usually responds by lack of indications in the display. Only discontinuities would result in indications. However some of these "discontinuities" may be nonrelevant or metallurgical. Such "metallurgical" indications occurring at the weld fusion zones are illustrated in View A of Figure 4-20. View B shows the same reflections for the fusion zones, but in this case a discontinuity is located in the weld.

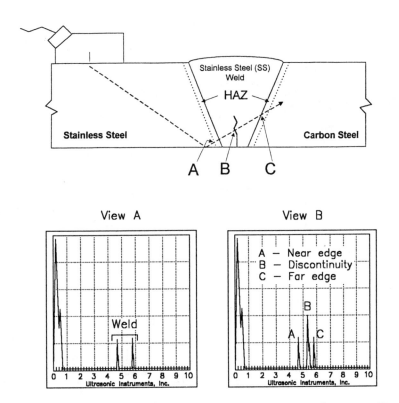

Figure 4-20. Weld Indications Using Angle Beam Contact Techniques

This most often occurs when there are different materials (carbon steel and stainless steel for example) welded together and there are different alloys used in the weldment. The weld itself is essentially a cast material embedded in a mold created by the base metal forming the joint. Therefore, this fusion or heat-affected zone (HAZ) between the base metal being joined and the weld area may cause indications due to reflections of the ultrasound. Depending on the material being joined and the weld filler material, such reflectors may be more pronounced or not present at all. The weld groove commonly has discontinuities such as porosity and slag that produce indications as illustrated in Figure 4-21.

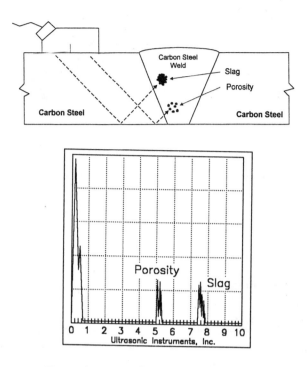

Figure 4-21. Porosity and Slag Indications in Weld Groove

Surface cracks are sometimes detected when testing with a shear wave produced by an angle beam transducer. Figure 4-22 shows a surface wave indication from a crack in the surface of the test specimen.

With pitch-and-catch testing using two transducers, the initial pulse does not interfere with reception as it does when using the single transducer. Figure 4-23 illustrates the indications received from a relatively thin test specimen using two transducers. Tandem or dual angle beam transducers are used to improve near surface resolution. The transit time of the ultrasound when passing through the Lucite wedge on which the transducers are mounted are given an additional advantage in that the initial pulse is moved to the left in the same way the water-path separation occurs in immersion testing. Figure 4-24 shows an indication from a discontinuity which lies only 0.02 inch (0.5 mm) below the sound entry surface of the material.

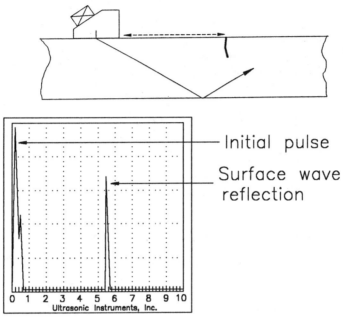

Figure 4-22. Surface Crack Indication Using Angle Beam Technique

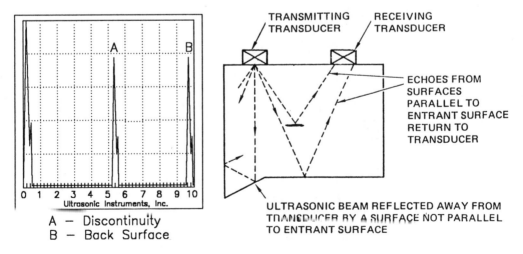

A — Discontinuity
B — Back Surface

Figure 4-23. Two-Transducer Indications

4-37

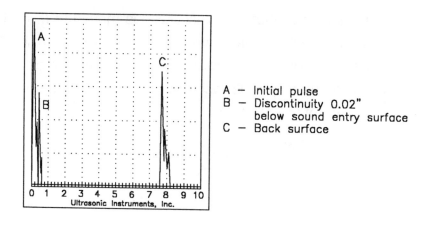

Figure 4-24. Indications of Near Surface Discontinuity

- Nonrelevant Indications

    Coarse-grained material causes reflections or "hash" across the width of the display (as illustrated in Figure 4-25) when the test is attempted at a high frequency. To eliminate or reduce the effect of these unwanted reflections, lower the frequency and change the direction of the sound beam by using an angle beam transducer with a shorter sound-path distance if possible.

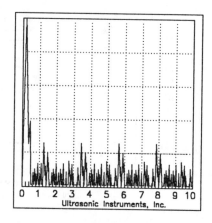

Figure 4-25. Coarse-Grained Indications

When testing cylindrical specimens (especially when the face of the transducer is not curved to fit the test surface), additional indications following the back surface indication will appear as in Figure 4-26.

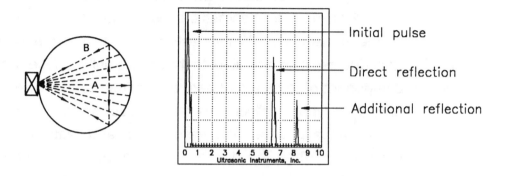

Figure 4-26. Nonrelevant Indication from Cylindrical Specimen

In testing long specimens, mode conversion occurs from the sound beam striking the sides of the test specimen and returning as reflected shear waves as illustrated in Figure 4-27. Changing to a larger diameter (narrower sound beam) transducer will lessen this problem.

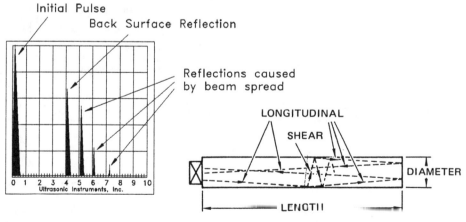

Figure 4-27. Nonrelevant Indication from Long Bar Specimen

4-39

Surface waves generated during straight beam testing also cause unwanted nonrelevant indications when they reflect from the edge of the test specimen as in Figure 4-28. This type of nonrelevant indication is easily identified since movement of the transducer will cause the indication from the surface wave to move across the display with the movement of the transducer. When testing with two angle beam transducers, it is possible to have a small surface wave component of the sound beam transmitted to the receiving unit. This type of reflection is readily recognized by varying the distance between the transducers and watching the indication. When the distance is increased, the apparent discontinuity indication moves away from the initial pulse.

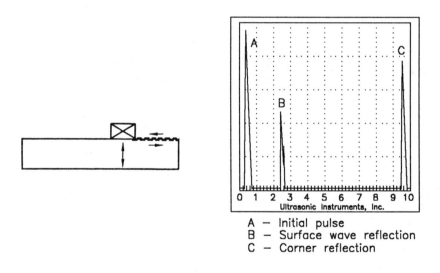

A — Initial pulse
B — Surface wave reflection
C — Corner reflection

Figure 4-28. Nonrelevant Surface Wave Edge Reflection

When using angle beam transducers, a certain amount of unwanted reflections are received from the wedge. These indications appear immediately following the initial pulse as illustrated in Figure 4-29. The reflections from within the wedge are easily identified because they are still present on the display when the transducer is lifted off the test specimen.

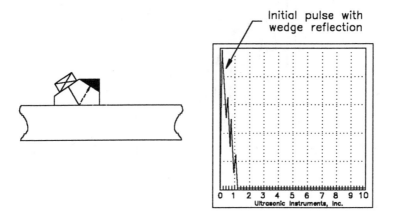

Figure 4-29. Nonrelevant Indication from Plastic Wedge

With continued use, the piezoelectric crystal in the transducer may become defective. When this happens, the indication may be characterized by a prolonged ringing which adds a "tail" to the initial pulse as in Figure 4-30. As the prolonged ringing effect results in a reduced capability of the system to detect discontinuities, the transducer is discarded or repaired.

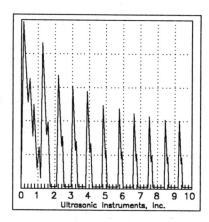

Figure 4-30. Nonrelevant Indications from Defective Transducer Crystal

Tip Diffraction

- General

    "Tip diffraction" is a technique useful for establishing the extremities of cracks that primarily propagate through the thickness of plates, pipes, and vessels. The concept is to take advantage of the signal generated from the tip of the flaw as it diffracts or bends the ultrasound beam passing over or under it. Notice in Figure 4-31 that the bending of the incident beam results in a wave that propagates from the top or tip of the notch. This signal locates the endpoint of the crack and defines its through-wall dimension. Tip diffracted signals can be generated by either the shear or longitudinal wave modes. The use of longitudinal waves complicates the interpretation, as both longitudinal waves and shear waves will be present in the test specimen. This testing approach is known as *bimodal* testing. We will concentrate on contact pulse-echo approaches of the tip diffraction technique.

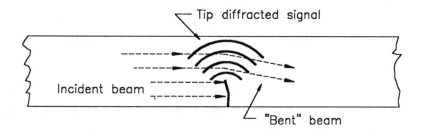

Figure 4-31. Tip Diffraction Concept

A typical application of the technique is in sizing or assessing a crack's height as measured from the inside of the component toward the outside or vice versa. From Figure 4-31 we can see that a signal will be reflected at the base of the crack. Likewise, a tip diffracted signal will be generated at the tip as long as the crack height does not exceed the beam spread. For deeper cracks, we simply move the search unit until we maximize the tip signal; however, characterizing the tip signal in this fashion can be a challenge, even for the most experienced technicians.

Applications of tip diffraction techniques include:

- sizing of under-clad cracking in pressure vessels and piping.
- determination of through-wall dimension of cracks in pipe that has been overlayed with multilayers of weld material.
- determination of the size or diameter of porosity in welds, composites, and ceramics.

Calibration for tip diffraction for through-wall dimensional assessment of inside diameter (ID) connected cracking requires a reference block that contains notches. These notches range from approximately 10 percent to 90 percent through-wall in 5 percent to

10 percent increments. The calibration is done in terms of depth as opposed to sound-path distance. Essentially, as the notch height increases, the time-of-flight for the tip signal decreases; therefore the tip signal arrives earlier in time or closer to the initial pulse.

You realize that the corner trap signal will always peak at the same location in the display, and that is the screen position representing the thickness of the plate or test article. The only indication that will change CRT positions will be the tip signal. It will move closer to the initial pulse as the crack height increases. UT system displays are presented in Figure 4-32 to illustrate the point.

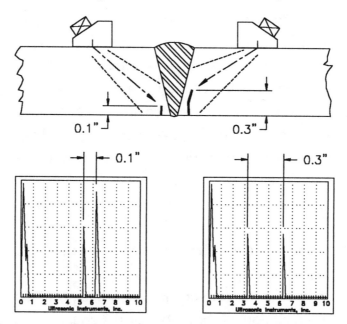

Figure 4-32. Corner Trap and Tip Diffracted Indications

Angle Beam Flaw Locator

It is possible to make a simple, but accurate, "flaw locator" by making a photocopy of the figures on the next few pages. This simple technique will help you understand the concept of angle beam inspection as shown in Figure 4-33. Photocopying Figures 4-34 through 4-39 for individual use is authorized by the publisher. Refer to Appendices E and F for samples of calibration sheets and data recording forms.

With the simple drawing program on your computer, you can generate similar sheets to cover any thickness of material or sound-path angle. Follow these steps when you have access to a calibrated ultrasonic instrument and a specimen with a known flaw.

STEP 1   Obtain maximum UT signal from indication and determine the **surface distance** from exit point of transducer to the weld centerline.

STEP 2   While maintaining the maximum UT signal, determine **sound-path distance** by reading the screen on a properly calibrated UT instrument.

STEP 3   Place transparency on top of the proper sound-path angle (45°, 60° or 70°) and move to match the surface distance in Step 1.

STEP 4   Use a grease pencil to mark the flaw location which will be at the sound-path distance in Step 2.

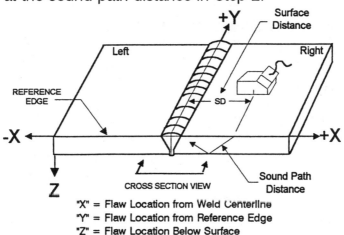

"X" = Flaw Location from Weld Centerline
"Y" = Flaw Location from Reference Edge
"Z" = Flaw Location Below Surface

Figure 4-33. Typical Angle Beam Inspection

4-45

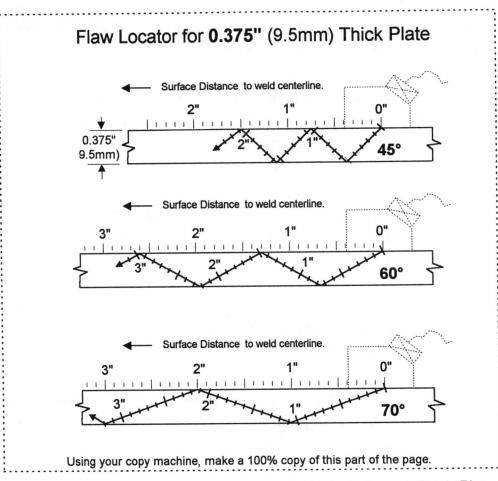

Figure 4-34. Sound-path Angles for 0.375-Inch (9.5 mm) Thick Plate

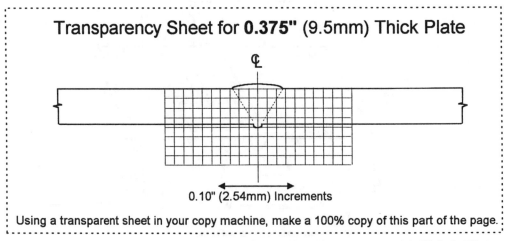

Figure 4-35. Transparency Sheet for 0.375-Inch (9.5 mm) Thick Plate

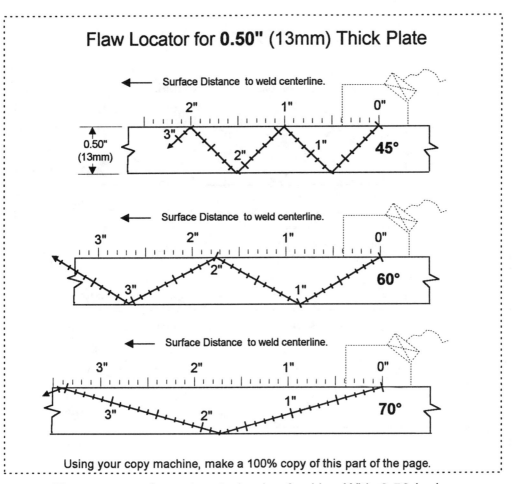

Figure 4-36. Sound-path Angles for Use With 0.50-Inch (13 mm) Thick Plate

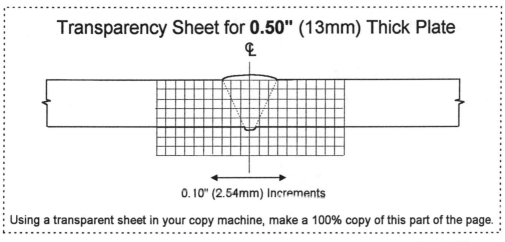

Figure 4-37. Transparency Sheet for 0.50-Inch (13 mm) Thick Plate

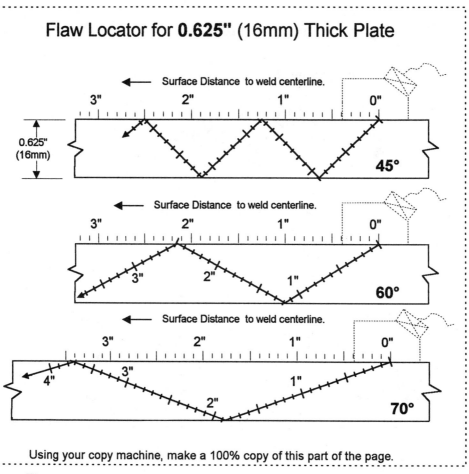

Figure 4-38. Sound-path Angles for Use With 0.625-Inch (16 mm) Thick Plate

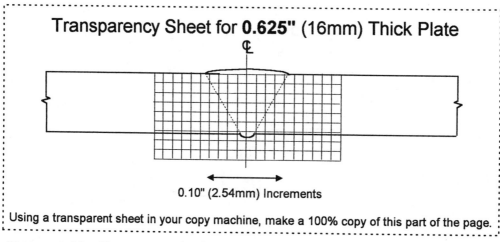

Figure 4-39. Transparency Sheet for 0.625-Inch (16 mm) Thick plate

CHAPTER 5: CALIBRATING TRANSDUCERS

TABLE OF CONTENTS

Page

GENERAL 5-1
GENERAL EQUIPMENT QUALIFICATIONS 5-1
GENERAL CALIBRATING TECHNIQUE 5-2
TRANSDUCER CALIBRATING EQUIPMENT 5-2
    General 5-2
    Test Setup 5-3
    Function 5-3
    Recording Method 5-3
    Manipulative Equipment 5-5
    Reflector Targets 5-5
    Pulser 5-6
    Wide Band Receiver 5-6
    Display System 5-7
RECORDING OF TRANSDUCER BEAM PROFILES 5-7
    General 5-7
    Flat-Disc Transducer Measurements 5-7
    Focused Transducer Measurements 5-9
    Cylindrically-Focused Transducer Measurements 5-10
ANALYSIS OF TRANSDUCER DATA 5-12
    General 5-12
    Waveform 5-12
    Frequency 5-12
    Damping Factor 5-12
    Sensitivity 5-13
    Focal Length 5-14
    Beam Amplitude Profiles 5-14
    Beam Width and Symmetry 5-15

LIST OF FIGURES

Figure   Page

5-1   Equipment Functional Diagram 5-4
5-2   Camera Recording Method 5-4
5-3   Typical Transducer Data Sheet 5-8
5-4   Flat-Disc Transducer Measurement 5-9
5-5   Focused Transducer Measurements 5-10
5-6   Cylindrically-Focused Transducer Measurements 5-11

CHAPTER 5

CALIBRATING TRANSDUCERS

General

Ultrasonic transducers, though identical in appearance and manufactured to the same specification, usually have individual characteristics. Acoustic anomalies may exist because of variations in crystal cutting, areas of poor bond to lens, or backing and misalignment of parts in the transducer assembly.

General Equipment Qualifications

Specialized wide-band transmitting and receiving equipment is required to accurately measure transducer variables. Much of these efforts are conducted via computer software programs available from a variety of sources. We will describe the overall process as performed via semi-automated equipment.

In analyzing transducer characteristics, the crystal is excited by a voltage spike that will not distort the natural mode of operation. The return signals received by the transducer are amplified without distortion and are displayed in a manner that will provide a permanent photographic record. The pages that follow describe the special instrumentation equipment and techniques required for measuring or calibrating the recording transducer characteristics such as frequency, sensitivity damping factor, beam size, beam symmetry, and beam focal distance.

General Calibrating Technique

In general, the transducer calibrating technique consists of scanning a small reflector (a ball bearing, a flat post, or a thin wire) in an immersion tank with the ultrasonic beam. As the transducer is moved over the reflector, a changing response that represents a distance amplitude plot of the beam in profile is produced on the oscilloscope. At the highest amplitude portion of the beam, the return signal waveform is photographically recorded while the transducer is held stationary. The waveform is then analyzed to obtain information relating to the frequency, damping ability, and sensitivity of the transducer unit. Using precision manipulative equipment, the transducer is moved over the target and dynamic recording of the beam symmetry is obtained by use of an open-shutter camera. When these recorded measurements are used in specifying or selecting transducers to be used for testing materials or articles, more uniform test results may be expected.

Transducer Calibrating Equipment

- General

    Equipment used to measure the send/receive characteristics of an ultrasonic transducer is capable of reproducing an exact indication on the oscilloscope of the signals sent and received by the transducer. The movement of the transducer over the reflector is accurately controlled. With data potentiometers coupled to sense the motion, a distance amplitude plot of the sound beam is produced on the oscilloscope. An open-shutter camera (or computer programming) is then used to record the beam profile.

- Test Setup

    The transducer is placed in a couplant tank made of Lucite or glass so that the immersed transducer and reflector can be viewed through the couplant. The reflector is scanned by the sound beam with accurate motion of the scanning transducer ensured through the use of milling table crossfeeds to move the transducer. Potentiometers coupled to the crossfeeds convert motion data into electrical signals which are fed into the horizontal position controls. The horizontal oscilloscope display indicates the distance in inches that the transducer traverses. Either X or Y directions of search unit movement are produced by switching from the output of one potentiometer to the output of the other.

- Function

    Figure 5-1 is a functional block diagram of instrumentation equipment. The equipment consists of a timer, a delay unit, a pulser and a wide-band receiver. The unit repeatedly pulses the transducer with a sharp spike and then amplifies the returned signals fed back through the transducer. During operation, the timer triggers both the delay unit and the pulser tube which, at an adjustable time, later triggers the oscilloscope.

- Recording Method

    Figure 5-2 shows how the response curve is recorded (possibly with an open-shutter camera). The sweep delay, as shown, is used to delay the presentation across the oscilloscope screen. By this method, a permanent record, calibrated in thousandths of an inch, (hundredths of a mm) of the response curve describing the uniformity of the sound beam is produced. This information is related to specific abnormalities in the transducer, such as variations in damping, crystal thickness, lens composition, and dimensional nonuniformity.

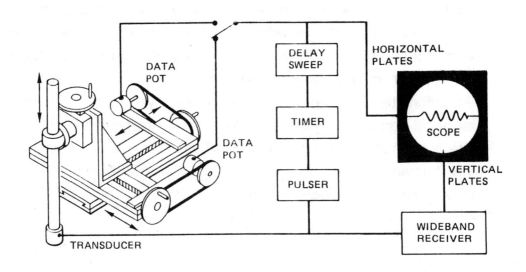

Figure 5-1. Equipment Functional Diagram

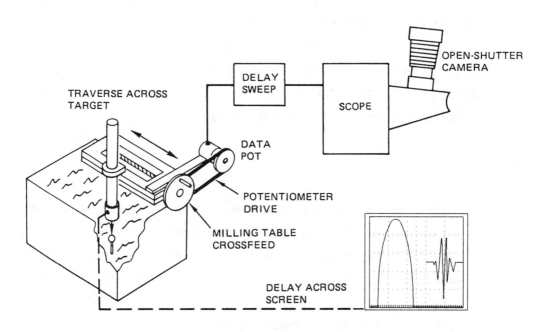

Figure 5-2. Camera Recording Method

- Manipulative Equipment

To obtain precise sound beam and focal length measurements, precision elevating and traversing mechanisms are required. Milling table crossfeeds consisting of heavy micrometer screw slides calibrated in thousandths of an inch (hundredths of a millimeter) are used. Two of the slide screws are fitted with sprocket and chain drives connected to data potentiometers that develop the sweep signal. By relating the micrometer reading to the distance the trace has moved across the oscilloscope screen, the recording is calibrated in inches (or millimeters) per oscilloscope division. The two data potentiometers, one on the transverse and one on the longitudinal movement, are provided so that one plot can be made across the target and then, by switching to the other potentiometer, a plot rotated 90° from the first plot can be made. Thus, two recordings of the beam profile can be made without turning or disturbing the mounting of the transducer.

- Reflector Targets

Reflector targets must be carefully chosen; a bad target will seriously distort the signal and will produce invalid information. In most cases, precision steel balls are used, particularly when calibrating focused transducers. The diameter of the ball must be as small as possible. The size of the effective reflecting surface of the ball is held to less than one-quarter wavelength of the transducer frequency to prevent frequency distortion and undue influence of the target on the measurement of the beam. When analyzing larger-diameter flat transducers, a ball target may not offer adequate return signal amplitude for profile recording. In that case, a flat-topped post as small in diameter as possible is used. The transducer must be held perpendicular to the flat-top surface while testing. Best results are obtained from the use of ball reflectors, since they eliminate the difficulty in holding the transducer normal to the flat surface.

Selection of ultrasonic reflectors varies with the geometry of each crystal and lens. Reflectors must be small compared with the beam size measured. For example, a flat, circular reflector of 12.5 percent of the crystal diameter is adequate for testing flat-disc transducers used to detect fairly large imperfections. Spherically-focused transducers, used to detect very small areas, produce sound beams much smaller than those produced by unfocused transducers; reflector size is small in proportion. In one experiment, performed by the AEC Hanford Laboratories, the sound beam traversed a 0.029-inch-diameter (0.74-mm) ball.

- Pulser

The transducer test requires a pulser with a short pulse capability. To analyze the natural frequency and the damping characteristics of the transducer, the transducer must be excited with a voltage pulse that will not drive the crystal into any abnormal oscillation. This requirement demands that the pulse duration be as short as possible; much less than one period of the natural resonant frequency of the crystal. For analysis of high-frequency (5 to 25 MHz) transducers, the recommended pulse duration is 0.025 microseconds with a rise time of 10 nanoseconds. (A microsecond is one-millionth of a second and a nanosecond is one-billionth of a second.)

- Wide Band Receiver

To prevent the received signal from becoming distorted, a receiver with a wide-band radio frequency amplifier is used. A recommended receiver is one with a bandwidth of 1.5 to 60 MHz, a rise time response of 10 nanoseconds, and a gain of about 40 dB.

- Display System

    An effective display system has sufficient bandwidth and rise time to present the information without distortion. Oscilloscopes and computer monitors (through appropriate software) offer combinations of delay and time base expansion features that are desirable for recording transducer beam profiles.

Recording of Transducer Beam Profiles

- General

    Transducer data sheets are prepared, as shown in Figure 5-3, for mounting of photographic records and recording of the transducer analysis factors. The following paragraphs describe various methods used to obtain transducer beam profiles.

- Flat-Disc Transducer Measurements

    Figure 5-4 shows a beam profile plot of responses picked up by a flat-disc transducer positioned in water over a reflector made from the butt end of a metal drill which was cut and polished flat. The flat end of the drill and the crystal face were held parallel while the transducer scanned over the reflector along the four parallel paths shown. These four beam amplitude profiles (taken with a moving transducer) plus a return signal waveform (taken with a stationary transducer) were recorded on photographs to provide a permanent record of individual transducer characteristics.

TRANSDUCER ACOUSTICAL ANALYSIS

FREQUENCY DAMPING FACTOR, BEAM WIDTH/DIAMETER, FOCAL LENGTH, BEAM SYMMETRY

SERIAL NO. _____   DATE _____
TYPE _____ FOCUSED _____   HOUSING STYLE _____
FREQUENCY ____ 15 MHz ____   CRYSTAL SIZE ____ 3/16 ____
CRYSTAL ___ LITHIUM SULFATE ___   LENS ____ 1/4 RAD. ____

REFLECTORS USED: INSPECTOR _____
                                      DRIVER PULSE:
A. ___ 0.039 DIA. STEEL BALL ___   DURATION ____ 0.065 ____ MICROSECONDS
B. ___ 0.039 DIA. STEEL BALL ___   AMPLITUDE ____ 150 ____ VOLTS
C. ___ 0.039 DIA. STEEL BALL ___

CONNECTING CABLES:

	LENGTH	TYPE
PULSER TO CRYSTAL	5'	RG 62
RECEIVER TO SCOPE	5'	RG 62

DRIVER PULSE: RECEIVED SIGNAL TRACE A
: _____ DB DOWN

TIME →

TRACE

A. (WAVEFORM)
  0.1 MICROSEC/CM
  2.0 VOLTS/CM

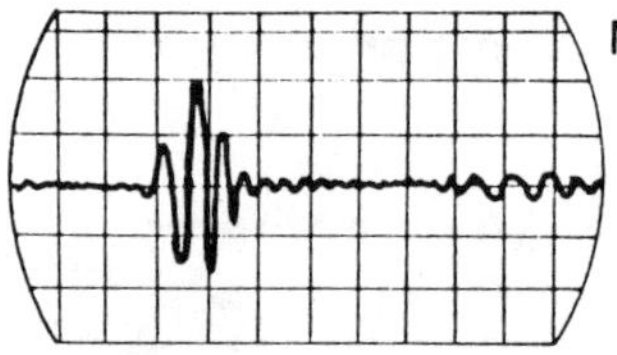

MEASURED FREQUENCY __17.5__ MHz TRACE A

FOCAL LENGTH __0.558__ INCHES

__20__ MICROSEC WATER PATH (ROUND TRIP TIME)

B. (Y DIRECTION)
  2 MICROSEC/CM
  0.5 VOLTS/CM
  0.004 INCH/DIV

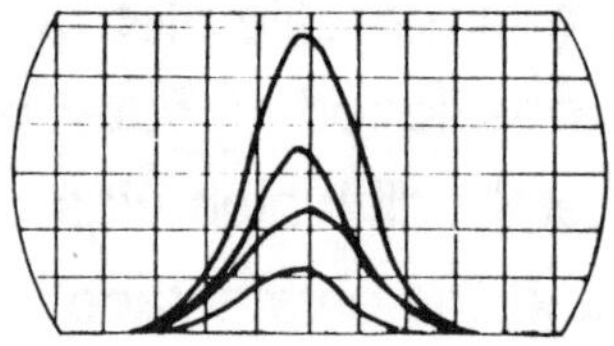

DAMPING FACTOR __3.0__ TRACE A

BEAM WIDTH OR DIA. __0.007__ INCHES 3dB AMP. POINTS TRACE B

C. (X DIRECTION)
  2 MICROSEC/CM
  0.5 VOLTS/CM
  0.004 INCH/DIV

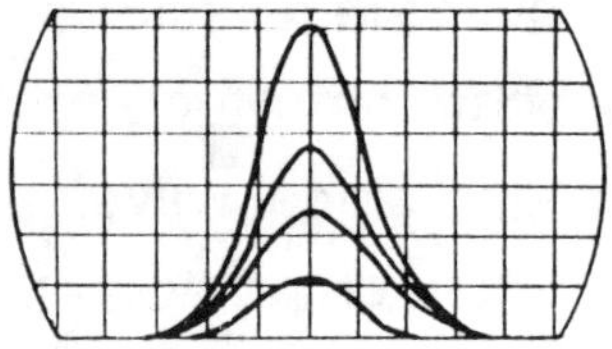

BEAM SYMMETRY:
LENGTH VS AMP. _____
270° – 90° _____ X
TRACE B
180° – 360° _____ X
TRACE C

COURTESY AUTOMATION INDUSTRIES

Figure 5-3. Typical Transducer Data Sheet

5-8

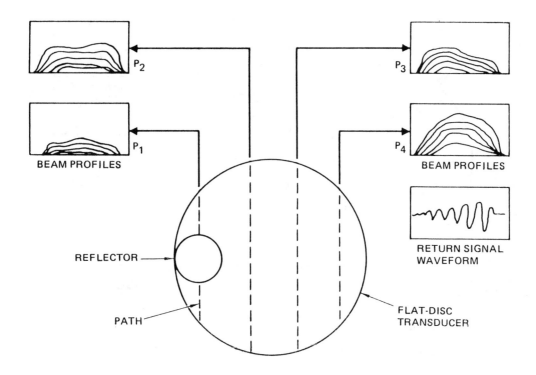

Figure 5-4. Flat-Disc Transducer Measurement

- Focused Transducer Measurements

Figure 5-5 shows the basic transducer measurements taken from a focused transducer. With the reflector stationary, a waveform was obtained. Two beam amplitude profile plots were taken with the transducer traversed in the X-axis and the Y-axis. If the depth of field for a focused transducer is required, beam profiles may be taken at points inside and outside the focal point.

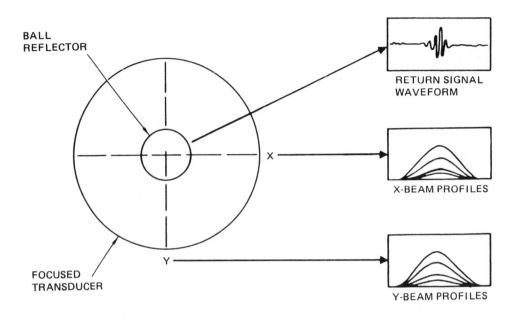

Figure 5-5. Focused Transducer Measurements

- Cylindrically-Focused Transducer Measurements

    A wedge-shaped sound beam, focused in width and unfocused along the length, is produced by a cylindrically-focused transducer which has a concave lens. Beam width is measured by traversing the immersed transducer across the length of the beam over a steel ball (or wire) reflector as shown in Figure 5-6. Ball diameter selected depends on the frequency, crystal size and lens radii of the transducer being tested. A rule of thumb is to select as small a reflector as possible that will still produce adequate signal levels for profiling.

    Since beam width is usually narrow, the problem of maintaining ball alignment while traversing along the beam length may be avoided by substituting a piece of wire for the ball. The wire diameter depends on the same factors that determine ball diameter selection, i.e., frequency, crystal size, and lens radii.

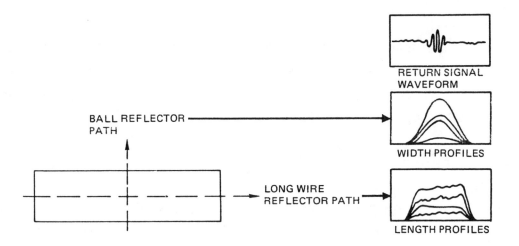

Figure 5-6. Cylindrically-Focused Transducer Measurements

Two beam amplitude profiles are produced by translating along the beam length over the wire, and then translating across the beam width over the ball reflector.

The point on the reflector selected for the beam width measurement is determined by the position of the transducer when the beam length measurement is at its point of highest amplitude. With the transducer held stationary, the waveform is also recorded at this point. If the depth of field for the focused area of the beam is required, the beam profile is taken with the transducer moved to points where the reflector is nearer than the focal point and beyond the focal point.

Analysis of Transducer Data

- General

    In the following paragraphs each of the main headings on the transducer data sheet are discussed. For each transducer tested, the waveform and beam profile plots are analyzed as outlined in the paragraphs that follow.

- Waveform

    The return signal waveform is calibrated in millivolts on the vertical scale and time on the horizontal scale to allow a determination of crystal frequency, damping factor, and sensitivity.

- Frequency

    In this analysis, the actual frequency of transducer operation is measured and compared to the design frequency. The actual frequency is a measurement of the acoustic wave in the water medium. As this is the frequency of the energy used when testing material, this is the frequency that is recorded. To record the acoustical frequency of the transducer, the first reflected signal from the ball target is analyzed. Trace A in Figure 5-3 illustrates this signal. The frequency may be calculated if the period (time base) is known, since frequency equals the number of complete cycles per unit of time.

- Damping Factor

    The damping factor is defined as the number of positive half-cycles within the pulse that are greater in amplitude than the first half-cycle. By counting the number of cycles generated by the crystal when reacting to the reflected pulse, a measure of the damping factor is reached. Trace A in Figure 5-3 illustrates this measurement. In this method, the damping factor is essentially a measurement of the time required for the crystal to return to a quiescent state after excitation. The resolution of the transducer is directly related to the damping factor. The smaller the damping factor, the better the ability of the transducer to resolve two signals arriving very close together in time.

- Sensitivity

Sensitivity is a measure of the ability of the transducer to detect the minute amount of sound energy reflected from a given size target at a given distance. The vertical amplitude of the signal received, as shown in trace A of Figure 5-3, calibrated in volts per centimeter (unit length) measures sensitivity. With the amplitude and duration of the pulse known and the amplification factor of the wide-band receiver known and held constant, sensitivity is measured in volts peak-to-peak or in decibels down with respect to the pulse voltage. The ultrasonic reflectors used in a test for sensitivity vary with the geometry of the crystal and lens. In general, the reflector is small compared to the beam size measured (roughly equal in size to actual defects the transducer is expected to detect). For flat, straight-beam transducers, a flat, circular reflector of 12.5 percent of the crystal diameter is adequate. Beam sizes of focused transducers, used to detect very small discontinuities, are much smaller than the beam sizes of flat transducers; therefore, the reflector is also smaller. Steel balls ranging in size from 0.030 to 0.050 inch (0.76 to 1.27 mm) in diameter have been used successfully for testing focused units. These tiny balls are also used for measuring the beam width of cylindrically-focused transducers. These units are focused in the width dimension and unfocused along the beam length. If difficulties are experienced in aligning the transducer with the ball while traversing in the beam length direction, a fine, small-diameter wire may be laid along the lengthwise path as a substitute for the ball.

- Focal Length

The focal length information for focused transducers is the water-path distance at which a maximum return signal is obtained. Focal length is the time base measurement on the oscilloscope between the excitation pulse and the point of maximum amplitude response. The transducer is held over the center of the ball target and moved toward or away from the ball until the maximum reflected signal is received. The focal length is then noted and recorded.

- Beam Amplitude Profiles

The beam amplitude profiles show amplitude envelopes of each half-cycle with the vertical scale calibrated in millivolts of transducer return signal and the horizontal scale calibrated in mils, or centimeters, of transducer travel. The motion of the transducer across the target drives a data potentiometer which in turn delays the composite rf (radio frequency) signal across the oscilloscope and a distance amplitude recording for each individual cycle is produced. The highest amplitude cycle records the major envelope, the next highest amplitude cycle records the next lower curve, and so on. This system of recording produces superimposed response curves from each individual cycle with respect to each other. The symmetry of these curves with respect to one another is indicative of uniformity of operation of the transducer in the send/receive modes. The symmetry of these curves is affected by variations in damping, crystal thickness, lens thickness, and bonding of the transducer components.

- Beam Width and Symmetry

The beam width is read directly from the width of the profile envelope displayed on the calibrated horizontal axis, or at the 3 dB down points on each side of the profile peak. Nonsymmetry is recognized as variations in the profile patterns of the propagated sound beam and, through critical analysis of these beam envelope variations, normal and abnormal conditions can be identified. Nonsymmetry may be caused by backing variations, lens centering or misalignment. Porosity in lenses and small imperfections in electrodes and bonding have also been linked to distortion in beam profiles.

APPENDIX A

COMPARISON AND SELECTION OF NDT PROCESSES

TABLE OF CONTENTS

      Page

GENERAL A-1
METHOD IDENTIFICATION A-1
NDT DISCONTINUITY SELECTION A-2
DISCONTINUITY CATEGORIES A-4
    Inherent Discontinuities A-4
    Processing Discontinuities A-5
    Service Discontinuities A-5
DISCONTINUITY CHARACTERISTICS AND
METALLURGICAL ANALYSIS A-5
NDT METHODS APPLICATION AND LIMITATIONS A-6
    General A-6
    Selection of the NDT Method A-7
    Limitations A-8
BURST A-8
COLD SHUTS A-11
FILLET CRACKS (BOLTS) A-15
GRINDING CRACKS A-18
CONVOLUTION CRACKS A-21
HEAT-AFFECTED ZONE CRACKING A-24
HEAT-TREAT CRACKS A-27
SURFACE SHRINK CRACKS A-30
THREAD CRACKS A-34
TUBING CRACKS A-37

HYDROGEN FLAKE A-40
HYDROGEN EMBRITTLEMENT A-43
INCLUSIONS A-46
INCLUSIONS A-49
LACK OF PENETRATION A-52
LAMINATIONS A-55
LAPS AND SEAMS A-59
LAPS AND SEAMS A-62
MICROSHRINKAGE A-65
GAS POROSITY A-68
UNFUSED POROSITY A-71
STRESS CORROSION A-74
HYDRAULIC TUBING A-77
MANDREL DRAG A-80
SEMICONDUCTORS A-83
HOT TEARS A-86
INTERGRANULAR CORROSION A-89

LIST OF FIGURES

Figure   Page

A-1   Liquid Penetrant Test A-2
A-2   Magnetic Particle Test A-2
A-3   Ultrasonic Test A-3
A-4   Eddy Current Test A-3
A-5   Radiographic Test A-4
A-6   Burst Discontinuities A-9
A-7   Cold Shut Discontinuities A-12
A-8   Fillet Crack Discontinuity A-16
A-9   Grinding Crack Discontinuity A-19
A-10 Convolution Crack Discontinuities A-22

A-11	Heat-Affected Zone Cracking Discontinuity	A-25
A-12	Heat-Treat Crack Discontinuities	A-28
A-13	Surface Shrink Crack Discontinuities	A-31
A-14	Thread Crack Discontinuities	A-35
A-15	Tubing Crack Discontinuity	A-37
A-16	Hydrogen Flake Discontinuity	A-41
A-17	Hydrogen Embrittlement Discontinuity	A-44
A-18	Weldment Inclusion Discontinuities	A-47
A-19	Wrought Inclusion Discontinuities	A-50
A-20	Lack of Penetration Discontinuities	A-53
A-21	Lamination Discontinuities	A-56
A-22	Lap and Seam Discontinuities in Rolled Threads	A-60
A-23	Lap and Seam Discontinuities in Wrought Materials	A-63
A-24	Microshrinkage Discontinuity	A-66
A-25	Gas Porosity Discontinuity	A-69
A-26	Unfused Porosity Discontinuity	A-72
A-27	Stress Corrosion Discontinuity	A-75
A-28	Hydraulic Tubing Discontinuities	A-78
A-29	Mandrel Drag Discontinuities	A-81
A-30	Semiconductor Discontinuities	A-84
A-31	Hot Tear Discontinuities	A-87
A-32	Intergranular Corrosion Discontinuity	A-90

APPENDIX A

COMPARISON AND SELECTION OF NDT PROCESSES

General

This appendix summarizes the characteristics of various types of discontinuities and lists the NDT methods that may be employed to detect each type of discontinuity.

The relationship between the various NDT methods and their capabilities and limitations when applied to the detection of a specific discontinuity is shown. Such variables as type of discontinuity (inherent, process, or service), manufacturing processes (heat treating, machining, welding, grinding, or plating), and limitations (metallurgical, structural, or processing) also help in determining the sequence of testing and the ultimate selection of one test method over another.

Method Identification

Figures A-1 through A-5 illustrate five NDT methods. Each illustration shows the three elements involved in all five tests, the different methods in each test category, and tasks that may be accomplished with a specific method.

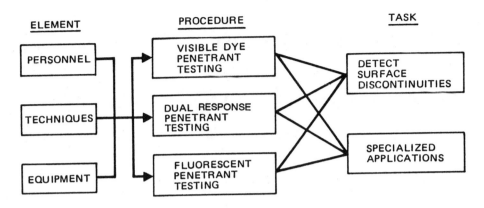

Figure A-1. Liquid Penetrant Test

NDT Discontinuity Selection

The discontinuities that are discussed in the following paragraphs are only some of the many hundreds that are associated with the various materials, processes, and products currently in use. During the selection of discontinuities for inclusion in this chapter, only those discontinuities which would not be radically changed under different conditions of design, configuration, standards, and environment were chosen.

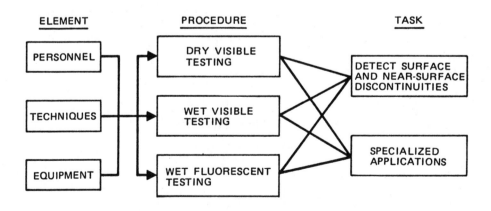

Figure A-2. Magnetic Particle Test

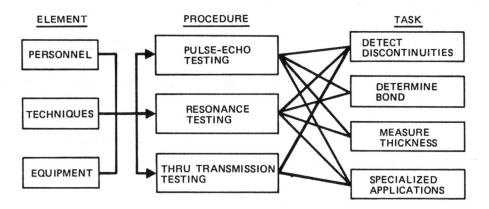

Figure A-3. Ultrasonic Test

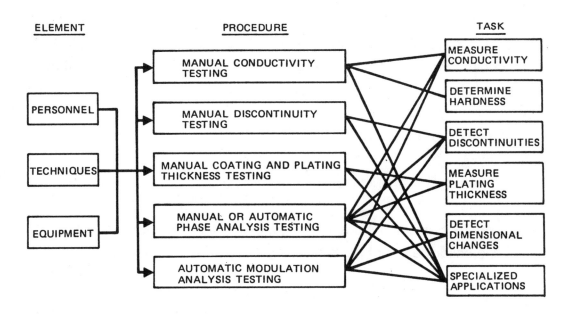

Figure A-4. Eddy Current Test

A-3

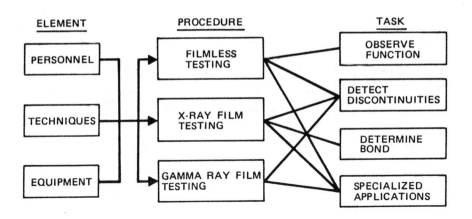

Figure A-5. Radiographic Test

Discontinuity Categories

Each of the specific discontinuities are divided into three general categories: inherent, processing, and service. Each of these categories is further classified as to whether the discontinuity is associated with ferrous or nonferrous materials, the specific material configuration, and the manufacturing processes, if applicable.

- Inherent Discontinuities

    Inherent discontinuities are those discontinuities that are related to the solidification of the molten metal. There are two types.

    - Wrought

        Inherent wrought discontinuities cover those discontinuities which are related to the melting and original solidification of the metal or ingot.

- Cast

    Inherent cast discontinuities are those discontinuities which are related to the melting, casting, and solidification of the cast article. It includes those discontinuities that would be inherent to manufacturing variables such as inadequate feeding, gating, excessively-high pouring temperature, entrapped gases, handling, and stacking.

- Processing Discontinuities

    Processing discontinuities are those discontinuities that are related to the various manufacturing processes such as machining, forming, extruding, rolling, welding, heat treating, and plating.

- Service Discontinuities

    Service discontinuities cover those discontinuities that are related to the various service conditions such as stress corrosion, fatigue, and wear.

Discontinuity Characteristics and Metallurgical Analysis

"Discontinuity characteristics," as used in this chapter, encompasses an analysis of specific discontinuities and references actual photos that illustrate examples of the discontinuity. The discussions cover the following.

- Origin and location of discontinuity (surface, near surface, or subsurface)

- Orientation (parallel or normal to the grain)

- Shape (flat, irregularly-shaped, or spiral)

- Photo (micrograph and/or typical overall view of the discontinuity)

- Metallurgical analysis (how the discontinuity is produced and at what stage of manufacture)

NDT Methods Application and Limitations

- General

    The technological accomplishments in the field of nondestructive testing have brought test reliability and reproductibility to a point where the design engineer may now selectively zone the specific article. Zoning is based upon the structural application of the end product and takes into consideration the environment as well as the loading characteristics of the article. Such an evaluation in no way reduces the end reliability of the product, but evaluation does reduce needless rejection of material that otherwise would have been acceptable. Keep in mind that the design engineer must design the most economical component(s), both in terms of cost and use of resources, that will meet the requirements of the application.

    Just as the structural application within the article varies, the allowable discontinuity size will vary, depending on the configuration and method of manufacture. For example, a die forging that has large masses of material and an extremely thin web section would not require the same level of acceptance over the entire forging. The forging can be zoned for rigid control of areas where the structural loads are higher, and less rigid for areas where the structural loads permit larger discontinuities.

The nondestructive testing specialist must also select the method which will satisfy the design objective of the specific article and not assume that all NDT methods can produce the same reliability for the same type of discontinuity.

- Selection of the NDT Method

    In selecting the NDT method for the evaluation of a specific discontinuity, keep in mind that NDT methods may supplement each other and that several NDT methods may be capable of performing the same task. The selection of one method over another is based upon such variables as those listed below.

    - Type and origin of discontinuity

    - Material manufacturing processes

    - Accessibility of article

    - Level of acceptability desired

    - Equipment available

    - Cost

    A planned analysis of the task must be made for each article requiring NDT testing.

    The NDT methods listed for each discontinuity in the following paragraphs are in order of preference for that particular discontinuity. However, when reviewing the discussions, it should be kept in mind that new techniques in the NDT field may alter the order of test preference. Literature is available from several resources that addresses many of these specialized NDT methods and techniques.

- Limitations

The limitations applicable to the various NDT methods will vary with the applicable standard, the material, and the service environment. Limitations not only affect the NDT method but, in many cases, they also affect the structural reliability of the test article. For these reasons, limitations that are listed for one discontinuity may also be applicable to other discontinuities under slightly different conditions of material or environment. In addition, the many combinations of environment, location, materials, and test capability do not permit mentioning all limitations that may be associated with the problems of locating a specific discontinuity. The intent of this chapter is fulfilled if you are made aware of the many factors that influence the selection of a valid NDT method.

Burst

- Category - Processing

- Material - Ferrous and Nonferrous Wrought Material

- Discontinuity Characteristics

Surface or internal. Straight or irregular cavities varying in size from wide open to very tight. Usually parallel with the grain. Found in wrought material that required forging, rolling, or extruding (Figure A-6).

A FORGING EXTERNAL BURST

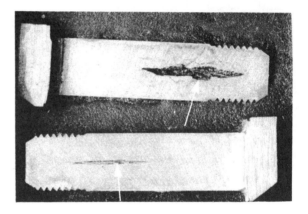

B BOLT INTERNAL BURST

C ROLLED BAR INTERNAL BURST

D FORGED BAR INTERNAL BURST

Figure A-6. Burst Discontinuities

- Metallurgical Analysis

    - Forging bursts are surface or internal ruptures caused by processing at too low a temperature, excessive working, or metal movement during the forging, rolling, or extruding operation.

    - A burst does not have a spongy appearance and is therefore distinguishable from a pipe, even when it occurs at the center.

    - Bursts are often large and are very seldom healed during subsequent working.

- NDT Methods Application and Limitations

    - Ultrasonic Testing Method

        - Normally used for the detection of internal bursts.

        - Bursts are definite breaks in the material and resemble a crack, producing a very sharp reflection on the scope.

        - Ultrasonic testing is capable of detecting varying degrees of burst, a condition not detectable by other NDT methods.

        - Nicks, gouges, raised areas, tool tears, foreign material, or gas bubbles on the article may produce adverse ultrasonic test results.

- Eddy Current Testing Method

    - Not normally used. Testing is restricted to wire, rod, and other articles under 0.250 inch (6.35 mm) in diameter.

- Magnetic Particle Testing Method

    - Usually used on wrought ferromagnetic material in which the burst is open to the surface or has been exposed to the surface.

    - Results are limited to surface and near surface evaluation.

- Liquid Penetrant Testing Method

    - Not normally used. When fluorescent penetrant is to be applied to an article previously dye penetrant tested, all traces of dye penetrant should first be removed by prolonged cleaning in applicable solvent.

- Radiographic Testing Method

    - Not normally used. Such variables as the direction of the burst, close interfaces, wrought material, discontinuity size, and material thickness restrict the capability of radiography.

Cold Shuts

- Category - Inherent

- Material - Ferrous and Nonferrous Cast Material

- Discontinuity Characteristics

    Surface and subsurface. Generally appear on the cast surface as smooth indentations which resemble a forging lap (Figure A-7).

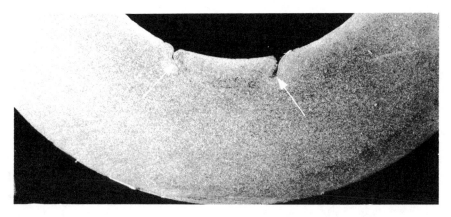

A SURFACE COLD SHUT

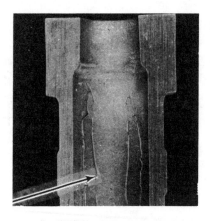

B INTERNAL COLD SHUT

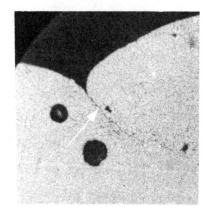

C SURFACE COLD SHUT MICROGRAPH

Figure A-7. Cold Shut Discontinuities

- Metallurgical Analysis

    Cold shuts are produced during casting of molten metal and may be caused by splashing, surging, interrupted pouring, or the meeting of two streams of metal coming from different directions. Cold shuts are also caused by the solidification of one surface before other metal flows over it, the presence of interposing surface films on cold, sluggish metal, or any factor that prevents fusion where two surfaces meet. Cold shuts are more prevalent in castings formed in a mold having several sprues or gates.

- NDT Methods Application and Limitations

    - Liquid Penetrant Testing Method

        - Normally used to evaluate surface cold shuts in both ferrous and nonferrous materials.

        - Indications appear as a smooth, regular, continuous, or intermittent line.

        - Liquid penetrants used to test nickel-based alloys, certain stainless steels, and titanium should not exceed one percent sulfur or chlorine.

        - Certain castings may have surfaces that are blind and from which removal of excess penetrant may be difficult.

        - The geometric configuration (recesses, orifices, and flanges) of a casting may permit buildup of wet developer, thereby masking any detection of a discontinuity.

- Magnetic Particle Testing Method

    - Normally used for the evaluation of ferromagnetic materials.

    - The metallurgical nature of some corrosion-resistant steel is such that, in some cases, magnetic particle testing indications are obtained which do not result from a crack or other harmful discontinuities. These indications arise from a duplex structure within the material, wherein one portion exhibits strong magnetic retentivity and the other does not.

- Radiographic Testing Method

    - Cold shuts are normally detectable by radiography while testing for other casting discontinuities.

    - Cold shuts appear as a distinct, dark line or band of variable length and width and a definite, smooth outline.

    - The casting configuration may have inaccessible areas that can only be tested by radiography.

- Ultrasonic Testing Method

    - Not recommended. As a general rule, cast structure and article configuration do not lend themselves to ultrasonic testing.

- Eddy Current Testing Method

    • Article configuration and inherent material variables require the use of specialized probes.

Fillet Cracks (Bolts)

- Category - Service

- Material - Ferrous and Nonferrous Wrought Material

- Discontinuity Characteristics

    Surface. Located at the junction of the fillet with the shank of the bolt and progressing inward (Figure A-8).

- Metallurgical Analysis

    Fillet cracks occur where a marked change in diameter occurs, such as at the head-to-shank junction where stress risers are created. During the service life of a bolt, repeated loading takes place whereby the tensile load fluctuates in magnitude due to the operation of the mechanism. These tensile loads can cause fatigue failure starting at the point where the stress risers occur. Fatigue failure, which is surface phenomenon, starts at the surface and propagates inward.

- NDT Methods Application and Limitations

    - Ultrasonic Testing Method

        • Used extensively for service-associated discontinuities of this type.

- A wide selection of transducers and equipment enable on-the-spot evaluation for fillet cracks.

- Since fillet cracks are a definite break in the material, the scope pattern will be a very sharp reflection. (Propagation can be monitored by using ultrasonics.)

- Ultrasonic equipment has extreme sensitivity, and established standards should be used to give reproducible and reliable results.

A FILLET FATIGUE FAILURE

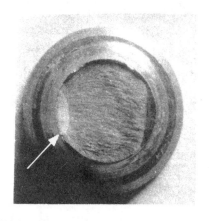
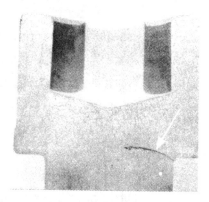

B FRACTURE AREA OF (A) SHOWING TANGENCY POINT OF FAILURE

C CROSS-SECTIONAL AREA OF FATIGUE CRACK IN FILLET SHOWING TANGENCY POINT IN RADIUS

Figure A-8. Fillet Crack Discontinuity

- Liquid Penetrant Testing Method

    • Normally used during inservice overhaul or troubleshooting.

    • May be used for both ferromagnetic and nonferromagnetic bolts, although usually confined to the nonferromagnetic.

    • Fillet cracks appear as sharp, clear indications.

    • Structural damage may result from exposure of high-strength steels to paint strippers, alkaline coating removers, deoxidizer solutions, etc.

    • Entrapment of penetrant under fasteners, in holes, under splices, and in similar areas may cause corrosion due to the penetrant's affinity for moisture.

- Magnetic Particle Testing Method

    • Only used on ferromagnetic bolts.

    • Fillet cracks appear as sharp, clear indications with a heavy buildup.

    • Sharp fillet areas may produce nonrelevant magnetic indications.

    • 16.6 pH steel is only slightly magnetic in the annealed condition; however, it becomes strongly magnetic after heat treatment and can then be magnetic particle tested.

- Eddy Current Testing Method

    • Not normally used for detection of fillet cracks. Other NDT methods are more compatible to the detection of this type of discontinuity.

- Radiographic Testing Method

    • Not normally used for detection of fillet cracks. Surface discontinuities of this type would be difficult to evaluate due to the size of the crack in relation to the thickness of the material.

Grinding Cracks

- Category - Processing

- Material - Ferrous and Nonferrous

- Discontinuity Characteristics

    Surface. Very shallow and sharp at the root. Similar to heat-treat cracks and usually, but not always, occur in groups. Grinding cracks generally occur at right angles to the direction of grinding. They are found in highly heat-treated articles, chrome-plated, case-hardened, and ceramic materials that are subjected to grinding operations (Figure A-9).

- Metallurgical Analysis

    Grinding of hardened surfaces frequently introduces cracks. These thermal cracks are caused by local overheating of the surface being ground. The overheating is usually caused by lack of coolant or poor coolant, a dull or improperly ground wheel, too rapid feed or too heavy cut.

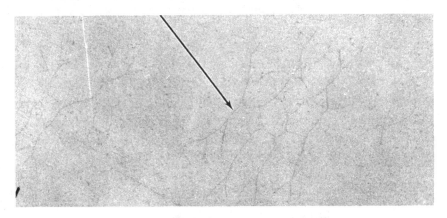

A TYPICAL CHECKED GRINDING CRACK PATTERN

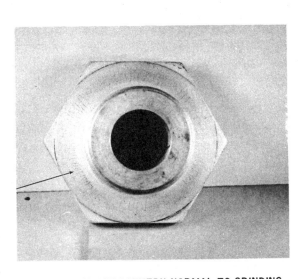

B GRINDING CRACK PATTERN NORMAL TO GRINDING   C MICROGRAPH OF GRINDING CRACK

Figure A-9. Grinding Crack Discontinuity

- NDT Methods Application and Limitations

    - Liquid Penetrant Testing Method

A-19

- Normally used on both ferrous and nonferrous materials for the detection of grinding cracks.

- Liquid penetrant indication will appear as an irregular, a checked, or a scattered pattern of fine lines.

- Grinding cracks are the most difficult discontinuities to detect and require the longest penetration time.

- Articles that have been degreased may still have solvent entrapped in the discontinuity and should be allowed sufficient time for evaporation prior to application of the penetrant.

– Magnetic Particle Testing Method

- Restricted to ferromagnetic materials.

- Grinding cracks generally occur at right angles to grinding direction, although in extreme cases a complete network of cracks may appear. In this case they may be parallel to the magnetic field.

- Magnetic sensitivity decreases as the size of the grinding crack decreases.

– Eddy Current Testing Method

- Although not normally used for detection of grinding cracks, eddy current equipment has the capability and can be developed for specific ferrous and nonferrous applications.

- Ultrasonic Testing Method

    • Not normally used for detection of grinding cracks. Other forms of NDT are more economical, faster, and better adapted to this type of discontinuity than ultrasonics.

- Radiographic Testing Method

    • Not recommended for detection of grinding cracks. Grinding cracks are too tight and too small. Other NDT methods are more suitable for detection of grinding cracks.

Convolution Cracks

- Category - Processing

- Material - Nonferrous

- Discontinuity Characteristics

    Surface. Range in size from microfractures to open fissures. Situated on the periphery of the convolutions and extend longitudinally in direction of rolling (Figure A-10).

- Metallurgical Analysis

    A rough "orange peel" effect of convolution cracks is the result of either a forming operation that stretches the material or from chemical attack, such as pickling treatment. The roughened surface contains small pits that form stress risers. Subsequent service application (vibration and flexing) may introduce stresses that act on these pits and form fatigue cracks as shown in Figure A-10.

A TYPICAL CONVOLUTION DUCTING

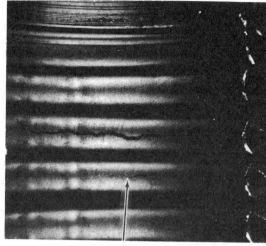

B CROSS-SECTION OF CRACKED CONVOLUTION

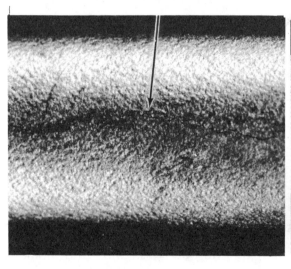

C HIGHER MAGNIFICATION OF CRACK SHOWING ORANGE PEEL

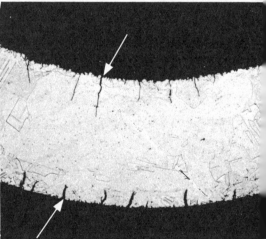

D MICROGRAPH OF CONVOLUTION WITH PARTIAL CRACKING ON SIDES

Figure A-10. Convolution Crack Discontinuities

- NDT Methods Application and Limitations

  - Radiographic Testing Method

A-22

- Used extensively for this type of failure.

- The configuration of the article and the location of the discontinuity limits detection almost exclusively to radiography.

- Orientation of convolutions to X-ray source is very critical since those discontinuities that are not normal to X-ray may not register on the film due to the small change in density.

- Liquid penetrant and magnetic particle testing may supplement, but not replace, radiographic and ultrasonic testing.

- The type of marking material (e.g., grease pencil on titanium) used to identify the area of discontinuities may affect the structure of the article.

- Ultrasonic Testing Method

  - Not normally used for the detection of convolution cracks. The configuration of the article (double-walled convolutions) and the presence of internal micro fractures are all factors that restrict the use of ultrasonics.

- Eddy Current Testing Method

  - Not normally used for the detection of convolution cracks. As in the case of ultrasonic testing, the configuration does not lend itself to this method of testing.

- Liquid Penetrant Testing Method

    • Not recommended for the detection of convolution cracks. Although the discontinuities are surface, they are internal and are superimposed over an exterior shell, which creates a serious problem of entrapment.

- Magnetic Particle Testing Method

    • Not applicable. Material is nonferrous.

Heat-Affected Zone Cracking

- Category - Processing (Weldments)

- Material - Ferrous and Nonferrous

- Discontinuity Characteristics

    Surface. Often quite deep and very tight. Usually run parallel with the weld in the heat-affected zone of the weldment (Figure A-11).

- Metallurgical Analysis

    Hot cracking or heat-affected zones of weldments increases in severity with increasing carbon content. Steels that contain more than 0.30 percent carbon are prone to this type of failure and require preheating prior to welding.

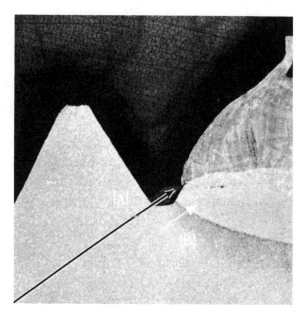

A MICROGRAPH OF WELD AND HEAT-AFFECTED ZONE SHOWING CRACK NOTE COLD LAP WHICH MASKS THE ENTRANCE TO THE CRACK

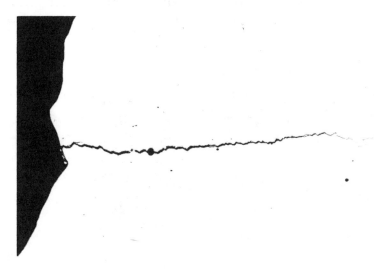

B MICROGRAPH OF CRACK SHOWN IN (A)

Figure A-11. Heat-Affected Zone Cracking Discontinuity

- NDT Methods Application and Limitations

  - Magnetic Particle Testing Method

    - Normally used for ferromagnetic weldments.

    - Prod burns are very detrimental, especially on highly-heat-treated articles. Burns may contribute to structural failure of article.

    - Demagnetization of highly-heat-treated articles can be very difficult due to metallurgical structure.

  - Liquid Penetrant Testing Method

    - Normally used for nonferrous weldments.

    - Material that has had its surface obliterated, blurred, or blended due to manufacturing processes should not be penetrant tested until the smeared surface has been removed.

    - Liquid penetrant testing after the application of certain types of chemical film coatings may be invalid due to the covering or filling of the discontinuities.

  - Radiographic Testing Method

    - Not normally used for the detection of heat-affected-zone cracking. Discontinuity orientation and surface origin make other NDT methods more suitable.

- Ultrasonic Testing Method

    • Used where specialized applications have been developed.

    • Rigid standards and procedures are required to develop valid tests.

    • The configuration of the surface roughness (i.e., sharp versus rounded root radii and the slope condition) is a major factor in deflecting the sound beam.

- Eddy Current Testing Method

    • Although not normally used for the detection of heat-affected-zone cracking, eddy current testing equipment has the capability of detecting ferrous and nonferrous surface discontinuities.

Heat-Treat Cracks

- Category - Processing

- Material - Ferrous and Nonferrous Wrought and Cast Material

- Discontinuity Characteristics

    Surface. Usually deep and forked. Seldom follow a definite pattern and can be in any direction on the part. Originate in areas with rapid change of material thickness, sharp machining marks, fillets, nicks, and discontinuities that have been exposed to the surface of the material (Figure A-12).

A FILLET AND MATERIAL THICKNESS CRACKS (TOP CENTER)
   RELIEF RADIUS CRACKING (LOWER LEFT)

B HEAT-TREAT CRACK DUE TO SHARP MACHINING MARKS

Figure A-12. Heat-Treat Crack Discontinuities

- Metallurgical Analysis

During the heating and cooling process, localized stresses may be set up by unequal heating or cooling, restricted movement of the article or unequal cross-sectional thickness. These stresses may exceed the tensile strength of the material, causing it to rupture. Where built-in stress risers occur (keyways or grooves), additional cracks may develop.

- NDT Methods Application and Limitations

    - Magnetic Particle Testing Method

        - For ferromagnetic materials, heat-treat cracks are normally detected by magnetic particle testing.

        - Indications normally appear as straight, forked, or curved indications.

        - Likely points of origin are areas that would develop stress risers, such as keyways, fillets, or areas with rapid changes in material thickness.

        - Metallurgical structure of age-hardenable and heat-treatable stainless steels may produce nonrelevant indications.

    - Liquid Penetrant Testing Method

        - Liquid penetrant testing is the recommended method for nonferrous materials.

        - Likely points of origin for heat-treat cracks are the same as those listed for magnetic particle testing.

- Materials or articles that will eventually be used in LOX systems must be tested with LOX-compatible penetrants.

- Eddy Current Testing Method

    - Although not normally used for the detection of heat-treat cracks, eddy current testing equipment has the capability of detecting ferrous and nonferrous surface discontinuities.

- Ultrasonic Testing Method

    - Not normally used for detection of heat-treat cracks. If used, the scope pattern will show a definite indication of a discontinuity. Recommended wave mode would be surface.

- Radiographic Testing Method

    - Not normally used for detection of heat-treat cracks. Surface discontinuities are more easily detected by other NDT methods designed for surface application.

Surface Shrink Cracks

- Category - Processing (Welding)

- Material - Ferrous and Nonferrous

- Discontinuity Characteristics

    Surface. Situated on the face of the weld, fusion zone and base metal. Range in size from very small, tight and shallow to open and deep. Cracks may run parallel or transverse to the direction of welding (Figure A-13).

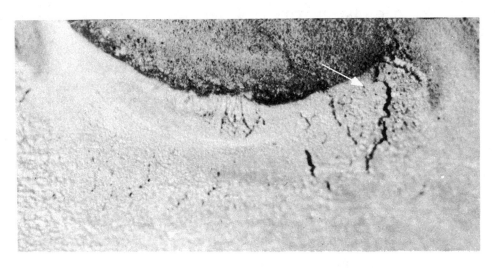

A TRANSVERSE CRACKS IN HEAT-AFFECTED ZONE

B TYPICAL STAR-SHAPED CRATER CRACK

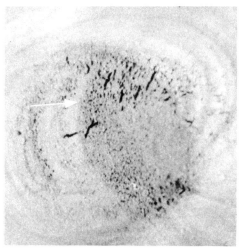

C SHRINKAGE CRACK AT WELD TERMINAL

Figure A-13. Surface Shrink Crack Discontinuities

- Metallurgical Analysis

    Surface shrink cracks are generally the result of improper heat application either in heating or welding of the article. Heating or cooling in a localized area may set up stresses that exceed the tensile strength of the material, causing the material to crack. Restriction of the movement (contraction or expansion) of the material during heating, cooling, or welding may also set up excessive stresses.

- NDT Methods Application and Limitations

    - Liquid Penetrant Testing Method

        - Surface shrink cracks in nonferrous materials are normally detected by use of liquid penetrants.

        - Liquid penetrant equipment is easily portable and can be used during in-process control for both ferrous and nonferrous weldments.

        - Assemblies that are joined by bolting, riveting, intermittent welding, or press fittings will retain the penetrant, which will seep out after developing and mask the adjoining surfaces.

        - When articles are dried in a hot air dryer or by similar means, excessive drying temperature should be avoided to prevent evaporation of penetrant.

- Magnetic Particle Testing Method

    • Ferromagnetic weldments are normally tested by magnetic particle method.

    • Surface discontinuities that are parallel to the magnetic field will not produce indications, since they do not interrupt or distort the magnetic field.

    • Areas such as grease fittings, bearing races, or other similar items that might be damaged or clogged by the bath or by the particles should be masked before testing.

- Eddy Current Testing Method

    • Ferrous and nonferrous welded sections can be inspected.

    • A probe or encircling coil could be used where article configuration permits.

- Radiographic Testing Method

    • Not normally used for the detection of surface discontinuities. During the radiographic testing of weldments for other types of discontinuities, surface indications may be detected.

- Ultrasonic Testing Method

    • Not normally used for detection of surface shrink cracks. Other forms of NDT (liquid penetrant and magnetic particle) give better results, are more economical, and are faster.

Thread Cracks

- Category - Service

- Material - Ferrous and Nonferrous Wrought Material

- Discontinuity Characteristics

   Surface. Cracks are transverse to the grain (transgranular) starting at the root of the thread (Figure A-14).

- Metallurgical Analysis

   Fatigue failures of this type are not uncommon. High cyclic stresses resulting from vibration and/or flexing act on the stress risers created by the thread roots to produce cracks. Fatigue cracks may start as fine submicroscopic discontinuities or cracks and propagate in the direction of applied stresses.

- NDT Methods Application and Limitations

   - Liquid Penetrant Testing Method

      - Fluorescent penetrant is recommended over non-fluorescent.

      - Low surface tension solvents, such as gasoline and kerosene, are not recommended cleaners.

- When applying liquid penetrant to components within an assembly or structure, the adjacent areas should be effectively masked to prevent overspraying.

A COMPLETE THREAD ROOT FAILURE

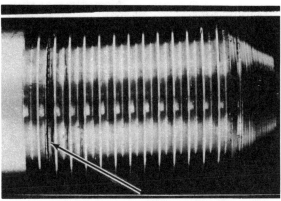

B TYPICAL THREAD ROOT FAILURE

C MICROGRAPH OF (A) SHOWING CRACK AT BASE OF ROOT

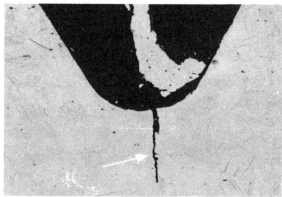

D MICROGRAPH OF (B) SHOWING TRANSGRANULAR CRACK AT THREAD ROOT

Figure A-14. Thread Crack Discontinuities

- Magnetic Particle Testing Method

    • Normally used to detect cracks at the threads on ferromagnetic materials.

    • Nonrelevant magnetic indications may result from the thread configuration.

    • Cleaning titanium and 440C stainless in halogenated hydrocarbons may result in structural damage to the material.

- Ultrasonic Testing Method

    • The article configuration can be examined utilizing the cylindrical guided-wave techniques. This method requires access to the article and poses interpretation difficulties.

- Eddy Current Testing Method

    • A specialized probe to fit thread size would be required.

- Radiographic Testing Method

    • Not recommended for detecting thread cracks. Surface discontinuities are best screened by NDT method designed for the specific condition. Fatigue cracks of this type are very tight and surface-connected. Detection by radiography would be extremely difficult.

Tubing Cracks

- Category - Inherent

- Material - Nonferrous

- Discontinuity Characteristics

Tubing cracks formed on the inner surface (ID), parallel to direction of grain flow (Figure A-15).

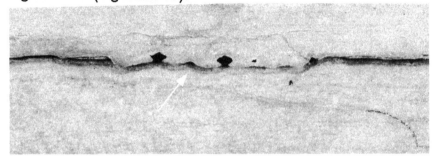

A TYPICAL CRACK ON INSIDE OF TUBING SHOWING COLD LAP

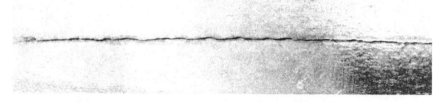

B ANOTHER PORTION OF SAME CRACK SHOWING CLEAN FRACTURE

C MICROGRAPH OF (B)

Figure A-15. Tubing Crack Discontinuity

- Metallurgical Analysis

    Tubing ID cracks may be attributed to one or a combination of the following:

    - Improper cold reduction of the tube during fabrication.

    - Foreign material may have been embedded on the inner surface of the tubes causing embrittlement and cracking when the cold-worked material was heated during the annealing operation.

    - Insufficient heating rate to the annealing temperature with possible cracking occurring in the 1200°F to 1400°F (649°C to 760°C) range.

- NDT Methods Application and Limitations

    - Eddy Current Testing Method

        - Normally used for detection of this type of discontinuity.

        - Tube diameters below 1 inch (2.54 cm) and wall thicknesses less than 0.150 inch (3.8 mm) are well within equipment capability.

        - Testing of ferromagnetic material may be difficult.

    - Ultrasonic Testing Method

        - Normally used on tubing.

- A wide variety of equipment and transducers are available for screening tubing for internal discontinuities of this type.

- Ultrasonic transducers have varying temperature limitations.

- Certain ultrasonic contact couplants may have high sulfur content which will have an adverse effect on high-nickel alloys.

- Radiographic Testing Method

    - Not normally used for detecting tubing cracks. Discontinuity orientation and thickness of material govern the radiographic sensitivity. Other forms of NDT (eddy current and ultrasonics) are more economical, faster, and more reliable.

- Liquid Penetrant Testing Method

    - Not recommended for detecting tubing cracks. Internal discontinuity would be difficult to process and interpret.

- Magnetic Particle Testing Method

    - Not applicable. Material is nonferrous under normal conditions.

Hydrogen Flake

- Category - Processing

- Material - Ferrous

- Discontinuity Characteristics

    Internal fissures in a fractured surface, flakes appear as bright, silvery areas. On an etched surface they appear as short discontinuities. Sometimes known as chrome checks and hairline cracks when revealed by machining. Flakes are extremely thin and generally align parallel with the grain. They are usually found in heavy steel forging, billets, and bars (Figure A-16).

- Metallurgical Analysis

    Flakes are internal fissures attributed to stresses produced by localized transformation and decreased solubility of hydrogen during cooling after hot working. Usually found only in heavy alloy steel forgings.

- NDT Methods Application and Limitations

    - Ultrasonic Testing Method

        - Used extensively for the detection of hydrogen flake.

        - Material in the wrought condition can be screened successfully using either the immersion or the contact method. The surface condition will determine the method most suited.

A 4340 CMS HAND FORGING REJECTED FOR HYDROGEN FLAKE

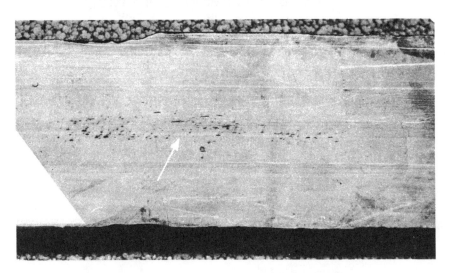

B CROSS SECTION OF (A) SHOWING FLAKE CONDITION IN CENTER OF MATERIAL

Figure A-16. Hydrogen Flake Discontinuity

- On the A-scan presentation, hydrogen flake will appear as hash on the screen or as loss of back reflection.

- All foreign materials (loose scale, dirt, oil, and grease) should be removed prior to any testing. Surface irregularities such as nicks, gouges, tool marks, and scarfing may cause loss of back reflection.

- Magnetic Particle Testing Method

    - Normally used on finish machined articles.

    - Flakes appear as short discontinuities and resemble chrome checks or hairline cracks.

    - Machined surfaces with deep tool marks may obliterate the detection of the flake.

    - Where the general direction of a discontinuity is questionable, it may be necessary to magnetize in two or more directions.

- Liquid Penetrant Testing Method

    - Not normally used for detecting flakes. Discontinuities are very small and tight and would be difficult to detect by liquid penetrants.

- Eddy Current Testing Method

    - Not recommended for detecting flakes. The metallurgical structure of ferrous materials limits their adaptability to the use of eddy current testing.

- Radiographic Testing Method

    • Not recommended for detecting flakes. The size of the discontinuity and its location and orientation with respect to the material surface restricts the application of radiography.

Hydrogen Embrittlement

- Category - Processing and Service

- Material - Ferrous

- Discontinuity Characteristics

    Surface. Small, nondimensional (interface) with no orientation or direction. Found in highly-heat-treated material that has been subjected to pickling and/or plating or in material exposed to free hydrogen (Figure A-17).

- Metallurgical Analysis

    Operations such as electroplating or pickling and cleaning prior to electroplating generate hydrogen at the surface of the material. This hydrogen penetrates the surface of the material, creating immediate or delayed embrittlement and cracking.

- NDT Methods Application and Limitations

    - Magnetic Particle Testing Method

        • Magnetic indications appear as a fractured pattern.

A   DETAILED CRACK PATTERN OF HYDROGEN EMBRITTLEMENT

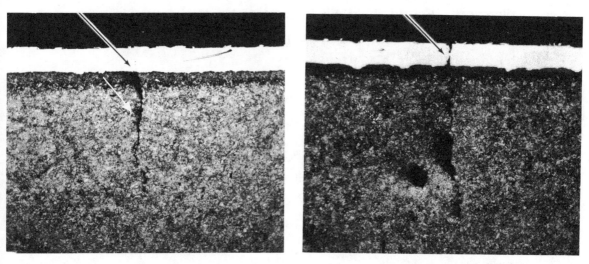

B   HYDROGEN EMBRITTLEMENT UNDER CHROME PLATE   C   HYDROGEN EMBRITTLEMENT PROPAGATED THROUGH CHROME PLATE

Figure A-17.   Hydrogen Embrittlement Discontinuity

- Hydrogen embrittlement cracks are randomly oriented and may be aligned with the magnetic field.

- Magnetic particle testing should be accomplished before and after plating.

- Care should be taken so as not to produce nonrelevant indications or cause damage to the article by overheating.

- Some alloys of corrosion-resistant steel are nonmagnetic in the annealed condition, but become magnetic with cold working.

- Liquid Penetrant Testing Method

  - Not normally used for detecting hydrogen embrittlement. Discontinuities on the surface are extremely tight, small, and difficult to detect. Subsequent plating deposit may mask the discontinuity.

- Ultrasonic Testing Method

  - Although ultrasonic equipment has the capability of detecting hydrogen embrittlement, this method is not normally used. Article configurations and size do not, in general, lend themselves to this method of testing. Surface wave and/or time-of-flight techniques are recommended.

- Eddy Current Testing Method

    • Not recommended for detecting hydrogen embrittlement. Many variables inherent in the specific material may produce conflicting patterns.

- Radiographic Testing Method

    • Not recommended for detecting hydrogen embrittlement. The sensitivity required to detect hydrogen embrittlement is, in most cases, in excess of radiographic capabilities.

Inclusions

- Category - Processing (Weldments)

- Material - Ferrous and Nonferrous Welded Material

- Discontinuity Characteristics

    Surface and subsurface. Inclusions may be any shape. They may be metallic or nonmetallic and may appear individually or be linearly distributed or scattered throughout the weldment (Figure A-18).

- Metallurgical Analysis

    Metallic inclusions are generally particles of metals of different density as compared to the density of the weld or base metal. Nonmetallic inclusions are oxides, sulphides, slag, or other nonmetallic foreign material entrapped in the weld or trapped between the weld metal and the base metal.

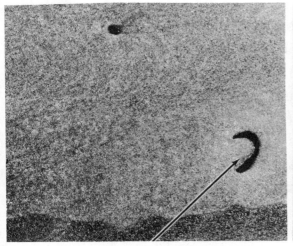

A METALLIC INCLUSIONS

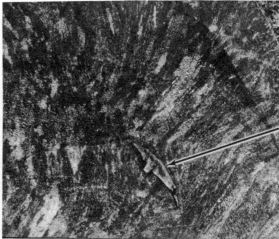

B INCLUSIONS TRAPPED IN WELD

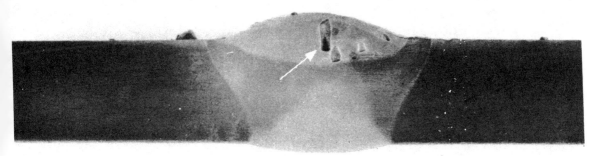

C CROSS SECTION OF WELD SHOWING INTERNAL INCLUSIONS

Figure A-18. Weldment Inclusion Discontinuities

- NDT Methods Application and Limitations

  - Radiographic Testing Method

    - This NDT method is universally used.

    - Metallic inclusions appear on the radiograph as sharply-defined, round, erratically-shaped or elongated white spots and may be isolated or in small linear or scattered groups.

- Nonmetallic inclusions will appear on the radiograph as shadows of round globules or elongated or irregularly-shaped contours occurring individually, linearly or scattered throughout the weldment. They will generally appear in the fusion zone or at the root of the weld. Less absorbent material is indicated by a greater film density and more absorbent materials by a lighter film density.

- Foreign material such as loose scales, splatter, or flux may invalidate test results.

- Eddy Current Testing Method

    - Normally confined to thin-walled welded tubing.

    - Established standards are required if valid results are to be obtained.

- Magnetic Particle Testing Method

    - Normally not used for detecting inclusions in weldments.

    - Confined to machined weldments where the discontinuities are surface or near surface.

    - The indications would appear jagged, irregularly-shaped, individually, or clustered and would not be too pronounced.

- • Discontinuities may go undetected when improper contact exists between the magnetic particles and the surface of the article.

- Ultrasonic Testing Method

  - • Not normally used for detecting inclusions. Specific applications of design or of article configuration, however, may require ultrasonic testing.

- Liquid Penetrant Testing Method

  - • Not applicable. Inclusions are normally not open fissures.

Inclusions

- Category - Processing

- Material - Ferrous and Nonferrous Wrought Material

- Discontinuity Characteristics

Subsurface (original bar) or surface (after machining). There are two types; one is nonmetallic with long, straight lines parallel to flow lines and quite tightly adherent. They are often short and likely to occur in groups. The other type is nonplastic, appearing as a comparatively large mass not parallel to flow lines. Found in forged, extruded, and rolled material (Figure A-19).

A TYPICAL INCLUSION PATTERN ON MACHINED SURFACES

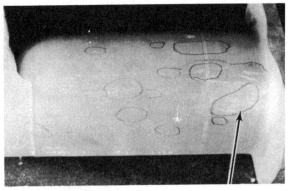
B STEEL FORGING SHOWING NUMEROUS INCLUSIONS

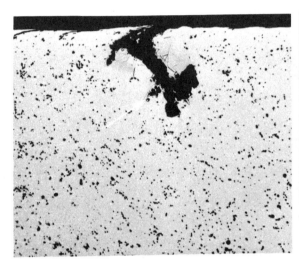
C MICROGRAPH OF TYPICAL INCLUSION

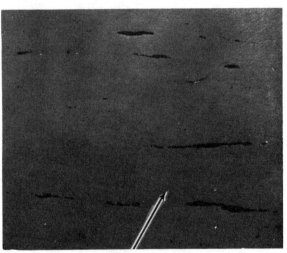
D LONGITUDINAL CROSS SECTION SHOWING ORIENTATION OF INCLUSIONS

Figure A-19. Wrought Inclusion Discontinuities

- Metallurgical Analysis

Nonmetallic inclusions (stringers) are caused by the existence of slag or oxides in the billet or ingot. Nonplastic inclusions are caused by particles remaining in the solid state during billet melting. Certain types of steels are more prone to inclusions than others.

- **NDT Methods Application and Limitations**

  - Ultrasonic Testing Method

    - Normally used to evaluate inclusions in wrought material.

    - Inclusions will appear as definite interfaces within the metal. Small, clustered condition or conditions on different planes cause a loss in back reflection. Numerous small, scattered conditions cause excessive "noise."

    - Inclusion orientation in relationship to ultrasonic beam is critical.

    - The direction of the ultrasonic beam should be perpendicular to the direction of the grain flow whenever possible.

  - Eddy Current Testing Method

    - Normally used for thin-walled tubing and small-diameter rods.

    - Eddy current testing of ferromagnetic materials can be difficult.

  - Magnetic Particle Testing Method

    - Normally used on machined surface.

- Inclusions will appear as a straight, intermittent, or a continuous indication. They may be individual or clustered.

- The magnetizing technique should be such that a surface or near surface inclusion can be satisfactorily detected when its axis is in any direction.

- A knowledge of the grain flow of the material is critical since inclusions will be parallel to that direction.

– Liquid Penetrant Testing Method

- Not normally used for detecting inclusions in wrought material. Inclusions are generally not openings in the material surface.

– Radiographic Testing Method

- Not recommended. NDT methods designed for surface testing are more suitable for detecting surface inclusions.

Lack of Penetration

- Category - Processing

- Material - Ferrous and Nonferrous Weldments

- Discontinuity Characteristics

    Internal or external. Generally irregular and filamentary occurring at the root and running parallel with the weld (Figure A-20).

A INADEQUATE ROOT PENETRATION

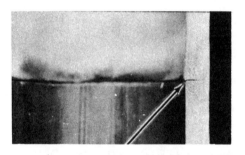

B INADEQUATE ROOT PENETRATION OF BUTT WELDED TUBE

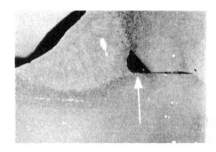

C INADEQUATE FILLET WELD PENETRATION KNOWN AS BRIDGING

Figure A-20. Lack of Penetration Discontinuities

- Metallurgical Analysis

    Caused by root face of joint not reaching fusion temperature before weld metal was deposited. Also caused by fast welding rate, too large a welding rod or too cold a bead.

A-53

- NDT Methods Application and Limitations

  - Radiographic Testing Method

    - Used extensively on a wide variety of welded articles to determine the lack of penetration.

    - Lack of penetration will appear on the radiograph as an elongated, dark area of varying length and width. Lack of penetration may be continuous or intermittent and may appear in the center of the weld at the junction of multipass beads.

    - Lack of penetration orientation in relationship to the radiographic source is critical.

    - Sensitivity levels govern the capability to detect small or tight discontinuities.

  - Ultrasonic Testing Method

    - Commonly used for specific applications.

    - Weldments make ultrasonic testing difficult.

    - Lack of penetration will appear on the scope as a definite break or discontinuity resembling a crack and will give a very sharp reflection.

  - Eddy Current Testing Method

    - Normally used to determine lack of penetration in nonferrous welded pipe and tubing.

- Eddy current testing can be used where other nonferrous articles can meet the configuration requirement of the equipment.

— Magnetic Particle Testing Method

- Normally used where back side of weld is visible.

- Lack of penetration appears as an irregular indication of varying width.

— Liquid Penetrant Testing Method

- Normally used where backside of weld is visible.

- Lack of penetration appears as an irregular indication of varying width.

- Residue left by the penetrant and the developer could contaminate any rewelding operation.

Laminations

- Category - Inherent

- Material - Ferrous and Nonferrous Wrought Material

- Discontinuity Characteristics

Surface and internal. Flat, extremely thin, generally aligned parallel to the work surface of the material. May contain a thin film of oxide between the surfaces. Found in forged, extruded and rolled material (Figure A-21).

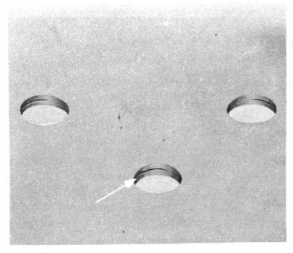

A LAMINATION IN 0.250 IN. PLATE

B LAMINATION IN 0.040 TITANIUM SHEET

C LAMINATION IN PLATE SHOWING SURFACE ORIENTATION

D LAMINATION IN 1 IN. BAR SHOWING SURFACE ORIENTATION

Figure A-21. Lamination Discontinuities

A-56

- Metallurgical Analysis

    Laminations are separations or weaknesses generally aligned parallel to the work surface of the material. They may be the result of pipe, blister, seam, inclusions, or segregations elongated and made directional by working. Laminations are flattened impurities that are extremely thin.

- NDT Methods Application and Limitations

    - Ultrasonic Testing Method

        - For heavier gauge material, the geometry and orientation of lamination (normal to the beam) makes their detection limited to ultrasonic testing.

        - Numerous wave modes may be used, depending upon the material thickness or method selected for testing. Automatic and manual contact or immersion methods are adaptable.

        - Laminations appear as a definite interface with a loss of back reflection.

        - Through-transmission and reflection techniques are applicable for very thin sections.

    - Magnetic Particle Testing Method

        - Articles fabricated from ferromagnetic materials are normally tested for lamination by magnetic particle testing methods.

- Magnetic indication will appear as a straight, intermittent indication.

- Magnetic particle testing is not capable of determining the overall size or depth of the lamination.

– Liquid Penetrant Testing Method

- Normally used on nonferrous materials.

- Machining, honing, lapping, or blasting may smear the surface of the material and thereby close or mask surface lamination.

- Acid and alkalines seriously limit the effectiveness of liquid penetrant testing. Thorough cleaning of the surface is essential.

– Eddy Current Testing Method

- Not normally used to detect laminations.

– Radiographic Testing Method

- Not recommended for detecting laminations. Laminations have very small thickness changes in the direction of the X-ray beam thereby making radiographic detection almost impossible.

Laps and Seams

- Category - Processing

- Material - Ferrous and Nonferrous Rolled Threads

- Discontinuity Characteristics

    Surface. Wavy lines, often quite deep and sometimes very tight, appearing as hairline cracks. Found in rolled threads in the minor pitch and major diameter of the thread and in direction of rolling (Figure A-22).

- Metallurgical Analysis

    During the rolling operation, faulty or oversized dies or an overfill of material may cause material to be folded over and flattened into the surface of the thread but not fused.

- NDT Methods Application and Limitations

    - Liquid Penetrant Testing Method

        • Compatibility with both ferrous and nonferrous materials makes fluorescent liquid penetrant the first choice.

        • Liquid penetrant indications will be circumferential, slightly curved, intermittent, or continuous indications. Laps and seams may occur individually or in clusters.

A TYPICAL AREAS OF FAILURE LAPS AND SEAMS

B FAILURE OCCURRING AT ROOT OF THREAD

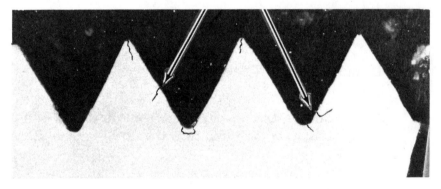

C AREAS WHERE LAPS AND SEAMS USUALLY OCCUR

Figure A-22. Lap and Seam Discontinuities in Rolled Threads

- Foreign material may not only interfere with the penetration of the penetrant into the discontinuity, but may cause an accumulation of penetrant in a nondefective area.

- Surface of threads may be smeared due to rolling operation, thereby sealing off laps and seams.

- Fluorescent and dye penetrants are not compatible. Dye penetrants tend to kill the fluorescent qualities in fluorescent penetrants.

– Magnetic Particle Testing Method

- Magnetic particle indications of laps and seams generally appear the same as liquid penetrant indications.

- Nonrelevant magnetic indications may result from threads.

- Questionable magnetic particle indications can be verified by liquid penetrant testing.

– Eddy Current Testing Method

- Probe coil design must match sample geometry.

– Ultrasonic Testing Method

- Not recommended for detecting laps and seams. Thread configurations restrict ultrasonic capability.

- Radiographic Testing Method

    • Not recommended for detecting laps and seams. Size and orientation of discontinuities restrict the capability of radiographic testing.

Laps and Seams

- Category - Processing

- Material - Ferrous and Nonferrous Wrought Material

- Discontinuity Characteristics

    - Lap Surface. Wavy lines which are usually not very pronounced nor tightly adherent since they usually enter the surface at a small angle. Laps may have surface openings which are smeared closed. Found in wrought forgings, plate, tubing, bar, and rod (Figure A-23).

    - Seam Surface. Lengthy, often quite deep, and sometimes very tight. Usually occur in fissures parallel with the grain, and, when associated with rolled rod and tubing, they may at times be spiral.

- Metallurgical Analysis

Seams originate from blowholes, cracks, splits, and tears introduced in earlier processing and elongated in the direction of rolling or forging. The distance between adjacent interfaces of the discontinuity is very small.

 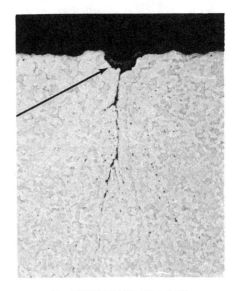

A TYPICAL FORGING LAP   B MICROGRAPH OF A LAP

Figure A-23. Lap and Seam Discontinuities in Wrought Material

Laps are similar to seams and may result from improper rolling, forging, or sizing operations. Corners may be folded over during the processing of the material or an overfill may exist during sizing that results in the material being flattened but not fused into the surface. Laps may occur on any part of the article.

- NDT Methods Application and Limitations

    - Magnetic Particle Testing Method

        • Magnetic particle testing is recommended for ferromagnetic material.

        • Surface and near-surface laps and seams may be detected by this method.

- Laps and seams may appear as straight, spiral, or slightly curved indications. They may be individual or clustered and continuous or intermittent.

- Magnetic buildup at laps and seams is very small; therefore, a magnetizing current greater than that used for the detection of cracks is necessary.

- Correct magnetizing technique should be used when examining for forging laps since the discontinuity may lie in a plane nearly parallel to the surface.

— Liquid Penetrant Testing Method

- Liquid penetrant testing is recommended for nonferrous material.

- Laps and seams may be very tight and difficult to detect, especially by liquid penetrant.

- Liquid penetrant testing of laps and seams can be improved slightly by heating the article before applying the penetrant.

— Ultrasonic Testing Method

- Normally used to test wrought material prior to machining.

- Surface wave and/or time-of-flight techniques permit accurate evaluation of the depth, length, and size of laps and seams.

- Ultrasonic indications of laps and seams will appear as definite interfaces within the metal.

— Eddy Current Testing Method

- Normally used for the evaluation of laps and seams in tubing and pipe.

- Other articles can be screened by eddy current where article configuration and size permit.

— Radiographic Testing Method

- Not recommended for detecting laps and seams in wrought material.

Microshrinkage

- Category - Processing

- Material - Magnesium Casting

- Discontinuity Characteristics

Internal. Small filamentary voids in the grain boundaries appear as concentrated porosity in cross section (Figure A-24).

- Metallurgical Analysis

Shrinkage occurs while the metal is in a plastic or semimolten state. If sufficient molten metal cannot flow into different areas as it cools, the shrinkage will leave a void. The void is identified by its appearance and by the time it occurs in the plastic range. Microshrinkage is caused by the withdrawal of the low melting point constituent from the grain boundaries.

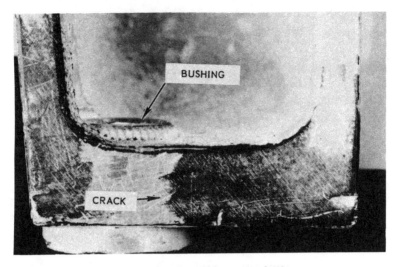

A CRACKED MAGNESIUM HOUSING

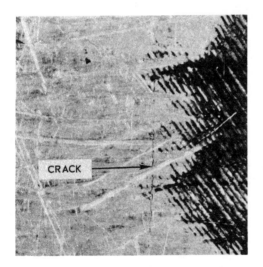

B CLOSE-UP VIEW OF (A)

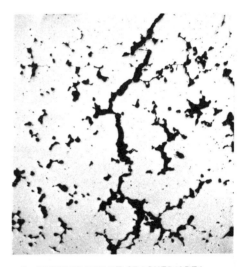

C MICROGRAPH OF CRACKED AREA

Figure A-24. Microshrinkage Discontinuity

- NDT Methods Application and Limitations

  - Radiographic Testing Method

    - Radiography is universally used to determine the acceptance level of microshrinkage.

    - Microshrinkage will appear on the radiograph as an elongated swirl resembling feathery streaks or as dark, irregular patches that are indicative of cavities in the grain boundaries.

  - Liquid Penetrant Testing Method

    - Normally used on finished machined surfaces.

    - Microshrinkage is not normally open to the surface; therefore, these conditions will be detected in machined areas.

    - The appearance of the indication depends on the plane through which the microshrinkage has been cut. The appearance varies from a continuous hairline to a massive porous indication.

    - Penetrant may act as a contaminant by saturating the microporous casting, affecting its ability to accept a surface treatment.

    - Serious structural or dimensional damage to the article can result from the improper use of acids or alkalies. They should never be used unless approval is obtained.

- Eddy Current Testing Method

    • Not recommended for detecting microshrinkage. Article configuration and type of discontinuity do not lend themselves to eddy current testing.

- Ultrasonic Testing Method

    • Not recommended for detecting microshrinkage. Cast structure and article configuration are restricting factors.

- Magnetic Particle Testing Method

    • Not applicable. Material is nonferrous.

Gas Porosity

- Category - Processing

- Material - Ferrous and Nonferrous Weldments

- Discontinuity Characteristics

    Surface or subsurface. Rounded or elongated, teardrop-shaped, with or without a sharp discontinuity at the point. Scattered uniformly throughout the weld or isolated in small groups. May also be concentrated at the root or toe (Figure A-25).

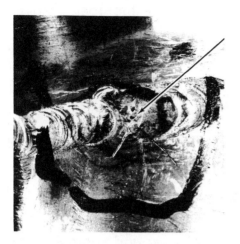

A TYPICAL SURFACE POROSITY

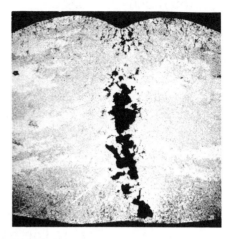

B CROSS SECTION OF (A) SHOWING EXTENT OF POROSITY

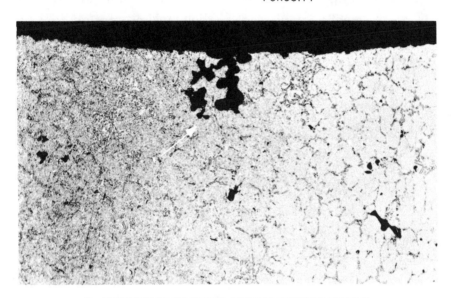

C MICROGRAPH OF CROSS SECTION SHOWING TYPICAL SHRINKAGE POROSITY

Figure A-25. Gas Porosity Discontinuity

- Metallurgical Analysis

Porosity in welds is caused by gas entrapment in the molten metal,

A-69

too much moisture on the base or filler metal, or improper cleaning or preheating.

- **NDT Methods Application and Limitations**

  - Radiography Testing Method

    - Radiography is the most universally used NDT method for the detection of gas porosity in weldments.

    - The radiographic image of a "round" porosity will appear as oval-shaped spots with smooth edges, while "elongated" porosity will appear as oval-shaped spots with the major axis sometimes several times longer than the minor axis.

    - Foreign material such as loose scale, flux, or splatter will affect validity of test results.

  - Ultrasonic Testing Method

    - Ultrasonic testing equipment is highly sensitive and is capable of detecting microseparations. Established standards should be used if valid test results are to be obtained.

    - Surface finish and grain size will affect the validity of the test results.

- Eddy Current Testing Method

    • Normally confined to thin-walled welded pipe and tube.

    • Penetration restricts testing to a depth of more than 0.25 inch (6.35 mm).

- Liquid Penetrant Testing Method

    • Normally confined to in-process control of ferrous and nonferrous weldments.

    • Liquid penetrant testing, like magnetic particle testing, is restricted to surface evaluation.

    • Extreme caution must be exercised to prevent any cleaning material, magnetic (iron oxide), and liquid penetrant materials from becoming entrapped and contaminating the rewelding operation.

- Magnetic Particle Testing Method

    • Not normally used to detect gas porosity. Only surface porosity would be evident. Near surface porosity would not be clearly defined since indications are neither strong nor pronounced.

Unfused Porosity

- Category - Processing

- Material - Aluminum

- Discontinuity Characteristics

    Internal. Wafer-thin fissures aligned parallel with the grain flow. Found in wrought aluminum that has been rolled, forged or extruded (Figure A-26).

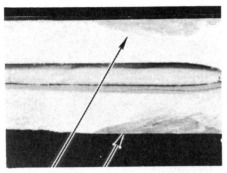

A. FRACTURED SPECIMEN SHOWING UNFUSED POROSITY

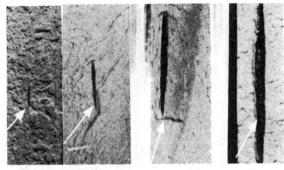

B. UNFUSED POROSITY EQUIVALENT TO 1/64 IN. (0.40 mm), 3/64 IN. (1.17 mm) 5/64 IN. (1.98 mm) AND 8/64 IN. (3.18 mm) (left to right)

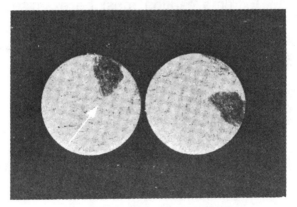

C. TYPICAL UNFUSED POROSITY

Figure A-26. Unfused Porosity Discontinuity

- Metallurgical Analysis

    Unfused porosity is attributed to porosity in the cast ingot. During the rolling, forging or extruding operations it is flattened into a wafer-thin shape. If the internal surface of these discontinuities is oxidized or is composed of a foreign material, they will not fuse during the subsequent processing. This results in an extremely thin interface or void.

- NDT Methods Application and Limitations

    - Ultrasonic Testing Method

        - Used extensively for the detection of unfused porosity.

        - Raw materials may be tested in the "as-received" configuration.

        - Ultrasonic testing fixes the location of the void in all three directions.

        - Where the general direction of the discontinuity is unknown, it may be necessary to test from several directions.

        - Method of manufacture and subsequent article configuration will determine the orientation of the unfused porosity to the material surface.

    - Liquid Penetrant Testing Method

        - Normally used on nonferrous machined articles.

        - Unfused porosity will appear as a straight line of varying lengths running parallel with the grain. Liquid penetrant testing is restricted to surface evaluation.

- Surface preparations such as vapor blasting, honing, grinding, or sanding may obliterate possible indications by masking the surface discontinuities and thereby restricting the reliability of liquid penetrant testing.

- Excessive agitation of penetrant materials may produce foaming.

— Eddy Current Testing Method

- Not normally used for detecting unfused porosity.

— Radiographic Testing Method

- Not normally used for detecting unfused porosity. Wafer-thin discontinuities are difficult to detect by a method that measures density or that requires that the discontinuity be perpendicular to the X-ray beam.

— Magnetic Particle Testing Method

- Not applicable. Material is nonferrous.

Stress Corrosion

- Category - Service

- Material - Ferrous and Nonferrous

- Discontinuity Characteristics

- Surface. Range from shallow to very deep, and usually follow the grain flow of the material; however, transverse cracks are also possible (Figure A-27).

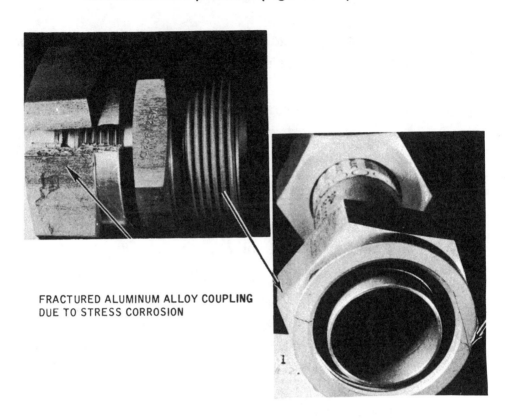

FRACTURED ALUMINUM ALLOY COUPLING
DUE TO STRESS CORROSION

Figure A-27. Stress Corrosion Discontinuity

- Metallurgical Analysis

The following three factors are necessary for the phenomenon of stress corrosion to occur: 1) a sustained static tensile stress, 2) the presence of a corrosive environment, and 3) the use of a material that is susceptible to this type of failure. Stress corrosion is much more likely to occur at high levels of stress than at low levels of stress. The type of stresses include residual (internal) as well as those from external (applied) loading.

A-75

- NDT Methods Application and Limitations

  - Liquid Penetrant Testing Method

    - Liquid penetrant is normally used for the detection of stress corrosion.

    - In the preparation, application, and final cleaning of articles, extreme care must be exercised to prevent overspraying and contamination of the surrounding articles.

    - Chemical cleaning immediately before the application of liquid penetrant may seriously affect the test results if the solvents are not given time to evaporate.

    - Service articles may contain moisture within the discontinuity which will dilute, contaminate, and invalidate results if the moisture is not removed.

  - Ultrasonic Testing Method

    - Advanced techniques have been successfully used to detect stress corrosion and stress corrosion cracking in the nuclear industry.

    - Indications appear in a variety of amplitudes, shapes, and characteristics.

    - Interpretation is often difficult, requiring highly-trained operators.

- Eddy Current Testing Method

    - Eddy current equipment is capable of resolving stress corrosion where article configuration is compatible with equipment limitations.

- Magnetic Particle Testing Method

    - Not normally used to detect stress corrosion. Configuration of article and usual nonferromagnetic condition exclude magnetic particle testing.

- Radiographic Testing Method

    - Not normally used to detect stress corrosion. Surface indications are best detected by NDT method designed for such applications; however, radiography can and has shown stress corrosion with the use of the proper technique.

Hydraulic Tubing

- Category - Processing and Service

- Material - Aluminum

- Discontinuity Characteristics

Surface and internal. Range in size from short to long, shallow to very tight and deep. Usually they will be found in the direction of the grain flow with the exception of stress corrosion which has no direction (Figure A-28).

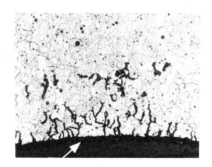
A INTERGRANULAR CORROSION

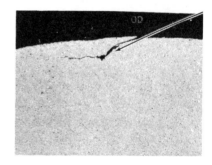
B LAP IN OUTER SURFACE OF TUBING

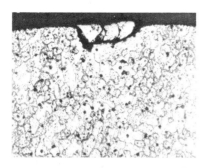
C EMBEDDED FOREIGN MATERIAL

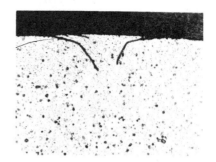
D TWIN LAPS IN OUTER SURFACE OF TUBING

Figure A-28. Hydraulic Tubing Discontinuities

- Metallurgical Analysis

    Hydraulic tubing discontinuities are usually one of the following.

    - Foreign material coming in contact with the tube material and being embedded into the surface of the tube.

    - Laps which are the result of material being folded over and not fused.

- Seams which originate from blowholes, cracks, splits, and tears introduced in the earlier processing, and then are elongated during rolling.

- Intergranular corrosion which is due to the presence of a corrosive environment.

- NDT Methods Application and Limitations

    - Eddy Current Testing Method

        - Universally used for testing of nonferrous tubing.

        - Heavier-walled tubing (0.25 inch or 6.35 mm and over) may not be successfully tested due to the penetration ability of the equipment.

        - The specific nature of various discontinuities may not be clearly defined.

        - Test results will not be valid unless controlled by known standards.

        - Testing of ferromagnetic material may be difficult.

        - All material should be free of any foreign material that would invalidate the test results.

    - Liquid Penetrant Testing Method

        - Not normally used for detecting tubing discontinuities. Eddy current is more economical, faster, and, with established standards, is more reliable.

- Ultrasonic Testing Method

    • Not normally used for detecting tubing discontinuities. Eddy current is recommended over ultrasonic testing since it is faster and more economical for this range of surface discontinuity and nonferrous material.

- Radiographic Testing Method

    • Not normally used for detecting tubing discontinuities. The size and type of discontinuity and the configuration of the article limit the use of radiography for screening of material for this group of discontinuities.

- Magnetic Particle Testing Method

    • Not applicable. Material is nonferrous.

Mandrel Drag

- Category - Processing

- Material - Nonferrous, Thick-walled Seamless Tubing

- Discontinuity Characteristics

    Internal surface of thick-walled tubing. Range from shallow, even gouges to ragged tears. Often a slug of the material will be embedded within the gouged area (Figure A-29).

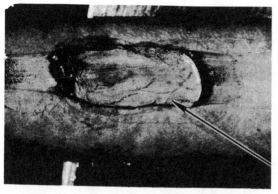
A. EMBEDDED SLUG SHOWING DEEP GOUGE MARKS

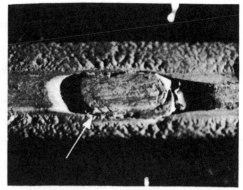
B. SLUG BROKEN LOOSE FROM TUBING WALL

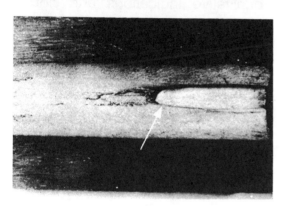
C. ANOTHER TYPE OF EMBEDDED SLUG

D. GOUGE ON INNER SURFACE OF PIPE

Figure A-29. Mandrel Drag Discontinuities

- Metallurgical Analysis

During the manufacture of thick-walled seamless tubing, the billet is ruptured as it passes through the offset rolls. As the piercing mandrel follows this fracture, a portion of the material may break loose and be forced over the mandrel. As it does, the surface of the tubing may be scored or have the slug embedded into the wall. Certain types of material are more prone to this type of failure than others.

- NDT Methods Application and Limitations

  - Eddy Current Testing Method

    - Normally used for the testing of thin-walled pipe or tube.

    - Eddy current testing may be confined to nonferrous materials.

    - Discontinuities are qualitative indications and not quantitative indications.

    - Several factors simultaneously affect output indications.

  - Ultrasonic Testing Method

    - Normally used for the screening of thick-walled pipe or tube for mandrel drag.

    - Can be used to test both ferrous and nonferrous pipe or tube.

    - May be used in support of production line, since it is adaptable for automatic instrumentation.

    - Configuration of mandrel drag or tear will produce very sharp and noticeable indications on the scope.

  - Radiographic Testing Method

    - Not normally used although it has been instrumental

in the detection of mandrel drag during examination of adjacent welds. Complete coverage requires several exposures around the circumference of the tube. This method is not designed for production support since it is very slow and costly for large volumes of pipe or tube. Radiograph will disclose only two dimensions and not the third.

- Liquid Penetrant Testing Method

    • Not recommended for detecting mandrel drag since discontinuity is internal and would not be detectable.

- Magnetic Particle Testing Method

    • Not recommended for detecting mandrel drag. Discontinuities are not close enough to the surface to be detectable by magnetic particles. Most mandrel drag will occur in seamless stainless steel.

Semiconductors

- Category - Processing and Service

- Material - Hardware

- Discontinuity Characteristics

    Internal. Appear in many sizes and shapes and various degrees of density. They may be misformed, misaligned, damaged, or may have broken internal hardware. Found in transistors, diodes, resistors, and capacitors (Figure A-30).

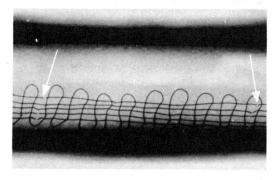
A. STRANDS BROKEN IN HEATER BLANKET

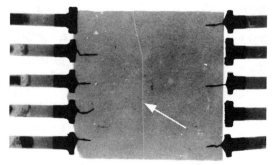
B. FINE CRACK IN PLASTIC CASING MATERIAL

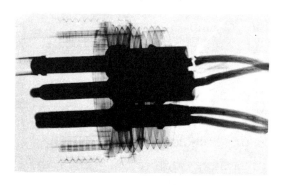
C. BROKEN ELECTRICAL CABLE

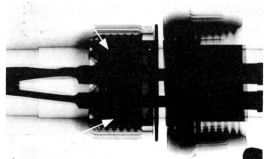
D. FOREIGN MATERIAL WITHIN SEMICONDUCTOR

Figure A-30. Semiconductor Discontinuities

- Metallurgical Analysis

    Semiconductor discontinuities such as loose wire, weld splash, flakes, solder balls, loose leads, inadequate clearance between internal elements and case and inclusions or voids in seals or around lead connections are the product of processing errors.

- NDT Methods Application and Limitations

    - Radiographic Testing Method

- Universally used as the NDT method for the detection of discontinuities in semiconductors.

- The configuration and internal structure of the various semiconductors limit the NDT method of radiography.

- Semiconductors that have copper heat sinks may require more than one technique due to the density of the copper.

- Internal wires in semiconductors are very fine and may be constructed from materials of different density such as copper, silver, gold, and aluminum. If the latter is used with the others, special techniques may be needed to resolve test reliability.

- Microparticles may require the highest sensitivity to resolve.

- The complexity of the internal structure of semiconductors may require additional views to exclude the possibility of nondetection of discontinuities due to masking by hardware.

- Positive positioning of each semiconductor will prevent invalid interpretation.

- Source angle should give minimum distortion.

- Preliminary examination of semiconductors may be accomplished using a vidicon system that would allow visual observation during 360° rotation of the article.

- Eddy Current Testing Method

    • Not recommended for detecting semiconductor discontinuities. Nature of discontinuity and method of construction of the article do not lend themselves to this form of NDT.

- Magnetic Particle Testing Method

    • Not recommended for detecting semiconductor discontinuities.

- Liquid Penetrant Testing Method

    • Not recommended for detecting semiconductor discontinuities.

- Ultrasonic Testing Method

    • Not recommended for detecting semiconductor discontinuities.

Hot Tears

- Category - Inherent

- Material - Ferrous Castings

- Discontinuity Characteristics

    Internal or near surface. Appear as ragged line of variable width and numerous branches. Occur individually or in groups (Figure A-31).

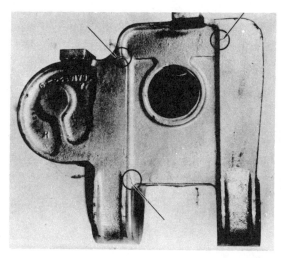

A. TYPICAL HOT TEARS IN CASTING

B. HOT TEARS IN FILLET OF CASTING

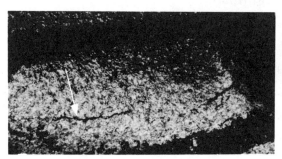

C. CLOSE-UP OF HOT TEARS IN (A)

D. CLOSE-UP OF HOT TEARS IN (B)

Figure A-31. Hot Tear Discontinuities

- Metallurgical Analysis

Hot cracks (tears) are caused by nonuniform cooling resulting in stresses which rupture the surface of the metal while its temperature is still in the brittle range. Tears may originate where stresses are set up by the more rapid cooling of thin sections that adjoin heavier masses of metal which are slower to cool.

- NDT Methods Application and Limitations

  - Radiographic Testing Method

    - Radiographic testing is the first choice since the material is cast structure and the discontinuities may be internal and surface.

    - Orientation of the hot tear in relation to the source may influence the test results.

    - The sensitivity level may not be sufficient to detect fine surface hot tears.

  - Magnetic Particle Testing Method

    - Hot tears that are exposed to the surface can be screened with magnetic particle method.

    - Article configuration and metallurgical composition may make demagnetization difficult.

    - Although magnetic particle testing can detect near surface hot tears, radiography should be used for final analysis.

    - Foreign material not removed prior to testing will cause an invalid test.

- Liquid Penetrant Testing Method

    - Liquid penetrant testing is recommended for nonferrous cast material.

    - Method is confined to surface evaluation.

    - The use of penetrants on castings may act as a contaminant by saturating the porous structure and thereby affecting the ability to apply surface finish.

    - Repeatability of indications may be poor.

- Ultrasonic Testing Method

    - Not recommended for detecting hot tears. Discontinuities of this type, when associated with cast structure, do not lend themselves to ultrasonic testing.

- Eddy Current Testing Method

    - Capable of detecting surface hot tears. Metallurgical structure, along with the complex configurations, may require specialized probes and techniques.

Intergranular Corrosion

- Category - Service

- Material - Nonferrous

- Discontinuity Characteristics

    Surface or internal. A series of small micro-openings with no definite pattern. May appear individually or in groups. The insidious nature of intergranular corrosion results from the fact that very little corrosion or corrosion product is visible on the surface. Intergranular corrosion may extend in any direction following the grain boundaries of the material (Figure A-32).

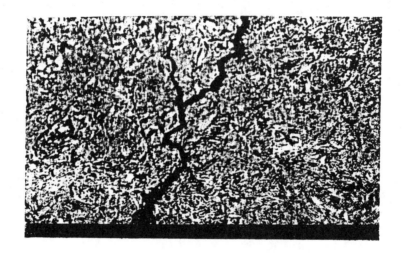

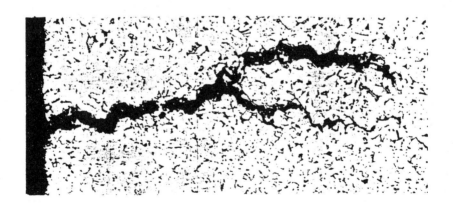

Figure A-32. Intergranular Corrosion Discontinuities

- Metallurgical Analysis

    Two factors that contribute to intergranular corrosion are:

    - Metallurgical structure of the material that is prone to intergranular corrosion, such as unstabilized 300 series stainless steel.

    - Improper stress relieving or heat treat may create the susceptibility to intergranular corrosion.

    Either of these conditions, coupled with a corrosive atmosphere, will result in intergranular attack.

- NDT Methods Application and Limitations

    - Liquid Penetrant Testing Method

        - Liquid Penetrant testing is the first choice due to the size and location of this type of discontinuity.

        - Chemical cleaning operations immediately before the application of liquid penetrant may contaminate the article and seriously affect test results.

        - Cleaning with solvents may release chlorine and accelerate intergranular corrosion.

        - Trapped penetrant solution may present a cleaning or removal problem.

- Ultrasonic Testing Method

    - Advanced techniques have been successfully used to detect stress corrosion and stress corrosion cracking in the nuclear industry.

    - Indications appear in a variety of amplitudes, shapes, and characteristics lending difficult interpretation.

- Eddy Current Testing Method

    - Eddy current can be used for the screening of intergranular corrosion.

    - Tube or pipe lend themselves readily to this method of NDT testing.

    - Metallurgical structure of the material may seriously affect the output indications.

- Radiographic Testing Method

    - Intergranular corrosion in the more advanced stages has been detected with radiography.

    - Sensitivity levels may prevent the detection of fine intergranular corrosion.

    - Radiography may not indicate the surface on which the intergranular corrosion occurs.

- Magnetic Particle Testing Method

    • Not recommended for detecting intergranular corrosion. Type of discontinuity and material restrict the use of magnetic particles.

APPENDIX B

GLOSSARY

A-scan Display A display in which the received signal is displayed as a vertical displacement from the horizontal sweep time trace, while the horizontal distance between any two signals represents the sound-path distance (or time of travel) between the two.

Absorption Coefficient, Linear The fractional decrease in transmitted intensity per unit of absorber thickness. It is usually designated by the symbol μ and expressed in units of cm^{-1}.

Acceptance Standard A control specimen containing natural or artificial discontinuities that are well defined and, in size or extent, similar to the maximum acceptable in the product. Also may refer to the document defining acceptable discontinuity size limits.

Acoustic Impedance The factor which controls the propagation of an ultrasonic wave at a boundary interface. It is the product of the material density and the acoustic wave velocity within that material.

Amplifier A device to increase or amplify electrical impulses.

Amplitude, Indication The vertical height of a received indication, measured from base-to-peak or peak-to-peak.

Angle Beam Testing A testing method in which transmission is at an angle to the sound entry surface.

Angle of Incidence The angle between the incident (transmitted) beam and a normal to the boundary interface.

Angle of Reflection The angle between the reflected beam and a normal to the boundary interface. The angle of reflection is equal to the angle of incidence.

Angle of Refraction The angle between the refracted rays of an ultrasonic beam and the normal (or perpendicular line) to the refracting surface.

Angle Transducer A transducer that transmits or receives the acoustic energy at an acute angle to the surface to achieve a specific effect such as the setting up of shear or surface waves in the part being inspected.

Anisotropic A condition in which properties of a medium (velocity, for example) vary according to the direction in which they are measured.

Array Transducer A transducer made up of several piezoelectric elements individually connected so that the signals they transmit or receive may be treated separately or combined as desired.

ANSI American National Standards Institute

API American Petroleum Institute

ASME American Society of Mechanical Engineers

ASNT American Society for Nondestructive Testing

ASTM American Society for Testing and Materials

Attenuation Coefficient A factor which is determined by the degree of scatter or absorption of ultrasound energy per unit distance traveled.

Attenuation The loss in acoustic energy which occurs between any two points of travel. This loss may be due to absorption, reflection, scattering, etc.

Attenuator A device for measuring attenuation, usually calibrated in decibels (dB).

B-scan Display A data presentation method that represents a cross-sectional or end view display of the test article.

Back Reflection The signal received from the back surface of a test object. Also referred to as back wall reflection.

Back Scatter Scattered signals that are directed back to the transmitter/receiver.

Background Noise Extraneous signals caused by signal sources within the ultrasonic testing system, including the material in test.

Baseline The horizontal line across the bottom of the CRT created by the sweep circuit.

Basic Calibration The procedure of standardizing an instrument using calibration reflectors described in an application document.

Beam Exit/Index Point The point on a transducer (primarily angle beam) indicating the physical location through which the emergent beam axis passes.

Beam Spread The divergence of the sound beam as it travels through a medium.

Bi-modal   The propagation of sound in a test article where at least a shear wave and a longitudinal wave exists. The operation of angle beam testing at less than first critical angle.

Boundary Indication A reflection of an ultrasonic beam from an interface.

Broad Banded   Having a relatively wide frequency bandwidth. Used to describe pulses which display a wide frequency spectrum and receivers capable of amplifying them.

C-scan A data presentation method yielding a plan (top) view through the scanned surface of the part. Through gating, only indications arising from the interior of the test object are indicated.

Calibration   To determine or mark the graduations of the ultrasonic system's display relative to a known standard or reference.

Calibration Reflector   A reflector with a known dimensioned surface established to provide an accurately reproducible reference.

Collimator   An attachment designed to reduce the ultrasonic beam spread.

Compensator   An electrical matching network to compensate for circuit impedance differences.

Compressional Wave   A wave in which the particle motion or vibration is in the same direction as the propagated wave (longitudinal wave).

Contact Testing A technique of testing in which the transducer contacts the test surface, either directly or through a thin layer of couplant.

Contact Transducer A transducer which is coupled to a test surface either directly or through a thin film of couplant.

Continuous Wave A wave that continues without interruption.

Contracted Sweep A contraction of the horizontal sweep on the viewing screen of the ultrasonic instrument. Contraction of this sweep permits viewing reflections occurring over a greater sound-path distance or duration of time.

Corner Effect The strong reflection obtained when an ultrasonic beam is directed toward the inner section of two or three mutually perpendicular surfaces.

Couplant A substance used between the face of the transducer and test surface to permit or improve transmission of ultrasonic energy across this boundary or interface. Primarily used to remove the air in the interface.

Critical Angle The incident angle of the sound beam beyond which a specific refracted mode of vibration no longer exists.

Cross Talk An unwanted condition in which acoustic energy is coupled from the transmitting crystal to the receiving crystal without propagating along the intended path through the material.

Damping (transducer) Limiting the duration of vibration in the search unit by either electrical or mechanical means.

Dead Zone The distance in a material from the sound entry surface to the nearest inspectable sound path.

Decibel (dB) The logarithmic expression of a ratio of two amplitudes or intensities of acoustic energy.

Defect/Flaw A material discontinuity whose size, shape, orientation, or location make it detrimental to the useful service of the test object or component.

Defect Indication The oscilloscope presentation of the energy returned by a rejectable flaw in the material.

Delamination A laminar discontinuity, generally an area of unbonded materials.

Delay Line A material (liquid or solid) placed in front of a transducer to cause a time delay between the initial pulse and the front surface reflection.

Delayed Sweep A means of delaying the start of horizontal sweep, thereby eliminating the presentation of early response data.

Delta Effect Acoustic energy re-radiated by a discontinuity.

Detectability The ability of the ultrasonic system to locate a discontinuity.

Diffraction The deflection, or "bending," of a wave front when passing the edge or edges of a discontinuity.

Diffuse Reflection Scattered, incoherent reflections caused by rough surfaces or associate interface reflection of ultrasonic waves from irregularities of the same order of magnitude or greater than the wavelength.

Discontinuity An interruption or change in the physical structure or characteristics of a material.

Dispersion, Sound Scattering of an ultrasonic beam as a result of diffused reflection from a highly-irregular surface.

Distance Amplitude Correction (DAC) Compensation of gain as a function of time for difference in amplitude of reflections from equal reflectors at different sound travel distances. Also referred to as time corrected gain (TCG), time variable gain (TVG) and sensitivity time control (STC).

Divergence Spreading of ultrasonic waves after leaving search unit, and is a function of diameter and frequency.

Dual-Element Technique The technique of ultrasonic testing using two transducers with one acting as the transmitter and one as the receiver.

Dual-Element Transducer A single transducer housing containing two piezoelectric elements, one for transmitting and one for receiving.

Echo See **Boundary Indication**.

Effective Penetration The maximum depth in a material at which the ultrasonic transmission is sufficient for proper detection of discontinuities.

Electrical Noise Extraneous signals caused by externally radiated electrical signals or from electrical interferences within the ultrasonic instrumentation.

Electromagnetic Acoustic Transducer (EMAT) A device using the magneto effect to generate and receive acoustic signals for ultrasonic nondestructive tests.

Evaluation The process of deciding the severity of a condition after an indication has been interpreted. Evaluation determines if the test object should be rejected, repaired, accepted, or replaced.

Far Field   The region beyond the near field in which areas of high and low acoustic intensity cease to occur.

First Leg   The sound path beginning at the exit point of the probe and extending to the point of contact opposite the examination surface when performing angle beam testing.

Focused Transducer   A transducer with a concave face which converges the acoustic beam to a focal point or line at a defined distance from the face.

Focusing   Concentration or convergence of energy into a smaller beam.

Fraunhofer Zone   See **Far Field**.

Frequency   Number of complete cycles of a wave motion passing a given point in a unit time (1 second); number of times a vibration is repeated at the same point in the same direction per unit time (usually per second).

Fresnel Field   See **Near Field**.

Gate   An electronic means to monitor an associated segment of time, distance, or impulse.

Ghost   An indication which has no direct relation to reflected pulses from discontinuities in the materials being tested.

Hertz (Hz)   One cycle per second.

Horizontal Sweep   See **Baseline**.

Horizontal Linearity A measure of the proportionality between the positions of the indications appearing on the baseline and the positions of their sources.

Immersion Testing A technique of testing, using a liquid as an ultrasonic couplant, in which the test part and at least the transducer face is immersed in the couplant and the transducer is not in contact with the test part.

Impedance (acoustic) A material characteristic defined as a product of particle velocity and material density.

Indication (ultrasonics) The signal displayed or read on the ultrasonic systems display.

Initial Pulse The first indication which may appear on the screen. This indication represents the emission of ultrasonic energy from the crystal face (main bang).

Interface The physical boundary between two adjacent acoustic mediums.

Insonification Irradiation with sound.

Interpretation The determination of the source and relevancy of an indication.

Isotropy A condition in which significant medium properties (velocity, for example) are the same in all directions.

Lamb Wave A type of ultrasonic vibration guided by parallel surfaces of thin mediums capable of propagation in different modes.

Linearity (area) A system response in which a linear relationship exists between amplitude of response and the discontinuity sizes being evaluated (necessarily limited by the size of the ultrasonic beam).

Linearity (depth) A system response where a linear relationship exists with varying depth for a constant size discontinuity.

Longitudinal Wave See **Compressional Wave**.

Longitudinal Wave Velocity The unit speed of propagation of a longitudinal (compressional) wave through a material.

Loss of Back Reflection Absence of or a significant reduction of an indication from the back surface of the article being inspected.

Major Screen Divisions The vertical graticule used to divide the CRT into 10 equal horizontal segments.

Manipulator A device used to orient the transducer assembly. As applied to immersion techniques, it provides either angular or normal incidence and fixes the transducer-to-part distance.

Material Noise Extraneous signals caused by the structure of the material being tested.

Miniature Angle Beam Block A specific type of reference standard used primarily for the angle beam method, but also used for straight beam and surface wave tests.

Minor Screen Divisions The vertical graticule used to divide the CRT into 50 equal segments. Each major screen division is divided into 5 equal segments or minor divisions.

Mode Conversion The change of ultrasonic wave propagation upon reflection or refraction at acute angles at an interface.

Mode The manner in which acoustic energy is propagated through a material as characterized by the particle motion of the wave.

Multiple Back Reflections Repetitive indications from the back surface of the material being examined.

Nanosecond One billionth (10^{-9}) of a second.

Narrow Banded A relative term denoting a restricted range of frequency response.

Near Field A distance immediately in front of a transducer composed of complex and changing wave front characteristics. Also known as the Fresnel field.

Node The point on the examination surface where the V-path begins or ends. (See **V-path**)

Noise Any undesired indications that tend to interfere with the interpretation or processing of the ultrasonic information; also referred to as "grass."

Nonrelevant Indication See **Ghost**.

Normal Incidence A condition where the angle of incidence is zero.

Orientation The angular relationship of a surface, plane, defect axis, etc., to a reference plane or sound entry surface.

Penetration (ultrasonic)   Propagation of ultrasonic energy through an article. See **Effective Penetration**.

Phased Array   A mosaic of probe elements in which the timing of the element's excitation can be individually controlled to produce certain desired effects, such as steering the beam axis or focusing the beam.

Piezoelectric Effect   The characteristic of certain materials to generate electrical charges when subjected to mechanical vibrations and, conversely, to generate mechanical vibrations when subjected to electrical pulses.

Pitch-Catch   See **Two-Probe Method**.

Polarized Ceramics   Ceramic materials that are sintered (pressed), heated (approximately 1000°C), and polarized by applying a direct voltage of a few thousand volts per centimeter of thickness. The polarization is the process that makes these ceramics piezoelectric. Includes sodium bismuth titanate, lead metaniobate, and several materials based on lead zirconate titanate (PZT).

Presentation   The method used to show ultrasonic information. This may include (among others) A-, B-, or C-scans displayed on various types of recorders, CRTs, LCDs or computerized displays.

Probe   Transducer or search unit.

Propagation   Advancement of a wave through a medium.

Pulse-Echo Technique   An ultrasonic test technique using equipment which transmits a series of pulses separated by a constant period of time; i.e., energy is not sent out continuously.

Pulse Length   Time duration of the pulse from the search unit.

Pulse Rate   For the pulse-echo technique, the number of pulses transmitted in a unit of time (also called pulse repetition rate).

Radio Frequency Display (RF)   The presentation of unrectified signals in a display.

Range   The maximum ultrasonic path length that is displayed.

Rarefaction   The thinning out or moving apart of the consistent particles in the propagating medium due to the relaxation phase of an ultrasonic cycle. Opposite in its effect to compression. The sound wave is composed of alternate compressions and refractions of the particles in a material.

Rayleigh Wave/Surface Wave   A wave that travels on or close to the surface and readily follows the curvature of the part being examined. Reflections occur only at sharp changes of direction of the surface.

Receiver   The section of the ultrasonic instrument that amplifies the electronic signals returning from the test specimen. Also, the probe that receives the reflected signals.

Reference Blocks   A block or series of blocks of material containing artificial or actual discontinuities of one or more reflecting areas at one or more distances from the sound entry surface. These are used for calibrating instruments and in defining the size and distance of discontinuous areas in materials.

Reflection   The characteristic of a surface to change the direction of propagating acoustic energy; the return of sound waves from surfaces.

Refraction A change in the direction and velocity of acoustic energy after it has passed at an acute angle through an interface between two different mediums.

Refractive Index The ratio of the velocity of a incident wave to the velocity of the refracted wave. It is a measure of the amount a wave will be refracted when it enters the second medium after leaving the first.

Reject/Suppression An instrument function or control used for reducing low amplitude signals. Use of this control may affect vertical linearity.

Relevant Indication In NDT, an indication from a discontinuity requiring evaluation.

Repetition Rate The rate at which the individual pulses of acoustic energy are generated; also **Pulse Rate**.

Resolving Power The capability measurement of an ultrasonic system to separate in time two closely-spaced discontinuities or to separate closely-spaced multiple reflections.

Resonance Technique A technique using the resonance principle for determining velocity, thickness or presence of laminar discontinuities.

Resonance The condition in which the frequency of a forcing vibration (ultrasonic wave) is the same as the natural vibration frequency of the propagation body (test object), resulting in large amplitude vibrations.

Saturation (scope) A term used to describe an indication of such a size as to exceed full screen height (100 percent).

Scanning (manual and automatic) The moving of the search unit or units along a test surface to obtain complete testing of a material.

Scattering Dispersion of ultrasonic waves in a medium due to causes other than absorption. See **Diffuse** and **Dispersion**.

Second Leg The sound path beginning at the point of contact on the opposite surface and extending to the point of contact on the examination surface when performing angle beam testing.

Sensitivity The ability to detect small discontinuities at given distances. The level of amplification at which the receiving circuit in an ultrasonic instrument is set.

Shear Wave The wave in which the particles of the medium vibrate in a direction perpendicular to the direction of propagation.

Signal-to-Noise Ratio (SNR) The ratio of amplitudes of indications from the smallest discontinuity considered significant and those caused by random factors, such as heterogeneity in grain size, etc.

Skip Distance In angle beam tests of plate, pipe, or welds, the linear or surface distance from the sound entry point to the first reflection point on the same surface.

Snell's Law The law that defines the relationship between the angle of incidence and the angle of refraction across an interface, based on a change in ultrasonic velocity.

Specific Acoustic Impedance A characteristic which acts to determine the amount of reflection which occurs at an interface and represents the wave velocity and the product of the density of the medium in which the wave is propagating.

Standardize See **Calibration**.

Straight Beam   An ultrasonic wave traveling normal to the test surface.

Surface Wave   See **Rayleigh Wave**.

Sweep   The uniform and repeated movement of a spot across the screen of a CRT to form the baseline.

Through-Transmission   A test technique using two transducers in which the ultrasonic vibrations are emitted by one and received by the other, usually on the opposite side of the part. The ratio of the magnitudes of vibrations transmitted and received is used as the criterion of soundness.

Tip Diffraction   The process by which a signal is generated from the tip (i.e., top of a fatigue crack) of a discontinuity through the interruption of an incident sound beam propagating through a material.

Transducer (search unit)   An assembly consisting basically of a housing, piezoelectric element, backing material, wear plate (optional) and electrical leads for converting electrical impulses into mechanical energy and vice versa.

Transmission Angle   The incident angle of the transmitted ultrasonic beam. It is zero degrees when the ultrasonic beam is perpendicular to the test surface.

Transmitter   The electrical circuit of an ultrasonic instrument that generates the pulses emitted to the search unit. Also the probe that emits ultrasonic signals.

Transverse Wave   See **Shear Wave**.

Two-Probe Method   Use of two transducers for sending and receiving. May be either send-receive or through-transmission.

Ultrasonic Absorption   A damping of ultrasonic vibrations that occurs when the wave transverses a medium.

Ultrasonic Spectrum   The frequency span of elastic waves greater than the highest audible frequency, generally regarded as being higher than 20,000 hertz, to approximately 1000 megahertz.

Ultrasonic System   The totality of components utilized to perform an ultrasonic test on a test article.

Ultrasonic Testing   A nondestructive method of inspecting materials by the use of high-frequency sound waves into or through them.

V-path   The path of the ultrasonic beam in the test object from the point of entry on the examination surface to the back surface and reflecting to the front surface again.

Velocity   The speed at which sound travels through a medium.

Video Presentation   A CRT presentation in which radio frequency signals have been rectified and usually filtered.

Water Path   The distance from the face of the search unit to the entry surface of the material under test in immersion testing.

Wavelength   The distance in the direction of propagation for a wave to go through one complete cycle.

Wedge/Shoe   A device used to adapt a straight beam probe for use in a specific type of testing, including angle beam or surface wave tests and tests on curved surfaces.

Wrap Around Nonrelevant indications that appear on the CRT as a result of a short pulse repetition rate in the pulser circuit of the test instrument.

APPENDIX C

THREE-PLACE VALUES OF TRIGONOMETRIC FUNCTIONS

Deg.	Sin	Tan	Sec	Csc	Cot	Cos	Deg.
0°	.000	.000	1.000	---	---	1.000	90°
1°	.017	.017	1.000	57.30	57.29	1.000	89°
2°	.035	.035	1.001	28.65	28.64	0.999	88°
3°	.052	.052	1.001	19.11	19.08	.999	87°
4°	.070	.070	1.002	14.34	14.30	.998	86°
5°	.087	.087	1.004	11.47	11.43	996	85°
6°	.105	.105	1.006	9.567	9.514	.995	84°
7°	.122	.123	1.008	8.206	8.144	.993	83°
8°	.139	.141	1.010	7.185	7.115	.990	82°
9°	.156	.158	1.012	6.392	6.314	.988	81°
10°	.174	.176	1.015	5.759	5.671	.985	80°
11°	.191	.194	1.019	5.241	5.145	.982	79°
12°	.208	.213	1.022	4.810	4.705	.978	78°
13°	.225	.231	1.026	4.445	4.331	.974	77°
14°	.242	.249	1.031	4.134	4.011	.970	76°
15°	.259	.268	1.035	3.864	3.732	.966	75°
16°	.276	.287	1.040	3.628	3.487	.961	74°
17°	.292	.306	1.046	3.420	3.271	.956	73°
18°	.309	.325	1.051	3.236	3.078	.951	72°
19°	.326	.344	1.058	3.072	2.904	.946	71°
20°	.342	.364	1.064	2.924	2.747	.940	70°
21°	.358	.384	1.071	2.790	2.605	.934	69°
22°	.375	.404	1.079	2.669	2.475	.927	68°
23°	.391	.424	1.086	2.559	2.356	.921	67°
24°	.407	.445	1.095	2.459	2.246	.914	66°
25°	.423	.466	1.103	2.366	2.145	.906	65°
26°	.438	.488	1.113	2.281	2.050	.899	64°
27°	.454	.510	1.122	2.203	1.963	.891	63°
28°	.469	.532	1.133	2.130	1.881	.883	62°
29°	.485	.554	1.143	2.063	1.804	.875	61°
30°	.500	.577	1.155	2.000	1.732	.866	60°
31°	.515	.601	1.167	1.942	1.664	.857	59°
32°	.530	.625	1.179	1.887	1.600	.848	58°
33°	.545	.649	1.192	1.836	1.540	.839	57°
34°	.559	.675	1.206	1.788	1.483	.829	56°
35°	.574	.700	1.221	1.743	1.428	.819	55°
36°	.588	.727	1.236	1.701	1.376	.809	54°
37°	.602	.754	1.252	1.662	1.327	.799	53°
38°	.616	.781	1.269	1.624	1.280	.788	52°
39°	.629	.810	1.287	1.589	1.235	.777	51°
40°	.643	.839	1.305	1.556	1.192	.766	50°
41°	.656	.869	1.325	1.524	1.150	.755	49°
42°	.669	.900	1.346	1.494	1.111	.743	48°
43°	.682	.933	1.367	1.466	1.072	.731	47°
44°	.695	0.966	1.390	1.440	1.036	.719	46°
45°	.707	1.000	1.414	1.414	1.000	.707	45°
Deg.	Cos	Cot	Csc	Sec	Tan	Sin	Deg.

APPENDIX D

Acoustic Properties of Materials

MATERIAL	DENSITY $\rho = gm/cm^3$	LONGITUDINAL WAVES		SHEAR (TRANSVERSE) WAVES		SURFACE (RAYLEIGH) WAVES	
		VELOCITY $V_L = cm/\mu s$	IMPEDANCE $Z_L = gm \times 10^3 / cm^2 \cdot s$	VELOCITY $V_T = cm/\mu s$	IMPEDANCE $Z_T = gm \times 10^3 / cm^2 \cdot s$	VELOCITY $V_R = cm/\mu s$	IMPEDANCE $Z_R = gm \times 10^3 / cm^2 \cdot s$
AIR	0.001	0.033	0.33	-	-	-	-
ALUMINUM 1100-O	2.71	0.635	1,720	0.310	840	0.290	788
ALUMINUM, ALLOY 2117-T4	2.80	0.625	1,750	0.310	868	0.279	780
BARIUM TITANATE	0.56	0.550	310	-	-	-	-
BERYLLIUM	1.82	1.280	2,330	0.871	1,600	0.787	1,420
BRASS (NAVAL)	8.1	0.443	3,610	0.212	1,720	0.195	1,580
BRONZE (P-5%)	8.86	0.353	3,120	0.223	1,980	0.201	1,780
CAST IRON	7.7	0.450	2,960	0.240	1,850	-	-
COPPER	8.9	0.466	4,180	0.226	2,010	0.193	1,720
CORK	0.24	0.051	12	-	-	-	-
GLASS, PLATE	2.51	0.577	1,450	0.343	865	0.314	765
GLASS, PYREX	2.23	0.557	1,240	0.344	765	0.313	698
GLYCERINE	1.261	0.192	242	-	-	-	-
GOLD	19.3	0.324	6,260	0.120	2,320	-	-
ICE	1.00	0.398	400	0.199	199	-	-
LEAD, PURE	11.4	0.216	2,460	0.070	798	0.063	717
MAGNESIUM, ALLOY M1-A	1.76	0.574	1,010	0.310	539	0.287	499
MOLYBDENUM	10.09	0.629	6,350	0.335	3,650	0.311	339
NICKEL	8.8	0.563	4,950	0.296	2,610	0.264	2,320
OIL, TRANSFORMER	0.92	0.138	127	-	-	-	-
PLASTIC (ACRYLIC RESIN-PLEXIGLASS)	1.18	0.267	315	0.112	132	-	-
POLYETHYLENE	-	0.153	-	-	-	-	-
QUARTZ, FUSED	2.20	0.593	1,300	0.375	825	0.339	745
SILVER	10.5	0.360	3,800	0.159	1,670	-	-
STEEL	7.8	0.585	4,560	0.323	2,530	0.279	2,180
STAINLESS 302	8.03	0.566	4,550	0.312	2,500	0.278	2,500
STAINLESS 410	7.67	0.739	5,670	0.299	2,290	0.216	2,290
TIN	7.3	0.332	2,420	0.167	1,235	-	-
TITANIUM (TI 150A)	4.54	0.610	2,770	0.312	1,420	0.279	1,420
TUNGSTEN	19.25	0.518	9,980	0.287	5,520	0.265	5,100
WATER	1.00	0.149	149	-	-	-	-
ZINC	7.1	0.417	2,960	0.241	1,710	-	-

APPENDIX E

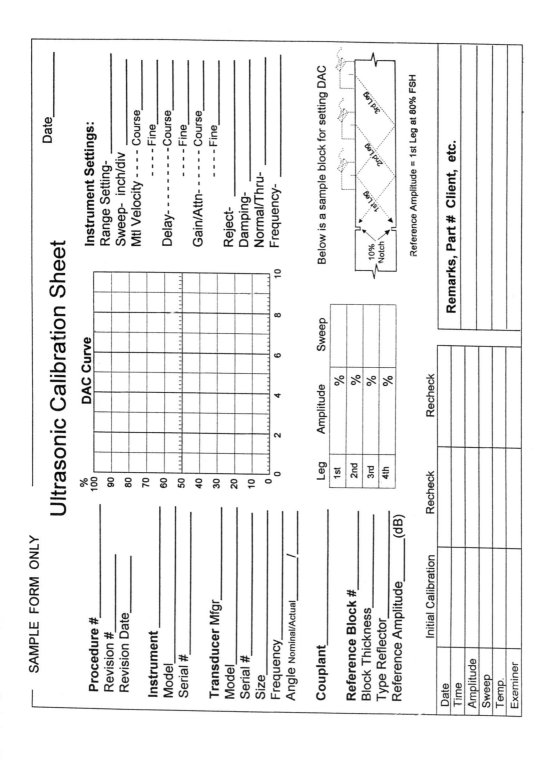

E-1

APPENDIX F

SAMPLE FORM ONLY

Angle Beam Examination Data Record

Job Number_____   Procedure Number_____

Drawing Number_____   Specification_____

Material_____   Reference Amplitude_____(dB)

Type_____   Scanning Amplitude _____

Average Part Thickness_____   Other_____

Indication Number	Scan Direction From − side From + side	Maximum % of DAC DAC From SDH or Notches	Sweep Reading Sound Path Distance	Surface Distance (+ or −) Exit Point to weld ℄	X Location (+ or −) ℄ to Indication	Y Location Inches from 0° Ref. and Total Length	Depth of Reflector Below the Surface

Notes:

−X ↑ Downstream
Flow →
↓ Upstream
+X

Sketch cross section of weld if needed.

Inspector_____ Level_____ Date_____

RESOLVING NATIONALITY CONFLICTS

The Role of Public Opinion Research

Edited by
W. Phillips Davison
Leon Gordenker

PRAEGER

PRAEGER SPECIAL STUDIES • PRAEGER SCIENTIFIC

Library of Congress Cataloging in Publication Data

Main entry under title:

Resolving nationality conflicts.

   Bibliography: p.
   Includes index.
   1. World politics--1945-   --Addresses, essays, lectures. 2. Self-determination, National--Public opinion--Addresses, essays, lectures. 3. Nationalism--Public opinion--Addresses, essays, lectures.
I. Davison, Walter Phillips, 1918-
II. Gordenker, Leon, 1923-   III. Princeton University. Center of International Studies.
D843.R435   323.1'1'0723   80-15128
ISBN 0-03-056229-5

Written under the auspices of the
Center of International Studies,
Princeton University

Published in 1980 by Praeger Publishers
CBS Educational and Professional Publishing
A Division of CBS, Inc.
521 Fifth Avenue, New York, New York 10017 U.S.A.

© 1980 by Praeger Publishers

All rights reserved

0123456789 145 987654321

Printed in the United States of America

FOREWORD

The concept of a new approach toward the lessening of armed hostilities through the role of the affected peoples was presented in January 1978 to Princeton University by the World War I Class of 1917, together with an offer of a grant for research.

The research covered a two-year period that included the centennial of Woodrow Wilson's graduation and the sixtieth anniversary of his efforts for peace at the Versailles conference. Appropriately it was Wilson's stress on self-determination that provided its basis. Timely also were three major world developments since his day that offer great usefulness for conflict resolution but are seldom taken into consideration:

First, the tremendous advancement in communication. Airlines, trains, and motor vehicles increasingly intermingle people and transport goods; and there are telephones, motion pictures, radio and television, and the print media to inform and entertain around the world.

Second, the rapid scientific growth of opinion surveys—the penetration by small samples into the minds and attitudes of people in many areas of the world. Such contacts, both direct and indirect, can be supplemented by trained observation of demonstrations and other mass actions.

Third, the very recent appearance of the computer, which can store any number of factors that may affect a situation, then can instantly relate those factors to each other, in order to reveal causes and indicate applicable solution methods.

Complementing all other types of peace efforts is an accurate current knowledge of the specific underlying attitudes and actions of the affected peoples. In their differing motives and reactions they hold the key to the several effective ways in which tensions and armed actions can be lessened.

During the summer of 1977, a working committee of the Class of 1917 marshaled from a variety of sources what was going on in the many kinds of efforts for peace, and became convinced that the approach under consideration was unique and had good potential. Continually since then the working committee has been engaged in the development of techniques. In December 1977, the Class Executive Committee approved a meeting with Princeton representatives, which was followed by a class referendum. As the approval of the Class became clear, together with pledges for more than the minimum funds that the university required, the project began, with the endorsement of Princeton's president. We are grateful to the university working team, which consisted of

Professor Cyril Black, director of the Center of International Studies; professor of politics Leon Gordenker; and professor (at Columbia University), W. Phillips Davison, a 1939 Princeton graduate.

Many '17ers, including the current class president, two vice-presidents, and three past presidents, helped to develop the research program. They held frequent meetings at Princeton, engaged in many telephone sessions, and collected information from many sources. Those involved included Lowell Turrentine of Palo Alto, California, former professor of law at Stanford University, who provided key subjects for investigation; S. Whitney Landon, former vice-president of the American Telephone and Telegraph Company, chairman of the American Arbitration Association's Executive Committee, and a director of the Association for many years; Kent G. Colwell, former vice-president of Guaranty Trust Company; Norris D. Jackson, of St. Paul, Minnesota, former director of labor relations for Northwest Orient Airlines; Percy H. Buchanan, formerly with Morgan Stanley and Company; and Mrs. Douglas Delanoy, former member of the Board of Trustees, Bryn Mawr College.

The major purpose of the Class of 1917 was to utilize the attitudes of the affected peoples as an added new dimension in the search for ways to reduce the human slaughter growing out of accelerating conflict. Rather than attempt to bring about the cessation of deep-seated major conflicts or of sporadic terrorism, the approach that was chosen centered on prevention of incipient wars and of the growth of small wars into greater ones. To accomplish this would involve, first, a means of receiving an early warning of an imminent conflict together with enough pertinent current information to describe it fully through some form of monitoring. It was quickly discovered that there are many sources from which to obtain monitors and observers. Needed next would be the services of those with expertise in determining all aspects of attitudes—initially to probe into the situation at its roots, and later to find opportunities to shift attitudes away from vehement hostility. Again, sources of such services were found to be readily obtainable.

The final stage would be to disseminate to the people whatever they needed to know in order to exercise their own ability to lessen armed hostilities. It was not the intention that this project should engage in mediation or other settlement methods. The job would eventually be one of assembling the various ways in which the people themselves could work toward the beginnings of a settlement—at least to keep a small fire from becoming a raging conflagration. This would be a cumulative matter of collecting and interrelating information, largely regarding the actions and attitudes of the people concerned. But again the services of experts in the public opinion field would be needed. It is considered practical that ultimately, through experience, a series of manuals to fit various situations will be prepared and made locally available to those who desire to organize measures to counteract armed action.

The Class of 1917, advancing in age and dwindling in numbers, must necessarily leave to others the pursuit of the ultimate goal. Its province has been to lay the groundwork to get something started (in the words of the dean of the Princeton Graduate School, "to light a candle"). For its part in the research plan, the Center of International Studies elected to handle all arrangements for three public forums in Princeton—two at June reunions and one two-day invitation conference in March 1979, with speakers from, or representing, a number of foreign countries. The March conference was opened by President William G. Bowen and Dean Donald E. Stokes at Princeton's Woodrow Wilson School, and included a dinner speech by Brian Urquhart, undersecretary-general of the United Nations. This book, prepared by the Center of International Studies, is based on the papers presented and some of the accompanying material submitted.

To its successors in carrying on the long-term objective, the Class of 1917 offers all other material it has collected, and any cooperation it is able to provide. Further, it is believed that those who took part in the conference or assisted in the gathering of information will generally hold themselves open to future cooperation. Those who have given valuable information and advice to the 1917 working group include Richard Baxter, vice-president of the Roper Organization; Ward B. Chamberlin, Jr., president of WETA (Public Broadcasting Service); Helen M. Crossley, research consultant to the U.S. Information Agency and past president of the World Association for Public Opinion Research; George H. Gallup, chairman of the American Institute of Public Opinion; George H. Gallup III, president of the American Institute of Public Opinion; Joseph E. Johnson, president emeritus, Carnegie Endowment for International Peace; William Rugh, public affairs officer, U.S. Information Agency; William H. Weathersby, former deputy director, U.S. Information Agency; Ralph K. White, professor emeritus, George Washington University; and Donald M. Wilson, vice-president of *Time*, Inc.

To all these and the many others who assisted, the 1917 committee and more than 40 contributing '17ers express very deep appreciation.

Archibald M. Crossley
President, Class of 1917
Princeton University

CONTENTS

	Page
FOREWORD	v
INTRODUCTION: ATTRACTIONS OF A FRESH APPROACH	xi

Chapter
1. SELF-DETERMINATION YESTERDAY AND TODAY
   Leon Gordenker — 1
   Self-Determination and the Seeds of Conflict — 2
   Selecting Cases for Study — 5
   Some Common Insights — 6

2. THE BASQUES IN SPAIN: NATIONALISM AND POLITICAL CONFLICT IN A NEW DEMOCRACY
   Juan J. Linz — 11
   Euskadi Today — 12
   Difficulty of Finding a Democratic Solution — 14
   The Basque Party System — 20
   Language and Politics — 30
   Natives and Immigrants — 34
   Class, Generation, and Politics — 37
   Religion—Another Cleavage — 41
   The Basque Country—A Society in Crisis — 43
   Notes — 51

3. COMMUNITY FRICTION IN BELGIUM: 1830–1980
   Paul Dabin — 53
   The Flemish Movement: From Linguistic Dualism (1870) to Regional Dualism (1970) — 54
   Birth of the Walloon Movement — 58
   Present State of the Flemish and Walloon Problems — 60
   The Situation of Brussels — 65
   Toward a Federalism Leading to Union or to Dislocation? — 66
   A Note on Public Opinion Research — 69
   Notes — 73

Chapter		Page
4.	PUBLIC OPINION AND THE SLIPPERY ROAD TO PEACE IN NORTHERN IRELAND	
	Gerald A. Fitzgerald	74
	The Legacy of Partition	75
	Recent Socioeconomic Changes	77
	Renewed "Troubles"	78
	Public Attitudes Toward Political Alternatives	79
5.	THE CONFRONTATION OF HARIJANS WITH INDIAN SOCIETY	
	Eric P. W. da Costa	83
	Four States Showing Rapid Progress	85
	Discrimination as Seen by Harijans	87
	The Importance of Family Planning and Education	94
	Perceptions of Progress	98
	The Utility of Attitude Surveys	103
6.	ARAB MINORITIES IN ISRAEL	
	Don Peretz	107
	Changing Social Structure	109
	The Land Question	111
	Growing Radicalism	113
	Palestinian Identity	115
	Polarization of Attitudes	119
	Recent Survey Data on Attitudes of and Toward Israeli Arabs	122
	Summary and Conclusions	129
	Note	131
7.	THE CASE OF THE MOROS IN THE PHILIPPINES	
	Linda S. Lichter	132
	Muslim-Christian Differences	134
	Negotiations Fail to Produce a Settlement	136
	A State of Neither War nor Peace	138
	Notes	139
8.	SELF-DETERMINATION IN QUEBEC: LOYALTIES, INCENTIVES, AND CONSTITUTIONAL OPTIONS AMONG FRENCH-SPEAKING QUEBECERS	
	Maurice Pinard	140
	Demographic Dimensions	141
	The Dual Loyalties of French Canadians in Quebec	142
	The Loyalties of Other Canadian Groups	147
	Determinants of Loyalty Patterns	149
	Trends in French Canadian Loyalties	150

Chapter		Page
	Sentiment Favoring Greater Autonomy for Quebec	151
	Trends in Support for Greater Autonomy	155
	The Independence of Quebec	156
	Trends in Separatist Support	159
	Sovereignty-Association	160
	A Mandate to Negotiate Sovereignty-Association	164
	Collective Incentives: Benefits and Costs of Various Options	167
	Loyalty, Incentives, and Independence	171
	Conclusion	172
	Postscript	173
	Notes	174
9.	PUERTO RICO'S STATUS DEBATE	
	Sonia Marrero	177
	Dissatisfaction with an Ambiguous Relationship	178
	Difficulties of Ascertaining Public Attitudes	180
	Can Surveys Measure Preferences for Something That Has Never Been Experienced?	187
	Notes	188
10.	APPLICATION OF OPINION RESEARCH TO CONFLICT RESOLUTION	
	W. Phillips Davison	190
	Two Scenarios From the Land of Imagination	191
	Considerations Affecting the Use of Opinion Research	196
	Possibilities of Conducting Public Opinion Research Worldwide	199
	What Information About Popular Attitudes Is Desirable?	202
	The Next Steps	208
	Notes	209
BIBLIOGRAPHY		211
LIST OF OTHER CENTER PUBLICATIONS		226
INDEX		233
ABOUT THE EDITORS AND CONTRIBUTORS		243

INTRODUCTION: ATTRACTIONS OF A FRESH APPROACH

The vigor and originality with which members of the Princeton Class of 1917 attacked the problem of peacemaking (see Foreword) impressed the editors of this volume. We were also somewhat startled. They looked back on the carnage of 1914-18, still very much alive in their memories, as well as on World War II, Korea, Vietnam, and a series of lesser conflicts, and said, in effect: "The killing has to stop." When we referred to the enormous efforts already devoted to problems of building a durable peace by countless generations of political leaders, scholars, and others, they were not impressed with the difficulties of finding new solutions. Their attitude was "Well, they didn't succeed; let's try again."

But the project that developed was not designed to attack all the causes of war at once. Initially it was to focus on threats to peace posed by the conflicting aspirations of national or communal groups within single states. Desires for self-determination had been one of the major issues of World War I, and have continued to cause conflict in many areas of the world. Could something be done to promote the peaceful realization of these national or communal aspirations? Was it possible that the relatively new techniques of opinion research could be used constructively in this endeavor?[1]

We were attracted to the Class of 1917's proposal for several reasons. The self-determination formula appeared to have become a more significant factor in causing conflicts than in settling them. Perhaps the time had come to search for a new formula, or at least for a new interpretation of the old one. Furthermore, the existing body of scholarly literature on nationality and nationality conflicts gave little attention to popular attitudes. As one scholar puts it, political scientists have tended to adopt a "view from the palace," to accept the existence of the nation-state as a given, and to see ethnic groups as "problems" that must be dealt with by central governments (Enloe 1979). It seemed likely that opinion research could help direct attention more to the aspirations and grievances of nationality groups as they are experienced by men and women in the streets and the fields. Supplementing the "view from the palace" might make solutions easier to find.

This proposal also interested us because it offered an opportunity to apply a comparative approach to the study of nationality problems. It appeared possible that the juxtaposition of individual, seemingly dissimilar, cases could lead to new insights that might suggest solutions to one or more of them. As the late Margaret Mead often pointed out, one way to solve a problem close at hand is to examine a similar one at a greater distance: those who wish to cure the ailments of New York City would do well to conduct their investigations in Tokyo. We thought that the formidable complexity of most situations involving conflict would make it difficult to discern the most important factors in any one case; comparison of several cases might help to reveal underlying patterns in each of them.

To anticipate a probable question, we should say at the outset that we have not found a formula for solving all nationality problems, or even any one of them. On the basis of the material presented in this volume, we do conclude, however, that opinion research offers a promising tool to those who are in a position to serve as peacemakers; the next step is to apply it experimentally in specific situations. Several common themes do emerge from the various case studies, and our attitude toward the concept of self-determination has certainly been affected. Indeed, we suspect self-determination may no longer be a useful term, in that it tends to be associated with conflict to achieve sovereignty or at least autonomy within a given territory. The term "self-realization" might be better, because some national, ethnic, or communal groups probably can best realize their aspirations through independence, some through limited autonomy, and some through closer relations with fellow citizens who belong to other groups and through greater integration into the larger society.

But to return to our narrative. After consulting recent literature on nationality problems, we composed an outline, indicating the kinds of information we thought would be useful in this study and suggesting ways of presenting it. This outline was sent to those who had agreed to prepare papers for a conference to be held at Princeton in March 1979. In accordance with well-established traditions of social science, contributors to this symposium declined to treat the outline as sacred; each proceeded in the manner he or she judged most suitable for the situation to be written about. With the benefit of hindsight, we believe that this assertion of intellectual independence was beneficial. An original approach to each case study facilitated the inclusion of insights that might not have been mentioned if a single outline had been rigidly followed. It also reduced the danger that common patterns might be seen where they did not in fact exist.

Because the case studies in this volume follow individual approaches, they are not arranged according to any analytic scheme. Rather, they are grouped geographically: the first three deal with Europe; the next three with countries in the Near East and Asia; and the final two with areas in North America and the Caribbean. An introductory chapter includes a brief discussion of the concept of self-determination and points out some of the common themes that run through several of the chapters. A final chapter suggests ways in which insights and information resulting from the project might be given practical application in the resolution of intrastate conflicts rooted in nationality or communal problems.

Throughout this volume the terms "nationality," "communal," and "ethnic" appear frequently. Sometimes they are used interchangeably, although their usually accepted meanings are not identical. Ethnicity generally refers to shared traits, often based on biological inheritance—bone structure, hair quality, shade of skin color, and so on, although traits that are not biologically determined are sometimes included. "Communal" refers to social organization, whether or not it corresponds to ethnicity. Thus, members of two distinct religious or social groups may constitute different communities, even though they represent the same or very similar biological stock.

"Nationality" is the most difficult term. Scholars are fond of pointing out the absence of a general agreement as to what a nationality is. Indeed, as many as 86 different definitions of nationality have been found in the literature (Jacob 1975). This is because it is largely a subjective phenomenon. No outsider can determine whether a particular group of people is a nationality; this can be done only by the people themselves. And, since there are many bases for nationalism—including ethnicity, language, religion, a shared cultural heritage or historical memory, and a common social and economic situation—it is possible for new nationalities to appear at any time, or for old nationalities to reappear. Members of some groups that often have been regarded as assimilated into larger cultures have in recent years put forth renewed national claims: the Welsh in England, the Bretons in France, American Indians in the United States. The nation of Pakistan was constituted following World War II on the basis of cultural self-determination, but almost immediately people in East Bengal began to demand autonomy and, ultimately, independence.

In spite of the many meanings that can be ascribed to the terms "ethnic," "communal," and "nationality," their usage in the following pages should not cause confusion if viewed in context. Furthermore, as used here, they all refer to a consciousness of whatever makes "us" different from "them." How this consciousness can be channeled into peaceful, productive paths, rather than toward conflict, is the major preoccupation of this book.

To the acknowledgments made in the Foreword, we would like to add several more. Our thanks go to Alec Gallup for making available materials from the Gallup-Kettering World Survey; to the Center of International Studies for its unfailing support of this venture; to Jane McDowall for the smooth accomplishment of administrative arrangements; to Gladys Starkey, Gail Wenrich, and June Garson for expert aid in preparing the manuscript; and especially to Cyril E. Black, long-time director of the Center, who well understands the use of both carrot and stick in bringing a research project to the point of publication. Our indebtedness to the distinguished contributors to this volume is obvious; we should add, moreover, that their tolerance in the face of editorial suggestions and good nature in meeting difficult deadlines made its preparation a pleasure rather than a burden. The assistance of J. Gérard-Libois, director-general of the Centre de Recherche et d'Information Socio-Politiques, Brussels; of Fernand Herman, who represented Belgium at the Princeton conference; and of Jan Logan, who facilitated the translation of various materials, is also gratefully acknowledged.

Finally, we must express particular gratitude to the Princeton Class of 1917 and its president, Archibald M. Crossley, without whose applied idealism, generous support, and intellectual contributions this project would not have been undertaken. We hope that the following chapters advance, at least to some degree, their vision of the future.

W. Phillips Davison
Leon Gordenker
December 1979

NOTE

1. It is not irrelevant that the current president of the Princeton Class of 1917 had devoted many years to public opinion research. Those who recall the U.S. presidential election of 1936 may remember that three previously little-known polling organizations—the Gallup, Roper, and Crossley polls—predicted the outcome correctly, in spite of a differing forecast by the prestigious *Literary Digest*. Archibald Crossley has been active in exploring new methods and applications of public opinion research ever since.

1

SELF-DETERMINATION YESTERDAY AND TODAY

Leon Gordenker

The idea of self-determination links directly to reorganizing the relationships of political authority. Woodrow Wilson made self-determination of peoples one of the Allied aims in World War I, thus deliberately encouraging the breakdown of the Austro-Hungarian and Ottoman empires. Challenges to the imperial authorities already existed in the form of Czech, Hungarian, Croat, and Slovenian nationalism. These movements, some of them with roots deep in the past, insisted on a territory in which the nation would rule. "One nation, one state" became a principle for the reorganization of Europe after World War I. Moreover, through the mandate arrangements of the League of Nations, it was introduced in a gingerly, implied fashion into the vision of a new international system. League mandates of indefinite duration for territories detached from the Ottoman and German empires suggested, especially in the Middle East, that the national claims of formerly subject peoples could not be overlooked, even if they were not immediately to be crowned with statehood.

National self-determination proved to be one of the most infectious doctrines of political change ever discovered. The aftermath of World War I indicated how quickly the idea could be applied and supported. The Russian Revolution was joined with the Versailles peace negotiations to map a new Europe. Independent states were founded by the Finns; Estonians; Latvians; Lithuanians; Poles; Czechs and Slovaks; Slovenes, Croats, and Serbs; and Hungarians. Greeks and Turks, among others, exchanged nationals. Few observers at that time would have dreamed that a large-scale, analogous process of self-determination would take place within the next two generations and would break up the vast colonial empires.

By 1950 many statesmen and observers understood that the breakdown of colonialism had begun. With the creation of Indonesia, the Dutch empire in Asia had been eliminated. In India the British had given way to independent India and Pakistan. Burma and the French colonies in Indochina also were independent or nearly so. Lebanon and Syria, which had been mandates under the League of Nations, became independent. Palestine turned into Israel and Jordan, and also provided an early indication of the difficulties that the new self-determination could create. From 1950 on, the wave of decolonization spread through Africa and the Caribbean. Nothing is left now of the old empires except a handful of minor island territories. All of this formal political change was carried out by anticolonial movements, together with the imperial powers, under the banner of national self-determination.

By 1960 the idea had so many firm adherents among governments that the United Nations (U.N.) General Assembly was able to adopt a resolution, by an overwhelming majority, characterizing self-determination as a right of all peoples. This declaration far exceeded the mention of self-determination in the U.N. Charter, where it was referred to as a principle. According to most governments in the world, the principle had turned into law and was binding on them. Since then, the doctrine that self-determination is a legal right of peoples has been extended so that its denial, it is argued, is a breach of the peace. Dozens of U.N. resolutions have reiterated the new doctrine.

SELF-DETERMINATION AND THE SEEDS OF CONFLICT

Such a triumph of national self-determination in both its applied and its doctrinal forms might be thought to obviate the disturbing effects of its implied challenge to political authority. But it has managed neither to eliminate some of the old issues nor to avoid creating new ones. In fact, applied self-determination has largely been a matter of setting up new states, rather than dealing with old national aspirations.

The territorial settlements after World War I had to do exclusively with the dissolution of the defeated empires. In the process some of the people who were separated from their clans by old territorial boundaries were reunited with them. But other divisions emerged from new territorial packages. As a result Germans lived in Poland and Czechoslovakia, and Hungarians found themselves in what is now Yugoslavia, where it was also possible to find Albanians and Bulgarians, let alone Macedonians whose cousins lived across the Bulgarian border. These were but a few of the situations in which national minorities remained excluded from the states set up in their names.

If it proved difficult to carve up the European map so as to put every person who might give loyalty a national state on the proper side of the border, what happened in Asia and Africa diverged even farther from the ideal of one nation, one state. With the major exception of India-Pakistan and a few other minor cases, old colonial boundaries simply became those of the new states. In the case of India, the attempt to separate peoples on the basis of religion, which became the fundamental idea of Pakistan, resulted in an expensive and frequently fatal forced migration of millions. The creation of Israel caused a similar but smaller migration of Arabs. Elsewhere, the retained colonial boundaries, established mainly for the convenience of European conquerors, slashed across clans and families, nations and tribes, with no regard for social existence. Keeping such boundaries as the basis of independence simply reinforced existing divisions.

In most of the new states, the governments quite consciously set out to promote a loyalty to the new sovereign unit. Such attachment had not existed earlier, for colonial authorities usually avoided welding divided peoples into a single, possibly very unmanageable, mass. Indeed, the indirect rule practiced by the British and Dutch colonial administrations tended to keep intact the existing national divisions within their overseas possessions.

The new efforts to create national loyalties had two special implications. First, they demonstrated that no government had any intention of reorganizing its territories along national lines if that meant the reduction of the realm. Because redrawing of borders almost invariably involved losses for someone, let alone the inconvenience of accustoming local people to new administrative and political ways, it never occurred. Furthermore, once governments embarked on nation-building policies, the exercise of self-determination by national or tribal groups violated state policy. Consequently, governments everywhere pushed aside any mention of self-determination in territories under their control.

The treatment of the right of self-determination as a one-time operation, taking place at the dissolution of the colonial empire, is now well-established in the practice of the U.N. and especially in the Organization for African Unity (OAU). The issue has been raised most frequently in the African context. In such rebellions of national groups as that of the Ibo in Nigeria, the OAU and the U.N. have pointedly avoided any but the most distant thought of intervention.

Such governmental attitudes toward self-determination may merely increase tension. They lead easily to coercive measures, for the distance between organizing to press for national self-determination and conspiring to commit treason may be perceived in official quarters as very short indeed.

Whether self-determination becomes an issue in a new country or in a well-established one, a fundamental characteristic of the concept implies a challenge. It involves autonomy or separateness of some degree. When full-blown, it leads to the creation, in some fashion, of a new nation-state to succeed an existing one. When autonomy is demanded by a group seeking self-determination, it implies a reduction in the authority of the existing government. Whether or not such autonomy is granted, seeking self-determination carries with it increasing separation and group coherence. Thus, national self-determination involves fundamental issues of governmental authority. Such issues almost always encounter strong ideological and behavioral resistance from existing elites. "Secession," writes Crawford Young, a student of cultural pluralism, "is a costly adventure, which can only be contemplated when perceived cultural threat reaches an extraordinary level of immediacy," because in nearly all situations the state will resist secession with force (Young 1976, p. 460).

Some governments try to accommodate national minorities without offering them the chance to set up a national state. These accommodations may include limited autonomy, as in some of the states of India, or symbolic deference to national identity, as in the Soviet Union. Or they may point in the direction of conciliatory integration, as in the Nigerian treatment of the movement that created the uprising in Biafra. Other governments try to eliminate legal and economic barriers to equal treatment within the unitary state.

Even if governments wish to accommodate the aspirations of minority nationalities, there frequently is no way in which an autonomous region, let alone a new sovereign state, can be established. Many nationalities are so geographically mixed with other peoples that there is no territorial area in which they predominate. Some are dependent on other nationalities for vital services, and live in a symbiotic relationship with them. Different peoples can sometimes be segregated into separate areas, as has been attempted in Cyprus, but at a cost of enormous suffering. No matter what is done, no matter how ingenious the arrangements for accommodating the aspirations of nationality groups, many cleavages will persist, and must be endured (Young 1976).

For all these reasons, most nationalities today do not live within the borders of their own national states. A recent survey of the larger ethnic groups, generally those of a million or more, found 91 major peoples without a national homeland. These ranged from the Achehnese of Indonesia to the Yoruba of Nigeria. In addition, 18 "major peoples" were separated from the national states in which they aspired to belong, such as the Catholic Irish of Northern Ireland and the Somalis of Ethiopia (Gastil 1978). If we include smaller groups that feel themselves to be nationalities, such as the Indian peoples of North and South America, it is clear that a large majority of national groups enjoy neither independence nor a significant degree of political autonomy.

With increasing frequency scholars have pointed to the danger that this situation poses to world peace. A study of ethnic conflict was prefaced with the following gloomy statement:

> Most countries are populated by several distinct ethnic groups; and as many as half of all countries have experienced or can expect substantial conflict among such groups in the second half of this century. Ethnic differences are the single most important source of large-scale

conflict within states; and they are frequently instrumental in wars between countries as well. (Heisler 1977, p. 1)

Another scholar expressed similar views in an article published in October 1978:

> Events of the past decade have now impressed upon even the most casual observer of world politics that ethnonationalism constitutes a major and growing threat to the political stability of most states. Rather than witnessing an evolution of stable state or superstate communities, the observer of global politics has viewed a succession of situations involving competing allegiances in which people have illustrated that an intuitive bond felt toward an informal and unstructured subdivision of mankind is far more profound and potent than are the ties that bind them to the formal and legalistic state structure in which they find themselves. (Connor 1978, p. 377)

The concept of self-determination, once regarded as a prescription for resolving conflict, has increasingly been used to justify violence and even war. It collides head-on with the principle that a sovereign state has the right to use all necessary means to preserve its integrity. New concepts that will allow for the peaceful fulfillment of national aspirations are clearly needed. Public opinion research, by directing attention to new dimensions of nationality problems, may be of assistance in defining new approaches.

SELECTING CASES FOR STUDY

Despite their large number, the complexity of situations in which aspirations for self-determination cause a threat to peace suggests that generalizations may be deceptive. In order to gain something approaching an overview of the problem, it seems desirable to select for study a number of cases with widely varying characteristics. These should involve national groups in different areas of the world. They should include claims for self-determination based on a variety of characteristics—language, religion, historical tradition, economic grievances, and others. They should represent some nationalities that aspire to their own national states and some that demand different political arrangements within an existing state. If a number of common themes emerge, or if certain elements are present in highly diverse cases, then it would appear possible that formulas for easing tensions and resolving conflicts among still other nationality groups would be suggested.

Furthermore, the situations selected should represent different stages of conflict or accommodation. One can scarcely hope to advance recommendations for resolving conflicts among nationality groups living within a single state unless one has some conception of how other contending nationalities have resolved their difficulties peaceably and why some multinational states enjoy domestic tranquillity. We therefore looked for some cases involving war and terrorism, and for others where discussions among nationality groups, even if heated, were carried on in a civilized manner.

Two more pedestrian considerations also affected our selection of cases for study. One was that interested scholars, who were already familiar with one or more situations involving nationality conflict, would be willing to prepare papers. The other was that data on mass attitudes would be accessible. Specifically, the existence of relevant information from public opinion surveys was a desideratum.

The five cases finally selected for principal attention satisfy these criteria fairly well. They are distributed among three continents. Two of them involve both linguistic and economic problems (Belgium and French-speaking Canada). One, while not concerning a nationality group in the sense that we are using the term here, nevertheless involves severe economic and social discrimination based on the accident of birth (the Harijans of India). One appears to be rooted mainly in an emotional and historical tradition (the Basques of Spain). And one involves religion, language, and history—all in abundant measure (the Arabs of Israel).

The cases are distributed along the continuum leading from armed violence to peace. Terrorism could be found in Spain and Israel; the Harijans of India were not infrequently involved in violent confrontations; the Canadians and Belgians seemed able to settle their differences more peacefully. Fairly extensive public opinion survey data that dealt with mass attitudes toward self-determination were available from all five countries, as were outstanding researchers who had for many years been concerned with each of the areas selected.

Three further cases were subsequently added to the original five, in order to give a broader basis for comparison. One of these, Northern Ireland, was a matter of serendipity. Gerald Fitzgerald, having recently returned from exploring the possibilities of integrated (Protestant-Catholic) education in that troubled region, agreed to review his data and cull information about self-determination in Northern Ireland and popular attitudes toward it. Briefer papers on sentiment in Puerto Rico and on nationalism among Muslims in the southern islands of the Philippines were prepared on the basis of materials immediately available in the New York area.

SOME COMMON INSIGHTS

Can common insights into issues concerning self-determination and the resolution of nationality conflicts be gained from examination of the case studies presented here? It appears that they can, even though at the present stage of this work they are unsystematic and do not form a theoretical whole. A few can be outlined here, anticipating the more detailed treatment in later chapters.

Ethnic divisions frequently involve dual loyalties. Individuals questioned may identify themselves with either the minority group or the broader national state, but they usually show some adherence to each center of loyalty. This splitting implies the existence of a middle ground for conciliation. In almost all the cases surveyed, the minorities who are struggling for greater self-determination feel that they have some ties with the state in which they live and with the other groups in that state. This is especially marked in Belgium and Canada. A speaker who described the situation in Belgium at the Princeton conference concluded:

> In the course of time, large sections of the population have become *more* Belgian, at the same time as they have become *more* Walloon or *more* Flemish. And so, there are two opposite but simultaneous trends. It is futile to state that one is more Walloon or Flemish than Belgian, or vice versa. All in all, there are many Belgians who prefer an arranged marriage to a judicial separation.

Similarly, in Canada a 1970 survey of French-speaking people in Quebec found that about 80 percent of them felt a "very strong" or "fairly strong" attachment to Canada—although this was counterbalanced by an even stronger loyalty to the Province of Quebec. When asked, in 1977, how they would define themselves, 45 percent of the French-speakers defined themselves as French Canadians; 41 percent considered themselves Quebecers. Another survey found that a majority of French-speakers in Quebec would like to have the provincial government exercise dominant powers in only three areas: education, cultural affairs, and natural resources. The armed forces, they felt, should be under the Canadian federal government's control; and other areas of government should be shared by the federal and provincial governments.

While the situation is less clear-cut, dual loyalties could also be found among a majority of Catholics in Northern Ireland and among the Arab minority in Israel. For example, when Israeli Arabs were asked in a 1974 survey whether they considered themselves "Palestinian" or "Israeli," a substantial number were willing to accept both terms. However, 29 percent preferred to be called "Israelis" and 41 percent preferred the term "Palestinian" (Tessler 1977).

Thus, a minority struggling for greater self-determination should not be thought of as a monolithic bloc. Many members of that minority probably feel loyalty not only to their own group but also to the state in which they live.

One of the most striking observations to come out of these case studies is that sustained violence can be carried on in the name of a nationality or religious group even when the overwhelming majority of its members oppose violence. In

Spain, for example, only about 23 percent of the population in the Basque region can speak the Basque language; and about 10 percent identify with the extremists who have resorted to force to gain their objectives. In other words, only a very small minority is willing to take armed action. But this minority suffices to throw the Spanish state into a turmoil.

Similarly, in Israel, according to a poll conducted in 1976, nearly two-thirds of the Arab minority who wanted greater self-determination advocated the use of nonviolent tactics. Only 19 percent were willing to support the use of force. The overwhelming majority wanted to keep the peace. And in Northern Ireland the existence of mass peace movements including both Protestants and Catholics indicates that most people on both sides oppose terrorism and favor the use of peaceful means to solve their problems.

The finding that many members of minority groups oppose the violence that is carried out in their name is, however, offset by another observation. This is that large numbers of people, even if they are against the use of arms, refuse to oppose that use actively. They indicate that they sympathize with the aims of the violent minority, even though they oppose the tactics that are used; or they say that they "understand" the terrorists and respect them. Thus, although most Israeli Arabs have made it clear that they disagree with the violent tactics of the Palestine Liberation Organization, they still refuse to denounce it. Similarly, a survey in 1977 in the Irish Republic showed that only 2 percent of the population approved of the campaign of violence carried on by the Provisional Wing of the Irish Republican Army, but 35 percent attributed idealistic motives to the Provisionals and expressed respect for them.

Although we do not have survey data from other areas on this question, it seems likely that a similar situation obtains in Canada, Puerto Rico, and elsewhere. Many more people have understanding for extremist tactics than are actually in favor of them.

Physical proximity does not necessarily promote favorable attitudes or even basic understanding among ethnic groups, as the data on India and Israel suggest. The French-Canadian, Basque, and Belgian cases involve either somewhat sharper or much sharper geographical separation than do the Israeli and Indian cases. The fact of distance, however, does not appear by itself to have a determining effect on the terms of the disputes.

Yet the research on attitudes in Israel suggests an untapped reservoir of goodwill, as both Arabs and Jews express the hope for improvement in relations with the other group. Although the data are less direct for Quebec, the unwillingness of much of the population to support complete severance from the rest of the country may have a similar meaning.

Rapid social change appears to have a close relationship to the development of nationalistic ideas. Certainly this has been the case for the Arabs in Israel, for the untouchables in India, and for both Flemish and Walloon communities in Belgium. It could be speculated that where social change causes uncertainty among a population, the members grasp for the certainty of ethnic identification as an element of stability and a means of defending newly won positions.

Economic expectations and satisfactions also link to ethnic identification and conflict. In Quebec clear differences of view about the desirability of various options for separation or autonomy relate to economic expectations. In Israel and India the minority resents economic inferiority. At the same time betterment of economic condition appears to be too gross an indicator of improvement or decline of group relationships to mean very much. It seems probable that the rate of change of economic status has an important bearing on mass attitudes. Furthermore, mere improvement of living standards, as has taken place among the Arabs in Israel, still leaves in place the important symbol of possession of land. The Arabs value the ownership of land highly, even if their economic return from it may not be as materially rewarding as urban employment. More broadly, symbolic values may frequently outweigh material advantages.

Research on Quebec suggests that the motivation for ethnic cohesion comes from internal beliefs, on the one hand, and from objective considerations, such as economic expectations, on the other. These objective considerations may include political alternatives. The Israeli case adds an external factor that is not constantly, or is hardly, present in other cases. This factor has to do with conditions and activities of Arab nationalists in the surrounding countries. Thus, actions external to the immediate dispute, such as the development of nationalistic movements among foreign groups and the export of their ideas, may affect a particular issue of self-determination.

Among many other observations that could be made on the basis of the research reported here, one may have special significance in the practical treatment of disputes. This is that survey data frequently enable one to determine the direction in which a situation is developing. Are demands for autonomy, or independence, growing? Or are they receding? Is approval for the use of violent tactics increasing or diminishing? To answer questions such as these, it would be necessary to have comparable surveys made from time to time over a period of months or years.

Sometimes the changes in attitude that can be detected by two or more successive surveys, using the same questions, suggest long-term, peaceful outcomes. For example, in 1967, 31 percent of the Arab minority in Israel said that Israel had a right to exist. By 1974 the proportion saying Israel had a right to exist had risen to 40 percent. As of 1971, 54 percent of the Arabs queried said they thought Arab youths had a future in Israel; by 1975 the percentage giving this reply had increased to 61.

Trends can also move in the other direction. Between 1971 and 1975 the proportion of Israeli Arabs who said that they had opportunity for frequent contacts with Jewish Israelis decreased by more than half—from 74 percent to 34 percent. The proportion of Arabs who considered such contacts essential also decreased, although by a lesser amount—from 80 percent to 61 percent.

This trend seems less auspicious for the establishment of peace in the Middle East.

No magic formula, or even a complicated but sure set of procedures, for defusing explosive national differences appears from these case studies. Had it done so, there would have been cause for high skepticism, in view of the complexities involved. A more modest, but still promising, conclusion is possible. Several lines for further exploration have been opened. These lines conceivably could have significance for diplomats and others concerned with governmental actions in issues raised by demands for self-determination.

The material presented here seems to support the basic hypothesis with which the venture began: Devoting more attention to mass attitudes can materially assist in resolving at least some conflicts based on demands for national self-determination. Opinion polls, of course, are not necessarily the key to peace. But they can provide a thermometer to show when the fevers that presage violence are starting to rise. They can lead those who must treat such situations, both persons abroad and those in the countries concerned, to a surer understanding of the nature of existing grievances. They can be combined with broader psychological and sociological propositions to indicate the boundaries of acceptable proposals. For example, it would be useless, or worse, for a mediator to suggest a course that the populations concerned, in opinions expressed through mass attitude surveys, had been rejecting by increasing majorities for many months.

The still rather blunt tool that seems to be taking shape in these pages needs additional refinement and more thought about its application. Its application may well require the creation of standing organizations or institutions that do not yet exist, either inside or outside of governments, to assemble data and follow developing cases. Some thoughts about these institutions will be presented in the final chapter of this book.

2

THE BASQUES IN SPAIN: NATIONALISM AND POLITICAL CONFLICT IN A NEW DEMOCRACY

Juan J. Linz

For the Basque nationalist a future Europe of nations rather than of states will include a new nation resulting from the separation of Basque territories from Spain and France, two states that for centuries have divided and oppressed the Euskal-Herria, or Basque people.[1]

In the view of most Spaniards, the Basques are one of the peoples that would fit Friedrich Engels' description of "those numerous small relics of people which after having figured for a longer or shorter period on the stage of history were finally absorbed as integral portions into one or the other of those more powerful nations whose greater vitality enabled them to overcome greater obstacles" (Rjasanoff 1916, pp. 215-16). They are one of those peoples who, while they had their own language, their own culture and style of life, and their own institutions, did not manage in the era of state-building to create an independent territorial political unit coinciding with the area of their culture.

During most of the Middle Ages, the Basques were divided among the Christian realms that emerged on the northern boundary of Spain during the struggle against Islam: Castile, Navarra, and the duchy of Gascony. The division of Navarra at the onset of modern times ultimately would transfer Basque lands

---

Support from a fellowship of the German Marshall Fund of the United States is gratefully acknowledged. Survey data used, unless otherwise indicated, are from studies designed and carried out by DATA, Madrid, under the direction of Francisco Andrés Orizo, Dario Vila Carro, and Manuel Gómez Reino. Their collaboration and the assistance of Rocío de Terán have been invaluable.

on the other side of the Pyrenees to France. None of the historical, political entities that have governed the Basques has ever been coterminous with the ethnic-linguistic Basque country. The kingdom of Navarra, saddled on the Pyrenees between Castile, Aragón, and France, also included extensive areas in which Basque was not spoken; and already at the end of the Middle Ages it did not include the core of the Basque homeland that was part of Castile. Navarra, which might have been the basis for the creation of the Basque state, ultimately was absorbed by its two great neighbors that were more successful in the process of state-building.

In a sense the Basques are one of those peoples without a history of statehood but with a clear identity that, in the age of nationalism, would become a national entity. Centuries of integration into Spain and France have, however, given many Basques another national identification. They have long felt, and still feel, that they are members of the French and Spanish nations. With the exception of some peasants and fishermen, they are bilingual. Others have not maintained the Basque language, even though they may feel the Basque identity strongly. The Basques are a linguistic minority but also something more than that.

EUSKADI TODAY

The present Basque territory includes, in the view of the nationalists, four Spanish provinces and three historical districts in France that, however, lack a political or administrative personality within the French state.[2] Three of the Spanish provinces, Alava, Guipúzcoa, and Vizcaya, have chosen in a referendum to constitute an autonomous community within the Spanish state, with the name of Euskadi. The constitution of 1978 and the autonomy statute of 1979 allow Navarra to accede to Euskadi, if it should choose to do so, by taking a series of political and legal steps.

I shall refer to the three provinces that now constitute an autonomous community within the Spanish state as Euskadi (traditionally called Vascongadas), but I also will speak sometimes of Euskadi Sur, in the sense used by the nationalists, as including all the territories they claim—that is, also the province of Navarra. The three territories of Labourd, Soule, and Lower Navarre, which are part of the French *département* of Basses Pyrénées (now Pyrénées Atlantiques), constitute for the nationalists a Euskadi Nord. The nationalists' dream is unification of all these territories in a new, independent Euskadi, symbolized by the slogan often painted on walls: 4 + 3 = 1.

The larger part, 86 percent of the territory and 91 percent of the population, of that greater Euskadi is in Spain. It is also in the Spanish Basque country that the nationalist movement was born at the turn of the century and that Basque nationalist parties have had their successes. It is therefore not surprising that the peaceful and violent struggle for autonomy and, ultimately, independence should have taken place in Spain. The French Basque country constitutes 39 percent of the area of the *département* of Pyrénées Atlantiques. Its 217,888 inhabitants in 1968 were less than 0.5 percent of the more than 50 million Frenchmen, while the 2.3 million inhabitants of the four Spanish provinces accounted for 6.9 percent, and the three provinces of Euskadi for 5.5 percent, of the Spanish population.

Not only demographic weight differentiates the French and the Spanish Basque countries. While the French area is largely rural, and most of its population is engaged in agriculture, fishing, and the tertiary sector linked with tourism, the Spanish Basque country is the most urbanized and industrial of the regions of Spain, and the provinces of Euskadi rank among the highest in per capita income. In contrast with most of the peripheral nationalities and linguistic groups in Italy, France, and northern Europe, the Basques are not a rural people living in a marginal area of underdevelopment, but are the inhabitants of one large metropolitan area (Bilbao), several large cities, and innumerable industrial towns almost continuous with each other. A Basque country of green hillsides and quaint villages lives on in travel posters, and there are many such places; but the inhabitants even of these work in industrial plants. Indeed, Bilbao, with a population density of 245 inhabitants per urbanized square kilometer, compares unfavorably with the 94 of New York and the 62 of Chicago, and its level of pollution can compete with that of any other great industrial metropolis.

Its wealth and rapid economic development, and the importance of its industries and financial institutions, have given the Basque country a unique position in the Spanish society and economy, as can be seen in Table 2.1. The Basque big business elite, particularly the Bilbao oligarchy, has played a dominant role in many sectors of the Spanish economy. While the Basque country has been politically and culturally dependent on the rest of Spain, and has been governed from Madrid, Basque businessmen have played a dominant role in the Spanish economy, together with their Catalan peers.

It makes little sense to speak of internal colonialism in attempting to explain in Marxist economic terms the sources and frustrations of national sentiment in the Basque country. There are within Spain regions that are economically subject to outside capitalist control; but these are far from the Basque country, and in some cases the metropolitan centers dominating them have been the banks and large corporate headquarters in Bilbao. In contrast with so many other areas in Europe that have risen in the defense of their traditional folk culture, the wealth and the opportunities created by industrialization have made the Basque country an area of immigration rather than of emigration. It has been the contact between the autochthonous population and the immigrants that has contributed decisively to the appearance of a nationalist sentiment; and immigrants provide one of the bases for internal stratification, and to some extent for the social and political cleavages within the population of the Basque country.

TABLE 2.1

Share of the Basque Country and Navarra in the Territory, Population, and Gross Product of Spain, 1970–71 (percent)

	Province				Total of Four Provinces
	Alava	Guipúzcoa	Vizcaya	Navarra	
Territory	.60	.40	.45	2.06	3.51
Population (1970)	.60	1.87	3.09	1.37	6.93
Production in agrarian sector	.62	.54	.90	2.25	4.31
Fishing sector		11.16	3.57		14.73
Industry and mining	1.44	4.03	6.18	1.85	13.50
Services	.62	2.21	3.67	1.34	7.84

Source: Azaola 1976, vol. II, p. 751.

The complexity of the Basque problem cannot be understood unless we keep in mind these basic facts, which deserve further elaboration (for which there is no room here). The Basque country, while peripheral geographically, occupies a central place in the Spanish economy; and its elites have played, and still do play, an important role in Spanish society. While there is a risk that radical nationalist terrorists might transform Euskadi into a Spanish Ulster, the Basque country of Spain is not like Ulster, separated by the sea from the rest of the United Kingdom and economically dependent on it.

DIFFICULTY OF FINDING A DEMOCRATIC SOLUTION

Robert Dahl, writing in *Polyarchy* on the problems caused for stable democracy by marked cultural pluralism, sets down unhappy and even tragic choices:

a) a polyarchy that provides mutual guarantees to its minorities but cannot respond to demands for solutions to major problems sufficiently well to hold the allegiance of the people, b) a hegemony that tries to meet these problems by coercing, if need be, members of one or more sub-cultures, or if the sub-cultures are also regional, c) separation into different countries. Only the last may enable polyarchy to survive among the dissenting minority. Thus, the price of polyarchy may be a break-up of the country and the price of territorial unity may be a hegemonic regime." (1971, pp. 120-21)

Spain, in the view of many, may be facing this tragic dilemma, and it depends on the skill of its political elites to avoid it, or at least to reduce the probability of any of those outcomes. Certainly the Spanish hegemonic model based on suppressing and ignoring the multinational character of Spain for 40 years has been proven totally unsuccessful. Some among the supporters of the peripheral nationalisms would argue that secession is the only alternative that would allow both the rest of Spain and a new Basque polity to be stable democracies. Independently of the opinion one might have about the value of maintaining the unity of Spain as a state, it needs to be emphasized that secession of the Basque country is not a viable alternative and would not create a new and stable democracy in that region. Therefore, only a third alternative is probable and viable: to accept the existence of the Spanish state and to find ways to accommodate the pluralism within that state. There are, however, many obstacles to a successful implementation of that solution in the case of the Basque country.

We do not need to spend much time arguing that the historical moment for assimilation of the Basque region to a Spanish nation-state, even to the degree that the French state was able to assimilate its linguistic and cultural minorities in the course of the nineteenth century, has passed—if it ever existed. The unity of the Spanish state might be maintained by a policy that would ignore the linguistic and cultural heterogeneity of Spain, and centralized government could probably be maintained by force, but at a high cost for those attempting to implement such a policy, since it would require a level of repression intolerable in a modern European country (although quite normal in other parts of the world). It certainly would not be possible within the framework of a democratic regime. There can, however, be little doubt that when faced with the alternatives of secession or unity at all costs, many Spaniards might still opt mistakenly for that second alternative.

Why not secession? First, let us stress that, today, secession of the four Basque provinces would not be a democratic alternative, since it would be the choice of at the most 43, and probably less than 40, percent of the electorate, assuming that all those supporting Basque parties would favor it—which is not now the case. In fact, the openly secessionist parties in 1979 had an even smaller vote, at the most 30 percent in one of the four provinces and considerably less in the other three. The new constitution was approved on December 6, 1978, by an absolute majority of the electorate in Navarra (and all those abstaining were not necessarily opposed to it), and 71 percent of those voting (42 percent of the electorate) in Alava gave it their support. In Vizcaya, 31 percent of the electorate supported it, and even in the most secessionist province, Guipúzcoa, 28 percent of the electorate voted for it. Those clearly saying "no" to the constitution were only 11 percent of the electorate; those saying "yes," 35 percent. Under democratic principles Navarra could under no circumstances be included in a new secessionist state; and in Alava such a decision would go against the clear will of 42 percent of the eligible population, and against the expressed desire of slightly less than one-third of the population in the other two provinces. If the unity of Spain as a state with autonomous regions, as provided by the new constitution, runs counter to the wishes of a significant minority of the population in two provinces of the Basque country, an independent Basque state would run counter to the democratically expressed will of even larger minorities.

Any discussion of self-determination faces the problem so well summarized by Sir Ivor Jennings when he wrote, "On the surface it seemed reasonable to let the people decide. It was in fact ridiculous because the people cannot decide until somebody decides who are the people" (quoted in Buchheit 1978). The people, for the Basque nationalists, are the inhabitants of the four provinces; exclusion of Navarra from any definition of the Basque people is totally unacceptable to many of the parties in the Basque country. From the results of the constitutional referendum and the last two elections, it is unlikely that "the people," including the Navarrese, would decide for secession. It could be argued that the territory of the Basque people could be defined as the three historical Basque provinces. Even assuming the unlikely situation that the majority in these provinces would favor secession, Navarra would then be for the Basque provinces an irredenta, and the conflict with the Spanish state would continue. The new Basque state would be torn by conflict between those not ready to give up the struggle for Navarra and those accepting the smaller Basque country. Navarra, on the other hand, would suffer from the demands of minorities wanting to join the new Euskadi. Secession would not have solved anything.

What is more, the new *Euskadi libre* would have a large minority identifying with Spain across the border; and it is likely that this minority would consider the new state as oppressive as, if not more oppressive than, the extreme Basque nationalists today see the Spanish state. Some of them would turn to violent actions, following the model of the Basque extremists who obtained independence for Euskadi.

It could be argued that the error lies in the use of the present and historical boundaries of the Basque country, and that one would have to adjust the political map to the ethnic-cultural-linguistic map. By segregating some parts of Navarra, some parts of Alava, and perhaps allowing some parts of Vizcaya to join Spain, one could create an ethnically-linguistically homogeneous Basque country. Such a solution, following pure Wilsonian principles and holding plebiscites at the communal level, would never satisfy the extreme Basque nationalists. In addition, the possibility of arranging a viable division of territory by this method does not exist. Plebiscites on independence at the municipal level would leave many enclaves, and by no means sparsely populated ones, within the new state. Undoubtedly people in large areas on the left bank of the Nervión River in metropolitan Bilbao, the capital of the new state, would never vote for such an alternative. Only a massive exchange of populations, a total disruption of the economy, and a coercively imposed option could create an internally homogeneous Basque state.

We do not need to invoke Spanish nationalism as an obstacle to the establishment of an independent Basque state. True, the nationalism of the Spanish army makes it highly unlikely that the army, at least today, would tolerate such an option, even if it were to be approved by a bare majority of the population. There is no need, therefore, to consider the enormous costs that the disruption of the economic, social, and cultural ties between the Basque people and the rest of Spain would have for Basques and Spaniards alike. In fact, it would be possible to argue that the costs might be greater for the Basques than for other Spaniards (except the working-class immigrants to Euskadi) if Spain would not accept the secession with good grace. Even assuming the continuing radicalization of Basque nationalists, only the composition of the population of Guipúzcoa makes secession a thinkable alternative; but for the Basque nationalists Euskadi reduced to Guipúzcoa would be an intolerable thought and could only provoke a violent irredentism among the Basques in other provinces, an irredentism that, as has been said, could not be appeased by modification of the political boundaries through a plebiscite.

It could be argued, and it certainly is being argued, by the proponents of the principle of self-determination that the Spanish state should, in view of the distribution of opinions and the likely outcome, allow a plebiscite on independence for the Basque country. But would such a plebiscite decide the question? Our guess is that it would not, since the partisans of independence would claim victory in parts of the region, would be resoundingly defeated in many others, and would represent a significant minority in others. Would such an outcome oblige them to accept the decision of the "misguided" people, of the people with a "false consciousness," "manipulated by the existing power structures and the media"? Or would their partial success be an incentive to continue the struggle with renewed energies? Our guess is that the latter would be the case. Certainly neither the elections nor the referendum on the constitution, with its ambiguous meaning due to the fundamental ambivalence in the position of the largest Basque political party, nor even the majority vote for the Autonomy Statute has led to a clarification of the issues. In fact, it could be argued: What is the point of allowing such a plebiscite on self-determination when consultations of the popular will have already given an answer?

If we add the costs to stability in the rest of the Spanish state deriving from a decision to hold a plebiscite on independence, the likely response of the Spanish nationalists to a democracy that is willing to risk even the loss of a part of the state—and for them of the nation—that has shared its history for centuries and has benefited from citizenship, it seems illogical even to consider such a possibility. Self-determination through democratic means is not a realistic option.

Once we have excluded hegemonic assimilation to a Spanish nation, as the present constitution has done, and the option of secession that it also has clearly rejected, the problem is to find a viable solution within a stable democratic Spanish state. Given the distribution of opinions within the Basque country, such a solution should not be impossible; but unfortunately there are serious obstacles to achieving it. Foremost among them is the willingness of a very small but very passionate active minority to turn to terrorism to achieve its goals. Terrorism of the ETA militants, in and of itself, is making civil life difficult in the Basque country and even in Spain as a whole, given the extension of their activities to Madrid.

Societies like the Federal Republic of Germany and Italy have learned to live with their terrorists. But we should be conscious of a very tragic fact: a significant minority of the population in the Basque country that is not ready to engage in terrorism is willing to give it a certain degree of moral support, or at least not to give such support to the representatives of the state that have to confront it. Another segment of the Basque population and leadership is willing to use the presence of terrorists to make demands on the central government, arguing that unless these demands are granted, terrorism is likely to continue—a position that has some elements of blackmail. In addition, those nationalists who are ready to settle for greater autonomy within the Spanish state are not ready to make a total break with those who favor independence, and consider the terrorists to be patriots striving for the same cause.

Many different motives enter into this fundamental ambivalence: unwillingness to give up in principle the ideal of independence, the need in some cases to collaborate with forces close to the extremists (for example, to overcome the weakness of Basque nationalism in Navarra at the electoral level by forming broad coalitions), the fear of being denounced as traitors to the national cause by the extremists, and the fear of losing support to them. The terrorism, above everything else, is a source of fear: fear that does not allow unambiguous statements, fear that obliges one to defend the extremists against the reprisals of the state, fear that prevents public stances that would involve serious risks for those taking them. The presence of extremists, the actions of the terrorists, the ambivalence of the moderates, and their ambiguity when it comes to taking a clear position also lead to a lack of trust on the other side. This has made the working out of stable autonomist solutions difficult. The question of whether the central government can devolve powers like the control of public order to a regional government controlled by moderate Basque parties has no easy answer.

Under such circumstances the central government might very well feel, as Eric Nordlinger (1972) has suggested, that the combination of territorially distinctive segments and a grant of partial autonomy will lead to further demands and, when these demands are refused, secession and civil war may follow. The combination of a small, active terrorist minority, a somewhat larger minority in latent sympathy with the terrorists, and a moderate plurality fundamentally unwilling to use repression against the terrorists and fearful of their actions (and at the same time unwilling to admit that its interests are closer to those of the central government than to those of the extremists, whose ultimate utopian hopes they share) makes collaboration on the basis of mutual trust difficult. The factor of terrorism and disloyal extremism unfortunately has not been built into the analysis by Arend Lijphart (1968) of the conditions making possible the establishment of consociational democracy. This is, however, a variable that we cannot ignore in the Spanish case.

Let us turn now to a more optimistic scenario. Let us assume that terrorism has been brought under control and that public opinion in the Basque country turns away from terrorism to search for viable solutions. The work of Lijphart would allow us to specify some of the factors that favor as well as oppose solutions similar to those achieved in countries like Switzerland and Belgium. The question is to what extent the factors that he considers conducive to a consociational solution are present today in the Basque country and in Spain. Foremost among them is a commitment to the maintenance of the system, as there is among the different ethnic groups and cantons of the Swiss confederation; among the immense majority of the Belgians, regardless of their views on the internal constitutional structure of Belgium; or among the Catalan political leadership in Spain. Certainly the rhetoric, perhaps more than the real thinking, of Basque nationalists seriously questions the existing system in Spain. It advocates secession from the Spanish state and never expresses sincere doubts about the ultimate viability of the separate Basque state that would be formed by the Basque provinces now under Spanish and French domination. The $4 + 3 = 1$ slogan reflects their strong sentiment.

The relative weakness, in demographic terms, of the Basque autonomous entity within the Spanish state, and even of all the nationalities with autonomy aspirations within the dominantly Castilian-speaking state, compounds the problem. The Spanish situation differs radically from that of Belgium, where the two linguistic groups are more or less equal. There is no real possibility of sharing power between equals or near equals. A commitment even to a federalized Spanish state would imply the abandonment of irredentism insofar as Euskadi Nord was concerned—unless the Spanish state would make this its cause, something that is impossible.

One factor emphasized in all the literature on consociational democracy is a feeling that the disaggregation of existing political entities—Switzerland, Belgium, for many years Lebanon, and perhaps even today Canada—would stimulate the hegemonic ambitions of other neighboring states, and that with all its defects the existing political unit has virtues that make it more valuable than the taking of that risk. In the case of the Basque country there is no such neighboring hegemonic power that would be perceived as such a threat, since it is inconceivable that France would have designs on an independent Basque country. Nor would it play the role that some of the great powers have played in the case of Belgium, and less visibly in the case of Switzerland, and seek to serve as protector of the new state. For the Basque nationalists both Spain and France are oppressors, and there is no sentiment that to be part of one or the other is a viable alternative that would justify consociational efforts.

True nationalists see the consociational solution within Spain as a tactic rather than an emotional question. In the highly loaded rhetoric of nationalism, the many pragmatic reasons for a principled commitment to the Spanish state are given little room. For the nationalists the question really is not autonomy within the Spanish state, nor a greater share in the making of decisions for the whole state, but the recovery of sovereignty and the negotiation of a federal structure on their own terms afterward. In the present situation the demands for ordered withdrawal of the police and armed forces of the state with a deadline, and the transfer of control of internal security to a regional government, are symbolic and threatening expressions of that ultimate aim of sovereignty.

It could be argued that these extreme formulations, the rhetoric of separatist nationalism, represent only minority positions in the whole region and that, therefore, they should be ignored and a compromise reached with the moderates. This, unfortunately, is only partly true, since the extremists have a significant plurality in one of the provinces and, given the inevitable devolution of power within the region in the absence of a true hegemonic center and the likely control of some local administrations by those advocating extreme alternatives, it is not clear how far the moderates would be able to control the actions of the plurality that bitterly rejects them and demands very extensive devolution, including control of internal order. It is not clear from the present record that Basque moderates have really given thought to the implications of some of the demands they have supported and to the ambivalences and ambiguities that in principle and practice contribute to making consociational solutions difficult.

THE BASQUE PARTY SYSTEM

The political party system of the Basque country and Navarra is, on the one hand, part of the Spanish party system and includes parties that function in all of Spain; on the other hand, it is a distinct system based on competition between these parties and the Basque nationalist parties (Linz 1978, 1980). In addition, various partisan organizations and coalitions compete for the vote of the Basque nationalist community.[3] Parties are therefore located on two axes: the left-right dimension and the national dimension, extending from Basque national independence to Spanish centralism.

The many cleavages of society in the Basque region are reflected in its politics. There are divisions along class lines, between natives and immigrants, between Basque-born who identify with Spain and those who support an independent Euskadi, between staunch defenders of the historical distinctness and Spanish identity of Navarra and those who favor its integration into Euskadi, and between social revolutionaries and reformist socialists.

The result of this social and cultural fragmentation is an extremely fragmented party system, sometimes complicated by electoral alliances between parties that compete with each other in other provinces. There are statewide parties—that is, Spanish parties called by the Basque nationalists *sucursalistas*, branch offices, so to speak, of the central parties. In some cases they adopt names including the word "Basque," and sometimes appear under a different name in the region. Some of these parties have, with greater or less enthusiasm, made aspirations to administrative or political autonomy of the region their own. Then there are the parties that present candidates in the region only; these we shall call Basque parties. They are not all of the same mind about the political future they desire for the Basque country. As elsewhere, some parties die and others are born. Coalitions between existing parties lead to new political groupings. In short, a very large number of political parties have been active in the Basque area in recent years. A list of the principal parties, with the initials by which they are often designated, follows:

AP	Alianza Popular (in the Basque Country in 1979 ran as Unión Foral del País Vasco, UFV)
AFN	Alianza Foral Navarra
ANV	Acción Nacionalista Vasca—Eusko Abertzale Ekintza
CC.OO	Comisiones Obreras
DCV	Democracia Cristiana Vasca
EE	Euskadiko Ezquerra (Basque Left)
ELA-STV	Euskal Langileen Alkartasuna-Solidaridad de Trabajadores Vascos
EMC-OIC	Movimiento Comunista de Euskadi-Organización de Izquierda Comunista
ESB	Euskal Sozialista Biltzarrea (Basque Socialist Party)
ETA	Euskadi ta Askatasuna (Euskadi and Freedom)
FE de las JONS(a)	Falange Española de las Juntas de Ofensiva Nacional-Sindicalista (auténtica)
HB	Herri Batasuna
IR	Izquierda Republicana
LAB	Langile Abertzale Batzordea (Committee of Patriotic Workers)
LCR	Liga Comunista Revolucionaria
ORT	Organización Revolucionaria de Trabajadores
PCE	Partido Comunista de Euskadi
PNV-EAJ	Partido Nacionalista Vasco-Euzko Alderdi Jeltzalea

PSOE	Partido Socialista Obrero Español. In the Basque Country it uses the name of PSE-PSOE, Partido Socialista de Euskadi-PSOE
PSOE(h)	Partido Socialista Obrero Español (sector histórico)
PTE	Partido del Trabajo de España
UAN	Unión Autonomista de Navarra
UCD	Unión de Centro Democrático
UFV	Unión Foral del País Vasco
UGT	Unión General de Trabajadores
UN	Unión Nacional
UNAI	Unión Navarra de Izquierda
UPN	Unión del Pueblo Navarro

These parties range widely from the center to the extreme left. They are also divided into those that would be satisfied with autonomy within the Spanish state and those that are explicitly loyal to that state; those that are ambivalent about the value of autonomy and those that are unequivocally committed to independence (some of them including the French Basque country in their sphere of action). Another point of disagreement is with regard to the incorporation of Navarra into Euskadi: some would leave it to the Navarrese electorate to decide; others consider Navarra to be ethnically and culturally part of the Basque country, and reject the possibility that Navarra might remain another autonomous region of Spain, defining the issue as nonnegotiable.

Some of these political groups have been unable to obtain any significant electoral strength in the region as a whole, but in some provinces they have gained a number of votes and in some local areas they have dominated elections.[4] The result has been a fractionalization of the Basque electorate (Núñez 1978; Cibrian unpublished paper, 1979).

It would be difficult to argue that any party can speak in the name of the Basques, and this obviously makes any consociational pattern of politics extremely difficult. In 1977 the largest plurality in the whole region of Euskadi Sur was obtained by a non-Basque party, the PSOE, with 24.7 percent of the vote, while the strongest Basque party, the PNV, obtained 28.5 percent. Adding the votes for the Unión Autonomista de Navarra (UAN), led by the PNV (6.81 percent in Navarra), its strength in Euskadi Sur would be 24.01 percent.

The complexity of Basque politics becomes even more apparent when we consider the largest plurality within each of the provinces. In Guipúzcoa the PNV obtained 30.49 percent of the vote, and in Vizcaya 29.86 percent. In Alava the largest plurality was that of the UCD, the party of Prime Minister Suárez, with 30.17 percent; in Navarra the UCD led with 28.60 percent. It is the weakness of the PNV in Alava (only 17.09 percent) and of the coalition led by the PNV in Navarra (6.81 percent), together with the absence of the UCD from the ballot in Guipúzcoa (in 1977) and its relative weakness in Vizcaya (15.89 percent), that accounted for the plurality of the PSOE in Euskadi Sur.

The problem is not only that none of the Basque parties can speak for a majority of the population in either the Vascongadas or in Euskadi Sur, but also that all the Basque parties added together in 1977 had no majority in the region. In fact, one could say that they barely had a majority in the most Basque of the provinces, Guipúzcoa, where the combined strength of the PNV with 30.49 percent, the EE with 9.11 percent, the ESB with 5.39 percent, and the ANV with 0.56 percent add up to 45.55 percent. Only by adding the moderate autonomist DCV (which since has joined UCD) with its 4.94 percent would one reach 50.49 percent. These figures, based on officially reported returns, could be modified by tenths or hundredths of a percent if we were to use those published in the Basque press, which probably include late returns from smaller communities that would increase slightly the vote of the EE and the PNV; but even then the added strength of all Basque parties in the Guipúzcoa electorate would be 51.61 percent. To this figure one would have to add the vote, for a candidacy that appears undefined, of Demócratas Independientes Vascos-Unión Foral para la Autonomía (4.63 percent).

In Vizcaya the situation is not very different. The total strength of the Basque parties—the PNV with 29.86 percent, the EE with 5.24 percent, the ESB with 2.62 percent, and the ANV with 0.81 percent—was 38.53 percent of the vote. To this one could add the 1.03 percent of the moderate autonomist DCV to reach a total of 39.56 percent. Therefore, in Vizcaya the plurality achieved by all Basque parties together in 1977 was a large one compared with the vote for the two most important statewide parties—the PSOE (24.49 percent) and the UCD (15.89 percent). However, the ideological heterogeneity of these parties cannot be ignored when we talk about the Basque vote.

In Alava the situation was even less favorable for the Basque parties, since the PNV was able to obtain only 17.09 percent of the vote; the added strength of the EE with 2.06 percent and the ESB with 2.17 percent would bring that figure up to 21.32, and the addition of the DCV with 2.71 percent would bring the total to 24.03 percent, representing less than one-fourth of the Alavese electorate.

The complex coalition politics in Navarra, the ambiguous position of some of these political groupings toward the incorporation of Navarra into Euskadi, and the alternative of a Navarrese autonomy within the Spanish state make it even more difficult to determine the strength of the pro-Euskadi vote there. The PNV-led UAN obtained 6.81 percent; and the Unión Navarra de Izquierda (UNAI), which would appear to be the equivalent in Navarra of the EE, obtained 9.28 percent. The total strength of both would therefore be 16.09 percent.

In the confrontation between Basque nationalism in all its variants and the Spanish state, we find significant minorities supporting not only statewide parties but also a party that can be defined as Spanish nationalist, Alianza Popular (AP), which under that name obtained 6.35 percent of the vote in Vizcaya and with that vote elected a deputy. In Alava it received 6.24 percent. In Navarra,

AP did not run candidates under its own name; but the Alianza Foral Navarra, perceived as its functional alternative, obtained 8.39 percent of the vote. And in Guipúzcoa a group appearing under the name Guipúzcoa Unida (including the extreme right) represents a similar position. It obtained 8.05 percent of the vote. In some respects the vote for these parties can be considered potentially ultraloyalist, since they would reject concessions to Basque nationalism going beyond a certain point defined by the Madrid government.

The UCD, in accordance with the policy of the present government, has accepted the autonomy aspirations of certain Spanish regions, but from the Basque point of view it is a *sucursalista* Spanish party. In 1977, it did quite well in the region, obtaining 30.17 percent of the vote in Alava, 28.60 percent in Navarra, and 15.89 percent in Vizcaya, where it had to compete for the Castilian-speaking electorate with the parties of the left, particularly the PSOE, which had the advantage of being a working-class party appealing to a largely working-class immigrant population. In addition the UCD had to divide the bourgeois Castilian-speaking vote with AP in 1977 and with the Unión Foral del País Vasco (UFV) in 1979.

If we consider the strength of AP and UCD, the two right-of-center statewide parties, we see that in 1977, in Alava and Navarra, more than one-third of the electorate identified with these Spanish parties. In Vizcaya one-fifth of the electorate voted for them.

The statewide parties on the left also present considerable fractionalization, even though in that camp the dominant party was and is the PSOE. Its competitor is not only the PCE, which in 1977 and in 1979 was disproportionately weak in the Basque country, with only 4.03 percent of the vote in 1977 compared with 24.75 percent for the PSOE. It also has to contend with the three main factions of the Marxist-Leninist or Marxist-Maoist Spanish political formations, which obtained 3.35 percent of the vote in the region. Without adequate survey data it is impossible to know to what extent the PSOE obtains the votes of the Basques, particularly the Basque-speaking working class, and how it divides that vote with the statewide revolutionary parties and with the Basque socialist formations. Our supposition is that the extreme left parties, particularly the Organización Revolucionaria de Trabajadores (ORT), had disproportionate strength among the native working class, and that the PSOE in 1977, and even more in 1979, was supported mainly by the immigrant working class.

One of the most striking features of the Basque party system is the weakness of the PCE. Its maximum strength in 1977 and 1979 was, respectively, 5.2 and 5.7 percent, and 4.8 percent in the 1980 election to the regional parliament, in Vizcaya, which is low, considering that this is an area of heavy metal industry, with very large plants and a concentrated working-class population on the left bank of the Nervión river on the outskirts of Bilbao. In some respects one could call this part of Vizcaya a Spanish Ruhr. In Guipúzcoa this weakness is even more visible. The PCE received 3.6 percent of the vote in 1977 and 3.1 percent in 1979, with a total vote for the extreme left parties of 2.3 percent in 1977 and 2.9 percent in 1979. The situation of the PCE was even worse in Navarra, where its 2.41 percent in 1977 and 2.18 percent in 1979 contrasted with the 4.98 percent obtained in 1977 and 4.26 percent in 1979 by a Maoist-supported candidacy. We do not know of any advanced industrial region of Europe where the Communists are followed so closely in their vote by parties to their left; and certainly in no province of Italy, Portugal, or (probably) France does any party to the left of the Communist party obtain twice its vote. In this sense the relative position of the extreme left, compared with the PCE, shows the strength of centrifugal and radical left tendencies in the Basque electorate. It is perhaps the clearest indication that the PCE, with its present strength in Spain, cannot ignore the competition of more leftist parties.

A good indication of the importance of this extreme left among the working class of Navarra is the fact that in the 1978 trade union elections the Sindicato Unitario of the ORT obtained (depending on the sources consulted) between 18 and 20 percent of the seats, compared with 10 to 16 percent for the UGT (Unión General de Trabajadores), linked with the PSOE, and 18 to 20 percent for the CC.OO (Comisiones Obreras), linked with the PCE, but including other parties in Navarra. To the fractionalization of the party system in the Basque country, we thus have to add a fractionalization of the trade union movement. In addition to the statewide union federations, there is a Basque trade union movement, which in Vascongadas, according to the Ministry of Labor data, obtained 12.5 percent of the seats and, according to the news agency EFE, 20.5 percent. In any case, the Basque labor movement is apparently weaker among the working-class electorate than the Basque nationalist parties are among the political electorate, an indication of the importance of nonworking-class voters for Basque nationalism, and also of the important presence of immigrants in the working class.

The distribution of seats in Parliament among the larger parties in 1977—nine for the PSOE, eight for the PNV, seven for the UCD, one for the EE, and one for the AP—can lead the observer to think that the Basque country has a moderate multiparty system. The reality, unfortunately, is infinitely more complex. Furthermore, while the Spanish party system at the national level consolidated and crystallized during the period between the legislative elections of June 15, 1977, and those of March 1, 1978, developments in the Basque country and Navarra presented quite a different picture.

A new political force, Herri Batasuna (HB), appeared on the scene. An electoral coalition of the *abertzale* (patriotic) left, whose aim is to struggle for national and social liberation, HB was formed by parties that had been illegal in 1977 (and mostly advocated abstention in the elections) and elements of ANV and ESB. It also enjoyed the support of the ETA. As a radical leftist and nationalist coalition, HB was in opposition to the PNV. It was committed to the ideal of independence, defending the integration of Navarra into Euskadi, demanding the immediate withdrawal of the Spanish police from the Basque country, and amnesty for all prisoners accused of political and terrorist acts. HB consequently advocated a "no" in the constitutional referendum of 1978 and active abstention from the referendum on the Autonomy Statute in 1979. Its three deputies, elected in 1979, have refused to take their seats in the Spanish Parliament. It has close connections with a nationalist radical labor organization, Langile Abertzale Batzordea (LAB; Committee of Patriotic Workers). The founding of the newspaper *Egin*, to compete with *Deia*, the organ close to the PNV, has been very important for the diffusion of its message. It is not always clear to what extent the main appeal of HB is extreme nationalism or its social revolutionary stance, nor is it always clear to what extent those who vote for it identify with the party and its program or are protesting the economic and social crisis in the Basque country.

Euskadiko Esquerra (EE), which before the founding of HB represented the left nationalist alternative at the electoral level, has suffered seriously from the competition of HB; and in 1979 its share of votes in the Basque nationalist camp was reduced, even though it increased its share in Euskadi from 6.05 to 7.81 percent (from some 61,000 to 80,000 votes). In the process it experienced an evolution toward more moderate positions, supporting the Autonomy Statute together with the PNV, although using very different arguments. There are indications that EE is perhaps more a national Communist party than an extreme nationalist party, and that its electorate is less heterogeneous and more ideologically committed than that of HB.

Between 1977 and 1979 the small DCV disappeared, with some of its leaders joining the UCD and its electorate probably dividing between that party and the PNV.

The struggle over the future of Navarra has reinforced the opposition to the UCD on the right as a result of the bolting of the party by one of its 1977 deputies, Jesús Aizpún, dissatisfied with the possibility foreseen in the constitution, and supported by the UCD, of allowing Navarra to decide whether it wants to join Euskadi. The new Unión del Pueblo Navarra (UPN) was able to obtain 10.97 percent of the Navarrese vote and one seat (compared with the Alianza Foral Navarra, its predecessor, which received 8.39 percent in 1977).

Some of the Spanish parties changed the E of España to the E of Euskadi in their names, and the forces represented in 1977 by AP appeared under the name Unión Foral del País Vasco (UFV).

The fragmentation on the left continued, even when the front names used in 1977 were replaced by the true names of the contending parties. Splinter Communist groups, like the Liga Comunista Revolucionaria (LCR) and the Movimiento Comunista de Euskadi-Organización de Izquierda Comunista (EMK-OIC in Basque), appeared on the ballot. The forces to the left of the PCE, the statewide Communist party, continued to confront it with greater competition than any Communist party in Europe experiences in an advanced industrial region. These extreme leftist parties, particularly EMK-OIC, opposed the Autonomy Statute and campaigned actively for the incorporation of Navarra into Euskadi.

One of the most important changes in the electoral game in 1979 was the presence of UCD on the ballot in Guipúzcoa, from which it had been absent in 1977. The governing party in Spain was able to make an honorable showing in that province with 15.06 percent of the vote, obtaining some of the support that in 1977 had gone to the DCV and probably many of the rightist Guipúzcoa Unida voters, even though some of that coalition must have gone to UFV. The nationwide extreme right coalition, Unión Nacional (UN), also appeared on the ballot in the three provinces of Euskadi.

The Carlists, who in 1977 had not been legalized, and therefore could not use their name, in 1979 appeared on the ballot in all four provinces as Partido Carlista. But they had little success except in Navarra, where, in spite of the change in ideology, family tradition must have contributed much to assure its 7.65 percent of the vote.[5]

In 1977 the PSOE had managed to obtain the largest plurality of any party in the Basque and Navarrese region, with 313,843 votes (24.75 percent of the ballots cast). In 1979 it was not able to repeat this feat. Its 247,785 votes represented only 19.22 percent, while the PNV had 23.15 percent and the UCD had 19.65 percent. The PSOE was the great loser in the region. Its defeat became even more apparent in the municipal elections less than two months later. It lost three of the nine parliamentary seats it had gained in 1977. It is not easy to explain this loss of votes from a sociological point of view, and divergent interpretations have been given within and outside of the party. Some argue that the votes must have gone to the Basque left, particularly HB, while others believe that many former PSOE voters became nonvoters—who constituted the largest "party" in the 1979 election in Euskadi. Postelection survey data do not seem to confirm the thesis that HB was the heir to some of the PSOE electorate; the analysis suggests the importance of abstention by PSOE voters in some of that party's strongholds.

After the 1979 election, therefore, the Basque party structure remains as fragmented as ever, with no party able to speak for the whole Basque electorate or clearly dominant in any of the provinces. This is a result of class and ideological divisions within the Basque nationalist camp and the fact that the statewide parties received some of the votes of the Basque autochthonous population and most of those of the numerous immigrants.

Fragmentation, found at the electoral level, in spite of the impact of the d'Hondt proportional system in districts with few deputies to be elected, like Alava and Navarra, can be found also at the parliamentary level. (The d'Hondt system provides a formula for determining the minimum number of votes a party must receive to be represented. It tends to favor the largest party and exclude very small parties.) The twenty-six seats assigned to the region are

TABLE 2.2

Vote in the Legislative Elections for the Lower House, Euskadi and Navarra, March 1979 (percent)

Party	Alava	Guipúzcoa	Vizcaya	Euskadi	Navarra	Euskadi and Navarra
HB	9.68	17.28	14.39	14.78	8.73	13.56
EE	4.54	12.58	5.72	7.81	–	6.24
PNV	22.28	25.86	28.45	26.89	8.30	23.15
Total Basque nationalists	36.50	55.73	48.56	49.48	17.03	42.95
Liga Comunista	–	–	–	–	.27	.06
LCR	.50	.64	.55	.58	.41	.54
EMK-OIC	1.02	1.42	1.31	1.19	1.14	1.18
ORT	1.01	.83	.54	.68	4.26	1.40
PTE	.64	–	–	.07	–	.06
PCE	3.21	3.07	5.67	4.54	2.18	4.07
PSOE	20.77	17.89	18.61	18.63	21.55	19.22
PSOE(h)	.57	–	.73	.47	–	.38
P. Carlista	.81	1.28	.25	.65	7.65	2.06
IR	–	.18	.22	.17	.20	.18
UCD	24.68	15.06	15.53	16.44	32.40	19.65
UPN	–	–	–	–	10.97	2.21
UFV	6.04	1.04	4.15	3.36	–	2.69
UN	.88	.68	1.32	1.07	–	.85
FE Jons(a)	–	.09	.14	.11	–	.08
Other	.16	.04	.10	1.16	.08	.14
Blank	.40	.23	.20	.23	.31	.24
Void	2.77	1.83	2.25	2.17	1.55	2.05
Total	99.96	100	100	100.08	100	100.01

distributed among six parties or coalitions, compared with five in 1977. In Euskadi proper, five parties shared twenty-one seats; the PNV had the largest number (seven), followed by UCD and the PSOE, with five each, HB with three, and EE with one.

Nevertheless, the 1979 election represented a considerable victory for Basque nationalism. The parties and coalitions that in 1977 had obtained 33.94 percent of the vote in the Basque-Navarrese region received 42.95 percent in 1979. Their gains were particularly impressive in Euskadi, where the same forces that in 1977 had obtained 38.53 percent were able to gain 49.48 percent of the ballots cast. Calculating the vote for HB, EE, and PNV in Euskadi, omitting the blank and void votes, gives the Basque nationalists a small majority in the region—50.70 percent of the valid votes—with the inclusion of Navarra the total would be reduced to 43.96 percent of the valid votes.

Basque nationalism made considerable progress in 1979, but it was still far from being a majority in the area claimed as Euskadi Sur, and has only a tiny majority in Euskadi. In addition, it should not be forgotten that the strength of the Basque parties is very unevenly distributed within the region. In the province of Alava, in spite of the great gains made by nationalists—from 21.25 percent in 1977 to 36.50 percent in 1979—this still represented only somewhat more than one-third of the voters. In Vizcaya the nationalists moved from 38.53 to 48.56 percent, just below an absolute majority of valid votes. In Navarra progress of the coalition formed by Basque parties has, by comparison, been minor. In only one of the provinces, Guipúzcoa, do the nationalist parties and coalitions have a clear majority, with 55.73 percent of the vote; but the PNV could claim a hegemonic position with its 25.86 percent.

The contrast with other countries in which there are nationalist movements or parties representing national linguistic minorites could not be greater. The nationalist movement until now has not been able to mobilize all of its potential constituency; it is even more unable to unite behind a single hegemonic party, and its penetration in the different provinces of Euskadi Sur is quite unequal. Further, nationalist strength varies sharply within provinces, particularly Navarra and Alava, as electoral maps would show; and the nationalists have had limited success in the metropolitan area of Bilbao, the largest city in the region and one of the great industrial centers of Spain.

It is difficult to predict whether this fractioned party system of the Basque country will be replaced by stronger formations. But it seems very doubtful that the major parties—the PNV, the PSOE, and the UCD—could completely dominate the sectors of the electorate potentially closest to them. Even if the PNV turns further to the left, the Basque leftist parties are unlikely to join with this party of Christian Democratic tradition that is willing to work within an autonomy granted by the Spanish state, even though holding on to the ideal of self-determination for a Basque nation. The Spanish nationalists supporting AP in 1977 and UFV and UN in 1979 are not likely to find the UCD an attractive alternative, and the many groups to the left of the PSOE are not likely to integrate into the largest socialist party in the region; nor are they, divided by deep ideological schisms, going to join a regionally weak PCE.

In addition, we should not forget that these different forces, while weak at the regional and even the provincial level, may have pluralities in some of the valleys of the Basque country and in the many small industrial communities that dot its landscape. Any tendency toward fusion or integration of these numerous parties is likely to be weakened by the importance that the municipality traditionally has had in Basque politics. Article 26 of the Statute of Autonomy of 1979 on the Basque regional parliament should contribute to freezing the present party system and perpetuating its fragmentation by assigning an equal number of representatives to each "historical territory" (irrespective of population), to be elected by proportional representation. In every respect the Basque country is internally less unitary and centralist than Catalonia, where the hegemony of Barcelona is long-standing.

The complexity of the party system becomes even more relevant when we consider the importance in Basque politics of militant minorities that have turned to terrorism in the struggle for national liberation and revolution, responding to the oppressive experience of the Franco period and taking advantage of the difficulties of maintaining order in a hostile area through the use of police forces alien to the society.

The political alignments in the Basque country are the result of linguistic-cultural and class cleavages. Both the Basque parties and the statewide parties are differentiated by their class composition, but the lines of cleavage are in no way simple. Some of the Basque-born and Basque-speakers vote for statewide parties, and some of those not born in the Basque country—Castilian-speaking immigrants—vote for the Basque parties. However, these deviants from the model pattern constitute a minority of the electorate of each party.

LANGUAGE AND POLITICS

As in many other nationalist movements, the affirmation, defense, and revival of the language has been a central goal of Basque nationalism.[6] Inspired by the success of the Catalan Renaixenca since its founding by Sabino Arana, the control of language policy and education has been one of the central tenets of the Basque movement in all the negotiations about autonomy. The recognition of both Basque and Castilian Spanish as official languages, the introduction of Basque-language teaching in the schools (which devote up to 25 percent of teaching time to it), and the right to control television have been some of the great victories of Basque nationalism after Franco.

The task confronting the Basques is not easy, however, and therefore they look at the Israeli introduction of Hebrew as a model. It is not certain that their success might not be closer to that of the Irish nationalists who wanted to make the Irish Republic a Gaelic-speaking country. In contrast with the Catalans, who have a literary tradition in their language and experienced a great cultural revival at the turn of the century, the Basques' language has had a less illustrious history. It was never a court language; Castilian was used even in local government institutions, where Latin was not used. The great Basque writers, Miguel de Unamuno and Pío Baroja in the early twentieth century, wrote in Castilian. Few books are published in Basque, and the older works were mostly religious books. Even today the two nationalist newspapers, *Deia* and *Egin*, publish only some pages in Basque. Many of the leaders of the nationalist movement learned the language late in life, moved by their political commitments.

A report prepared for the Basque Academy calculated the Basque-speaking population in Great Euskadi as 632,000 out of 2,784,000 living north and south of the border. This figure would mean that 23 percent speak the language. The same report calculates that 45 percent of the population of Guipúzcoa, 15 percent of that of Vizcaya, 8 percent of that of Alava, 11 percent of that of Navarra, and 34 percent of that of the French Basque country speak the language. The nationalists certainly will have a hard task to achieve their goal of making the population speak Euskera. The parties and affiliated organizations have devoted great effort to creating *ikastolas* (schools) that serve to transmit the language and the nationalistic spirit; and the decree of August 3, 1979, creates an opportunity to educate the younger generation even in those areas where Basque was never spoken—the children of immigrants or residents who might not wish to learn the language will be obliged to do so. However, the same report speaks of the need for 12,000 teachers and 17 billion pesetas in a ten-year program. Significantly, during the debate on the Autonomy Statute in the Madrid Parliament, one of the Spanish parties (Partido Socialista Andaluz) proposed an amendment to defend the linguistic rights of the immigrants.

Language is, therefore, an element in the conflict between the Basques and the Spanish state; but the Basques themselves face a situation somewhat similar to that of the Welsh, in that an effort to impose the language strictly and rapidly could alienate many Basques who do not speak it.

It will not be easy to assimilate the immigrants linguistically, particularly since Euskera is a difficult language and not a Romance language, as Catalan is. It is therefore difficult for Castilian-speaking immigrants to learn, although they may be willing and even eager to have their children learn the language so as to assure them a better future in the region. An immigrant from Estremadura asked his son in a television documentary about the life of a family in the past decade: "Do you want to stay here?" When the son answered in the affirmative, his father told him that he "better do well in school in Basque."

Let us turn to the relation between language use and political behavior and attitudes. (See Tables 2.3 and 2.4.)

TABLE 2.3

Party Preference in 1978 and Vote in 1979, by Reported Knowledge of the Basque Language (percent)

Knowledge of Basque	HB	EE		PNV		PCE		PSOE		UCD		AP		None		Don't Know/ No Answer		Total	
	1979	1978	1979	1978	1979	1978	1979	1978	1979	1978	1979	1978	1979	1978	1979	1978	1979	1978	1979
Reads	42	38	30	38	44	18		10	5	5	15	8		19	17	13	18	16	25
Speaks	42	48	30	40	47	17		12	5	9	17	8		25	18	15	21	19	26
Understands	53	69	35	59	55	27	8	20	11	22	22	8		47	23	24	25	32	33
None of above	45	29	65	37	44	66	92	78	85	77	76	92	100	52	77	72	75	66	66
No answer			3	1	1	5		1	2	1	2			2		3		2	1
	(55)	(76)	(23)	(143)	(77)	(55)	(12)	(302)	(55)	(153)	(41)	(26)	(3)	(117)	(77)	(251)	(28)	(1,140)	(389)

Note: In 1978, among the 13 "extreme Left" respondents, 75 percent did "none of above"; the percentage was zero for the 3 "extreme Right" respondents. In 1979, among the 14 "extreme Left" respondents, 57 percent did "none of above"; the figure was 80 percent for the 5 "extreme Right" respondents. These two groups are included in the total.

Source: Surveys by DATA, Madrid.

TABLE 2.4

Knowledge of Basque Language and Party Preference, 1978 (percent)

Knowledge of Basque	EE	PNV	PCE	PSOE	UCD	AP	None	Don't Know/ No Answer	Total (No.)
Reads	15	29	5	16	4	1	12	19	(188)
Speaks	16	26	4	16	9	1	13	17	(220)
Understands	14	23	4	17	6	1	15	17	(365)
None of above	3	7	5	32	16	3	8	24	(747)
No answer	11	7	14	12	10		11	31	(21)

Source: Surveys by DATA, Madrid.

Only a small number of those who know Basque give their vote to UCD, and practically none admit supporting AP, even though it is possible that a number of those not expressing a party preference in a 1978 survey might be giving their votes to one of these parties. Among those speaking the language, only 13 percent sympathized with the UCD. The PSOE, on the other hand, received around 23 percent of their votes. The overwhelming majority of the Basque-speakers, who are a minority in the Basque-born population (particularly if we include the natives of Navarra), gave their support to Basque parties: 37 percent to the PNV and 21 percent to the EE. This means that in 1978 the PNV received some 40 percent of its vote from Basque-speakers and the EE 48 percent, compared with the 12 percent of the PSOE vote coming from them, 17 percent of the PCE vote, and less than 10 percent of the UCD and AP. These figures can be compared with the 19 percent who, in the four provinces, say that they speak Basque.

The support for the Basque parties is not limited to those speaking Basque, since a very large proportion of the population of Basque origin that identifies as Basque does not speak the language. The most nationalist of the parties in 1978, the EE (which supports the independence of the Basque country), received 29 percent of its vote from those who do not even understand Basque, and the PNV, 37 percent. The Communist party received most of its votes from those who do not understand Basque—66 percent, a figure that rises to 78 percent among the supporters of the PSOE. The bourgeois statewide party, the UCD, received some votes from Basque-speakers, but 77 percent of its voters do not understand Basque, a figure that rises to 92 percent among voters for AP.

If we turn to the vote of those who do not understand Basque, both natives of the region and immigrants, the proportion supporting the PNV was 10 percent, and of those expressing a preference for EE, 4 percent. The bulk of the vote of the non-Basque-speakers went to the PSOE (47 percent), a result of the appeal of the Socialist party to the natives of Alava and Navarra, the two provinces where the nationalist sentiment is weakest, and of the weight of the immigrant working class in the whole region. Among those monolingual in Castilian who expressed a preference, one-quarter chose the UCD. To this one would probably have to add some of the votes of those not expressing a party preference.

NATIVES AND IMMIGRANTS

One of the great differences between the politics of multilingualism and multinationalism in Spain and in other countries of Western Europe is the fact that the Basque country and Catalonia are both regions of immigration rather than, like Brittany, Corsica, Sicily, Macedonia, and others, of emigration (Barbancho 1967; Miguel 1974). In contrast with Scotland and Wales, where only a small proportion of the population, mainly Irish immigrants, was not born in the region, these two highly developed industrial centers of Spain have received massive numbers of immigrants, particularly in the 1970s. The stability of multilingual polities like Belgium (with the exception of the Brussels area) and Switzerland has been due to the linguistic and cultural homogeneity of each area, if we omit the foreign immigrants. Conflicts arising from linguistic and cultural assimilation are therefore less salient, and the possibility of a homogeneous linguistic environment is greater.

Immigration, in addition, is to a considerable extent linked with the socioeconomic stratification system, since most of the immigrants are manual workers and form a disproportionate percentage of the unskilled laborers. They tend to concentrate in the industrial cities, where they constitute close to half of the population. The native/immigrant cleavage line in part coincides with the class cleavage and the urban/rural division. In Euskadi Sur, according to the 1970 census, 30 percent of the population were immigrants, although there were great differences between the provinces: from a low of 16 percent in Navarra to close to 37 percent in Vizcaya and 31 percent in Alava. While a significant number of those immigrants came from the overpopulated and poorer regions of Spain, a very large proportion came from Castile and León, areas of relatively high levels of education, independent property-owning peasants, and strong Spanish identification. For these people, discrimination aside, assimilation and integration into Basque society would be perhaps more difficult than for the immigrants from Andalucía and other poor regions.

Basque parties, while not advocating any discrimination against immigrants, and occasionally making gestures toward them, are basically ambivalent about the threats to Basque identity represented by immigration.[7] In fact, this theme was very salient in Basque nationalism at the turn of the century, and accounted for considerable tension between socialists and nationalists until the mid-1930s. Today the Basque left underlines the common class position and attempts to integrate immigrants into its movement on the basis of shared hostility to, if not hatred for, the capitalists—particularly big businesses whose identification with Spain (for various reasons) makes the Bilbao business elite, or at least an important sector of it, traitors to the national cause and class enemies.[8] It remains to be seen to what extent, particularly when the language policy of the nationalists begins to be implemented effectively, the common class interest will prove to be stronger than cultural identity.

It is an open question whether the PSOE, whose strongholds always have been communities with large numbers of immigrants, will be able to retain both natives and immigrants in its fold. There are indications that nationalist parties on the left have been able to compete effectively with the PSOE, and particularly with the PCE; and there are also indications that the PSOE leadership identification with the Basque cause may have contributed to a loss of votes among immigrants, in this case to the large "party" of nonvoters. Certainly the PSOE

TABLE 2.5

Place of Birth of Those Expressing a Preference for Various Parties in the Basque Country and Navarra, 1978 (percent)

Province	Party Preference						No Answer/ Don't Know	Total Sample
	EE	PCE	PSOE	PNV	UCD	None		
Alava	23	16	12	28	17	13	9	15
Guipúzcoa	29	13	12	21	11	32	15	17
Vizcaya	28	20	18	35	11	20	13	19
3 provinces	80	49	42	84	39	65	37	51
Navarra	3	23	16	7	26	12	53	25
Total	83	72	58	91	65	77	90	76
Outside region	17	28	42	9	35	23	10	24
	(76)	(55)	(302)	(143)	(153)	(117)	(251)	(1,140)
Father born in region	72	56	48	67	57	57	78	61

Note: The total includes the supporters of extreme left and extreme right parties.
Source: Survey by DATA, Madrid.

today receives a large share of its vote from immigrants, and by 1979 might have lost a significant number of its Basque voters to the socialist parties in the nationalist camp. While 80 percent of the EE supporters were born in the three provinces, this was true of only 49 percent of PCE adherents, and 41 percent of those of the PSOE (see Table 2.5). Around the center, leaning either to the left, like the PNV, or to the right, like the UCD, one of the dividing lines is clearly the native constituency of the PNV and the large proportion of those from outside the region in the UCD electorate, respectively 9 and 35 percent in a 1978 survey. Among those born in the region a certain proportion, particularly in Vizcaya and Guipúzcoa, are second generation immigrants (as the number of those whose parents were not born in the region shows). This poses all the well known problems of identification, including overidentification, in the process of integration into Basque society.

CLASS, GENERATION, AND POLITICS

Basque society is an advanced industrial capitalist society, deeply divided along class lines that cut across the lines of ethnic identification.[9] In strictly Marxist terms most Basques are "proletarians," since they are not the owners of the means of production. Given the small number of persons active in agriculture, peasant property owners do not constitute a significant proportion of the population, even though many Basque (but not immigrant) workers are also part-time farmer-landowners. On the other hand, the concentration of big industry, particularly in Vizcaya, makes a high bourgeoisie a key element in both society and politics. The coexistence of the Bilbao "oligarchy" with a large number of medium and small family owners and entrepreneurs (many of working-class background), particularly in Guipúzcoa, is another source of cleavage and tension not without political significance.

The managerial, professional, and white-collar sectors in this advanced industrial society play an increasingly important role, and for them the language and culture have particular emotional significance. The Basque language for them also represents a potential advantage in relation to those among the immigrant professionals and civil servants who know only Spanish. It is no accident that Carlos Garaikoetxea, the PNV leader, in appealing for a "yes" vote on the Autonomy Statute on the eve of the referendum, alluded to the opportunities autonomy would offer Basque youth in the new regional government. Pressures to force out the nonnative civil servants, university professors, and others, including real and bogus threats, undoubtedly are opening new opportunities to the Basque middle class.

If we turn to the class composition of the electorate of different parties (see Table 2.6), we will discover that the cleavage lines are far from clearly defined. If we take as a basis the subjective class identification chosen by the

TABLE 2.6

Party Preference and Class Identification, Basque Country and Navarra, 1978 (percent)

	Party Preference								
Subjective Social Class	EE	PCE	PSOE	PNV	UCD	AP	None	Don't Know/No Answer	Total
Upper or upper middle	8	6	9	18	19	36	16	6	12
Lower middle	41	38	32	35	38	23	40	31	34
Working	49	56	58	41	40	21	42	55	50
No answer	2	—	—	5	2	20	2	7	3
	(76)	(55)	(302)	(143)	(153)	(26)	(112)	(251)	(1,140)

Source: Survey by DATA, Madrid.

respondent for his family, we discover that the parties of the working-class left, the PCE and the PSOE, received over 50 percent of their support in 1978 from those who identified as working class, but even the UCD received 40 percent of its support from that class and AP, 21 percent. In the Basque electorate the revolutionary EE received 49 percent from the working class and the PNV, which has always considered itself an interclass party, 41 percent. The larger proportion of the PNV vote coming from the upper or upper-middle class, 18 percent, distinguished it from the EE (8 percent). In terms of class composition, the Basque nationalist PNV and the Spanish UCD were almost identical. The dividing line between those two parties is ethnic identification rather than class identity, even though those parties are placed by the electorate and by their own voters quite differently on the left-right scale.

Among the non-Basque parties, AP stands out clearly as the one receiving a large proportion of its support from the upper or upper-middle class, more than one-third. The AP electorate, both in class composition and in language, is clearly differentiated from that of the UCD. The two statewide working-class parties, the PCE and the PSOE, are not clearly differentiated in their class basis, although there is a slightly larger proportion of PSOE voters of more privileged socioeconomic position; and they are to only a small extent differentiated in terms of language, perhaps slightly more than in terms of socioeconomic status. The two parties on the left of the spectrum, EE and PCE, are differentiated by their linguistic composition rather than by their socioeconomic classes. In fact, the less radical PCE has more working-class voters, and the revolutionary EE has more middle-class voters.

Differences in age, as well as class, characterize the adherents of various parties (see Table 2.7). In the days of Francisco Franco, a division between fathers and sons among supporters of the PNV over the tactics to use in fighting the dictatorship led to the transformation of the youth organization of the party into the ETA, and also to the emergence of new, more radical parties. Today, also, one of the great differences between the electorate of EE (in 1978, and HB and EE in 1979) and that of the PNV is age. Radical nationalist and Marxist parties have attracted disproportionate support among the younger age cohorts. The granting of suffrage in 1979 to those 18 to 21 years old was one of the factors in the success of HB, which received a large part of its vote—22 percent, according to one 1979 survey—from those voting for the first time.

It is an open question whether those young Basques who voted EE and HB, and who were dominant in the constant street agitation in this period, will continue to hold those views as they grow older. The identification of sons and daughters of many PNV followers with more radical groups helps to account for the sympathy, or at least ambivalence, of moderate Basques toward the radicals and even toward their terrorist activities.

TABLE 2.7

Age and Marital Status of Male Supporters of Parties in the Basque Country and Navarra, 1978 (percent)

Age	EE	PCE	PSOE	PNV	UCD	AP	None	Total Sample
18–20	23	9	6	3	2		14	7
21–24	22	5	10	6	5		15	9
25–29	15	9	14	10	6	5	15	11
30–44	25	29	37	34	30	29	28	32
45–59	16	25	22	28	26	15	16	23
60+		24	10	17	30	51	11	18
Unknown			1	1			1	1
Marital status:								
single	54	31	25	21	14	20	43	26

Source: Survey by DATA, Madrid.

RELIGION—ANOTHER CLEAVAGE

Religion plays an important role in Spanish politics, although for various reasons no Christian Democratic party emerged after Franco. The Roman Catholic Church, at least initially, defined itself as neutral in relation to the parties (Linz 1979). Obviously, however, in matters like education, divorce, and abortion, differences between the positions of the parties and of the church become politically significant (see Table 2.8). Despite efforts of the PCE to overcome Catholic distrust, many Catholics, particularly in the Basque-Navarrese region, think that it is incompatible to be a good Catholic and a good Communist. The UCD in Spain and the PNV in the Basque country have long-term links with the Christian democratic movement, are perceived as defending Christian values, and therefore disproportionately attract the votes of practicing Catholics. AP has been even more inflexible than UCD on issues of religious significance, while the Marxist and socialist left has advocated a laicization of state, law, and education.

The Basques and Navarrese have been, at least until recently, the Spaniards most identified with the Catholic Church, giving a large number of their sons and daughters to the cloth. Basque bishops were disproportionately represented in the Spanish hierarchy, and the clergy enjoyed enormous influence and prestige in the smaller communities (Duocastella 1975; Núñez 1977). As in many other minority culture areas in Europe—Slovakia, Slovenia and Croatia, Flanders, Brittany, Wales—the clergy was identified with the autochthonous culture, was the intellectual elite using the vernacular, and as a result was a standard bearer of nationalism even where the hierarchy might object to its nationalist activities. The PNV, since its founding, often had the support of the local clergy (Larronde 1977, pp. 79-97). The winds of change after Vatican II led to the emergence of a left-oriented radical critique of capitalist society by many clergymen. This trend was particularly important in the Basque country, where social and national minority protest fused. The Franco regime, to comply with the Vatican Concordat, had to establish a special prison for opposition priests, mostly Basques. One of the consequences has been the moral support given to the nationalist struggle, and even indirectly to the ETA fighters, by members of the clergy. Many priests and former priests have become political leaders, including those of parties of the left and even extreme left.[10]

However, the mass electorates of parties of the left, the center, and the right are still differentiated along religious lines (see Table 2.8). Thus, the greater radicalism of the EE than the PCE is congruent with the very small proportion of good or practicing Catholics among the EE supporters and the large number of indifferents and atheists. In this latter respect it is not very different from the PCE. Religiosity, however, very clearly differentiated the electorate of the EE from that of the PNV: while 47 percent of the PNV supporters declared themselves good Catholics or practicing Catholics, only 10 percent of those of EE did so. While 16 percent of the PNV declared themselves indifferent,

TABLE 2.8

Party Preference and Religion in the Basque Country and Navarra, 1978 (percent)

Religion (Self-rated)	Party Preference						Don't	Total
	EE	PCE	PSOE	PNV	UCD	AP	None	Know/No Answer	
Good Catholic	1	–	3	9	6	22	2	13	6
Practicing	9	16	28	38	55	69	23	46	35
Not very practicing	14	9	24	25	24	6	23	19	21
Not practicing	25	16	20	11	10	–	12	7	13
Indifferent	35	39	22	16	5	–	22	9	17
Atheist	12	11	4	–	–	3	12	4	5
Other religion	1	6	0	–	–	–	1	–	1
	(76)	(55)	(302)	(143)	(153)	(26)	(117)	(251)	(1,140)

Source: Survey by DATA, Madrid.

35 percent of the EE did so, and an additional 12 percent defined themselves as atheists. More than language or class, religion was one of the dividing lines between the two main Basque parties in 1978.

Among the statewide parties religion also is a differentiating factor. Many more PCE supporters declared themselves indifferent (39 percent), or atheist (11 percent) compared with the PSOE (22 percent and 4 percent). None of the Communists claimed to be a good Catholic, while 3 percent of the PSOE voters did so; and the proportion of those declaring themselves practicing was 16 percent among the PCE supporters and 28 percent among those of the PSOE. In the religious Basque country, as in other parts of Spain, one of the key differentiating factors between the two Marxist parties is religion, in spite of the efforts of the PCE to extend a hand to the Catholics and the parading of priests and Catholic laymen by the party, which contrasts with the much more reticent position of the PSOE toward the church. More surprising is the fact that the PNV and the UCD electorate should be distinguished by their different levels of religiosity. Given the religious tradition of the Basque country and the long-standing identification of the PNV with the Christian democratic international, one could have expected that its electorate would be more religious than that of the UCD, but this was not the case in 1978. Although the minority that declared itself made up of very good Catholics was slightly larger among the PNV voters, the number of those indifferent was appreciably larger than in the UCD. The electorate of AP was clearly different from that of the UCD, with a much larger proportion of those claiming to be "good Catholics" (22 percent) and practicing Catholics (69 percent).

THE BASQUE COUNTRY—A SOCIETY IN CRISIS

A number of circumstances, not all derived from the political situation, converged at the time of the restoration of democracy to create a deep feeling of crisis in the Basque country. The frustrations and even hatred created by the indiscriminate response of the Franco regime and its police to ETA terrorist actions and the opposition the regime encountered in the region have left a bitter legacy.[11] There was, and still is, hope that the coming of democracy, the amnesty for those arrested for anti-regime activities, free elections, the preautonomy phase, an even more extended amnesty, the enactment of a constitution that includes provisions for devolution, and finally the approval of the Autonomy Statute will reverse the downward spiral of frustration, polarization, conflict, and violence. Certainly each of those measures did not immediately have that effect, and it remains to be seen whether the devolution after the election of the first Basque parliament and the investiture of an autonomous government will be able to reverse the trend. It will be always argued and be arguable whether some of those measures could have been taken earlier, faster, with less debate, and could have been more generous; but those who hold that view forget that those policies inevitably encountered opposition in broad sectors of the population, particularly in the armed forces, which suffered undeniable provocation and perceived them as a threat to the authority of the state and the unity of Spain.

The Basque situation is unique in the sense that consociational arrangements had to be worked out in the face of continuous terrorist activity, attacks on the legitimacy of the participants in the political process, and domination of the street by activist groups and violent youth—far from the model of slow accommodation among elites respecting and trusting each other and capable of committing their followers, as described by Arend Lijphart and others. In all the negotiations, including those between Suárez and Garaikoetxea, the leader of the PNV, in the summer of 1979, the threat of violence loomed in the background, and many statements linking demands and the need for compromise to the risk of continuous tension and violence could sound like blackmail. Whatever criticism of details can be offered by either side, there has been an extraordinary effort by those actively concerned with politics, and the two top leaders in particular, to reach for solutions.

The deep political crisis has been aggravated and compounded by a deep and growing economic crisis. For structural reasons, important sectors of Basque industry in the 1970s started to face serious difficulties. Shipbuilding, steel, and some consumer durables were experiencing a structural crisis linked in part to the international economic turnaround. These difficulties were compounded by the presence of a very militant working class and constant political strikes to support demands for amnesty or to protest the death of ETA activists or innocent bystanders. The imposition of the so-called revolutionary tax by the ETA, effectively collected with the help of information provided even by bank clerks about the financial status of individuals and enterprises, helped to discourage the business community from investing, and led to a flight of capital and industrial plants.

This tragic situation is reflected in the figures on violence during the period after Franco's death, the inevitable arrest of persons accused of politically related criminal acts, and the feeling of personal insecurity and lack of political freedom resulting from terrorist threats.

The resulting situation is reflected in the public opinion data collected in 1978 and 1979 on the political and the economic situation, and on the problem of terrorism. To probe attitudes about the political and economic situations, we used the following question: When asked about the economic (or political) situation by a friend from outside Spain—for an opinion on the present situation—would you say that many problems remain to be solved . . . that in general we cannot complain . . . or that the situation is growing more serious every day and that things cannot go on like that? In the whole of Spain, more than half of the respondents were pessimistic about the economic situation, but as of the summer of 1978, 57 percent were relatively optimistic on the political front.

At that time, in the Basque/Navarrese region, pessimism on the economy was 7 percent higher than the national average. In politics optimism was 16 percent lower and pessimism 14 percent higher. Even more significant is the fact that pessimism in all four provinces increased between the summer of 1978 and the summer of 1979, rising from 61 to 78 percent with respect to the economy and from 53 to 68 percent as far as politics was concerned. As in all other areas of opinion, there were very significant differences within the Basque region. Economic pessimism in Guipúzcoa, for example, rose from 76 to 93 percent during the period in question, while in Navarra it rose from 47 to 66 percent. The contrast is even more marked in regard to the political situation; in Guipúzcoa the proportion of those saying it was becoming more serious every day rose from 65 to 90 percent, while in Navarra the proportion of pessimists remained stable or may even have declined, the corresponding figures being 46 and 43 percent.

These responses undoubtedly reflect the actual economic crisis and the growing number of unemployed, but they are also colored by nationalist views, as shown by a comparison between responses of the supporters of the PSOE, the party of the immigrant working class, and those given by adherents of the PVN and, above all, of the EE and HB. In 1978, 62 percent of the PSOE supporters saw the economic situation as growing more serious every day, compared with 83 percent of the EE and 74 percent of the PNV. In 1979 the pessimistic proportion among the PSOE had risen to 75 percent, was 78 percent among the EE, and reached 91 percent among HB supporters.

Differences are even more marked with respect to the political situation, with 57 percent among the PSOE feeling in 1978 that "things cannot go on like that," compared with 75 percent among the EE. By 1979 the pessimists were 62 percent among PSOE voters, 65 percent among those of the PNV, 91 percent among the EE, and 85 percent among the HB.

The responses on both the economic situation and the political situation were closely linked with attitudes on centralism, autonomy, federalism, and independence in a linear way, indicating that it is not only the objective situation but also the ideological commitments that account for the perception of the crisis, even when we cannot leave out the possibility that the economic crisis might have contributed to the support for independence as a solution to the problem. If the latter were true, the not too hopeful prospects of the economy would be an additional factor causing polarization and instability for the new autonomous government.

The violence and bloodshed in the Basque country, the political fragmentations, and the lack of firm support for the constitution and the devolution statute have given rise to the specter of a Spanish Ulster. A comparison of the number of casualties should help to dispel it. In Northern Ireland, from 1969 to 1975, there were 1,391 deaths, including 131 policemen and 246 British soldiers. If we were to apply the ratio of casualties to population in Ulster to the Basque country and Navarra, there would have been 2,310 victims, far more than the 153 ETA victims and almost 40 victims in demonstrations and encounters with the police from 1976 to December 1979 in all of Spain (*El País*, December 14, 1979).

Nevertheless, the level of violence in Spain is sufficiently high to cause grave political repercussions, and the armed struggle of the ETA has continued. As of January 4, 1979, the newspaper *Deia* reported that there were 243 persons imprisoned in Spanish jails for actions having a political significance. Of these, 111 identified with ETA; the rest were mainly from the terrorist organization GRAPO and the MPAIAC, the liberation movement of the Canary Islands.

Since the process of transition to democracy started, and after the first elections, ETA violence has increased, and continues. A PSOE member was assassinated a few days after the Autonomy Statute referendum, and a foreign affairs spokesman and UCD deputy was held hostage for over a month in November 1979. The main targets have been the police forces and Basques who held public office, opposed the ETA, or refused to pay the revolutionary tax. An accounting received from military intelligence sources and published in the newspaper *Diario 16* on November 19, 1979, shows that 64 deaths were caused by actions of the ETA in 1978 and 23 during the first three months of 1979. In addition, ETA is reported to have carried out 23 kidnappings, 17 of them since 1977 (that is, after democracy had returned to Spain).

Without having reached the gravity of the violence in Ulster, terrorism and repressive responses to it, and the popular reactions to these responses, are part of daily life in the Basque country. We have explored the attitudes of the population toward terrorism by asking the following question: Thinking now of the persons involved in terrorism, which of the following expressions fits them best in your opinion: patriots, idealists, manipulated people, madmen, criminals?

In spite of the tension in the area, only 10 percent in 1978 and 15 percent in 1979 refused to answer. The question was, in itself, somewhat objectionable to those who consider the terrorists to be freedom fighters; but to use that term would have been even more objectionable to many others, and could have been construed as approval of terrorist actions by the researchers.

In the four provinces somewhat over 10 percent perceived terrorists as patriots. As we shall see from the party identification of those holding that opinion, this was largely an expression of support. More than one-third saw them as idealists in 1978, but this was reduced to one-fourth in 1979. These responses differentiate the position of the ETA terrorists from similar groups in West Germany or Italy, and bring them closer to the position the IRA holds in the minds of the Ulster Catholics.

One-third of the respondents perceived the terrorists as being manipulated by someone else, even though the number holding that opinion seemed to diminish slightly between 1978 and 1979. This belief is congruent with a widespread feeling in the world today that many events can be explained by the role of the secret services, by international conspiracies to destabilize governments, or by support given to terrorist organizations by radical third world leaders.

The proportion of those seeing "persons involved in terrorism" as madmen or as common criminals has increased, although these negative and delegitimizing opinions were held by only a small fraction of the population in the Basque country in 1978 and by one-fourth of the respondents in the summer of 1979.

Different subgroups of the population hold sharply differing opinions. Among people born in the three provinces, the proportion of those perceiving "persons involved in terrorism" as patriots was 19 percent in 1978, and among the minority born in the provinces *and* speaking the Basque language, the proportion rose to 27 percent. In addition, 44 percent of those born in Euskadi and 49 percent of those speaking Basque saw them as idealists. As a result, only 45 percent of the native-born and 31 percent of those speaking the language expressed one of the three negative opinions. Their attitudes contrast sharply with those of the immigrants in the three provinces, including the working class, among whom the proportion saying "patriots" was only 10 percent and "idealists" about 30 percent.

Opinions on terrorism vary directly with the party voted for in 1979. Fully 49 percent of HB voters see "persons involved in terrorism" as patriots, and another 47 percent as idealists. The EE supporters were more restrained, dividing equally between patriots and idealists (39 percent each), with a significant minority (16 percent) seeing them as common criminals. This indicates a slow drift toward more responsible positions in that party, which by 1979 had lost its more extreme nationalist supporters to the HB.

The most serious finding, however, is the not insignificant minority of PNV supporters who say "patriots," the significant number saying "idealists," and the refusal to define those involved in terrorism as madmen or criminals. The contrast between the moderate nationalists who support the PNV and the socialists who support the PSOE clearly shows the latent sympathy of the former for the terrorists. While 34 percent of PSOE voters were ready to call them common criminals, and another 13 percent madmen, the corresponding proportions for the PNV were 8 and 12 percent.

Memory of the common struggle against Francoism, which might explain why a minority of PCE and PSOE supporters see the terrorists as idealists, is, naturally, totally absent in the UCD constituency.

We also asked which groups were responsible for disorder and violence in Spain. This question obviously covers incidents of violence and terrorism not linked with the Basque situation.

Reflecting the climate of opinion in the region, we found that in spite of actions for which the ETA claimed responsibility, few people blamed the nationalists for the violence, although a significant number (38 percent) attributed responsibility to the extreme left. Even so, in 1978, still reflecting the leftist political climate in the region, 52 percent of the respondents attributed the violence to the extreme right. This figure rose to 59 percent among those who spoke Basque, with only 8 percent blaming the nationalists and 33 percent the extreme left. (Multiple answers to the question were allowed, so that some people mentioned more than one group.)

In the Basque country a large proportion attributed blame to the government (33 percent) and/or the police (28 percent). The importance of the legacy of conflict left by Franco is reflected by the fact that in the whole population 38 percent blamed the past dictatorship for the violence; and among those in the three provinces who spoke Basque, this figure rose to 49 percent.

Once more the difference in opinion between the native population and the immigrants is apparent. Among immigrant workers 17 percent attributed responsibility for the violence to the nationalists, and 35 percent to the extreme left, even though they were equally, if not more, prone to blame the government and the police. Among immigrants of the middle class, and in Navarra, where the UCD has considerable support, the proportion blaming the state and its agents was reduced to 20 percent of the population.

What kind of response would people support if terrorism were to continue? In the total Spanish population, our surveys show that only 1 percent favored accepting the demands of the terrorists, while 23 percent in 1978 and 16 percent in 1979 favored negotiation and dialogue with them. Half of the population advocated a policy affirming order and authority while respecting basic human rights, and 18 percent wanted to go further and declare all-out war against terrorism, using "all possible means." Only 5 percent favored declaring a state of emergency and adopting military measures.

The drop in willingness to negotiate with the terrorists appeared among the Communist, socialist, and UCD constituencies, but particularly among the supporters of the two parties of the left, whose leaders have been increasingly outspoken against terrorism. Even among UCD supporters only 6 percent were for military measures; the proportion ready to "use any means" (whatever the respondent understood by that) was about 25 percent, and did not increase by the time of the second survey. Even among the supporters of the right-wing AP, often perceived by others as an authoritarian threat to democracy, 33 percent opted for maintaining authority and order while respecting basic human rights.

In the Basque country the climate of opinion was, and still is, quite different from that in the rest of Spain. In 1978 a minority of 5 percent favored accepting the demands of the terrorists. Negotiation was advocated by 43 percent, maintenance of authority by 38 percent, and 7 percent favored stronger responses. In 1979 the basic response pattern had not changed in the same direction as in the rest of Spain. It remained stable.

It is this divergence in the response to the frustrations created by terrorism among the population of Spain as a whole and in the Basque country that is particularly dangerous. Demands for action by the government may increase among the general population, while these same actions may be perceived as rash and illegitimate in the Basque country.

In considering our analysis the reader should, however, not lose sight of the fact that responses in Euskadi might well have been distorted to a greater or lesser extent by the climate of fear created by terrorist actions.

On October 25, 1979, the Autonomy Statute was submitted to a referendum in the three provinces.[12] This time the PNV advocated a "yes" vote, rather than the abstention it had supported in the constitutional referendum in 1978, while the *abertzale* patriotic camp that in 1978 had advocated a "no" vote was divided between EE, which supported a "yes" vote, and HB, which advocated abstention. Under the circumstances prevailing in the Basque country, especially in many smaller communities, it required some courage to go to the polls and be counted as supporting the new statute. A "yes" vote was also supported by the UCD, the PSOE, and the PCE. AP left its voters free to support the statute or not, even though it had publicly expressed its misgivings about it. It is likely that most of the "no" votes came from its supporters and from those of UN and other Spanish right-wing groups.

The returns show that 54 percent of those eligible to vote (90.3 percent of those voting) supported the Autonomy Statute. There were some differences in the participation in the various provinces and areas of the Basque country, but less than one would have expected (see Table 2.9).

While far from elating the supporters of the Autonomy Statute, and not providing a solid basis of support and legitimacy for the new Basque institutions, the returns represented a victory for moderation and gave hope for the institutionalization of devolution. They were more favorable than one would have expected on the basis of the electoral returns for the parties advocating one or the other position in the March legislative election. It is difficult to say how that result came about, to what extent some of the PSOE sympathizers who did not vote in the legislative elections followed the party in its support of the statute, or to what extent HB was unable to convince some of its voters to reject the statute by abstaining.

Nevertheless, the complexities of the Basque question, some of which we have attempted to describe, may condemn it to become one of those "unsolvable" problems—like the difficultes in Ulster or the Arab-Israeli conflict. No easy solutions are available. Neither self-determination and secession, nor suppression, are viable. The moderate alternative of autonomy within Spain is threatened by violence—violence that is supported by a minority of the popula-

TABLE 2.9
Vote in the Referendum on the Basque Autonomy Statute, October 25, 1979 (percent of voters and of eligible voters)

Province	Yes	No	Blank	Void	Total Voters	Abstentions	Total Eligible
Alava	83.66	9.06	5.75	1.53	110,604	36.77	174,930
	52.90	5.73	3.64	.96	63.23		
Guipúzcoa	91.94	4.05	2.97	1.03	303,659	40.15	507,002
	55.03	2.42	1.78	.62	59.85		
Vizcaya	90.75	4.94	3.15	1.16	507,487	40.98	859,843
	53.56	2.92	1.86	.68	59.02		
Euskadi	90.29	5.14	3.41	1.16	921,560*	40.23	1,541,775
	53.97	3.07	2.04	.69	59.77		

*There is a slight discrepancy between the total reported for Euskadi and the sum of the figures for the three provinces.

Source: Ministerio del Interior, Dirección General de Político Interior. Referendum Estatuto Autonomía. País Vasco. Información. Avance de resultados por orden alfabético de municipios, 25 de octubre 1979.

tion but toward which many are ambivalent. The disloyal opposition to the Spanish state can count on the semiloyalty of the moderates who are not ready to give up the ultimate goal of independence.

In this context, and the climate of opinion created by it, the "consociational" efforts of Basque and Spanish politicians may be condemned to failure. The institutions of an autonomous Euskadi may fail to function satisfactorily, the economy may deteriorate further, and ultimately the Basque crisis may threaten the future of the new Spanish democracy.

NOTES

1. For information on the historical, political, social, linguistic, and economic background of the problem, see Linz 1973 and 1975. There is a growing literature on Basque nationalism. In English see Payne 1975; Medhurst 1972; González Blasco 1973. From a Spanish perspective the only overview for many years was García Venero 1968. An extreme nationalist and left perspective is given in Ortzi 1977. For the early history see Larronde 1977. A scholarly study of the evolution and social bases of Basque nationalism is Elorza 1978. On the Basque problem in the Second Republic (1931-36), see Escudero and Villanueva 1976; Fusi Aizpurua 1979; and Castells Arteche 1976. For the more recent history, with a chapter on Basque nationalism in France, see Apalategui 1979.

2. The best work on the geography and the social, demographic, and economic structure of the Spanish and French Basque country is Azaola 1976. For a description of the crisscrossing of political, administrative, historical, and linguistic boundaries, and some useful maps, see Rawkins 1972.

3. On political parties in the Basque country at the onset of democratization, see Pérez Calvo 1977. Sarrailh Ihartza 1962 reprints the platforms of several Basque groups, some now defunct, and provides a wealth of information from a nationalist perspective, including maps reflecting Basque "imperialist" ambitions. Pastor 1977 includes interviews with Basque leaders, ranging from the president of the provincial *diputación* (assassinated by the ETA) to ETA leaders, that reveal political positions during the period of transition to democracy in Spain.

4. There is no official publication of the post-1976 returns from elections and referenda. The Ministerio de la Gobernación has made available computer printouts of the returns by province and municipality, which we have used in our analysis. These data do not always coincide exactly with those reported in the press and those calculated by the parties on the basis of reports from polling places. All sources of information have gaps and errors, and figures for participation and total votes cast often do not add exactly.

In our analysis we have generally calculated the percentage of votes cast, including blanks and void ballots, rather than valid votes. (Blank and void ballots were .28 and 1.45 percent, respectively, in 1977; and .32 and 1.47 in 1978. In some cases these have a political significance.) This sometimes leads to small differences from results obtained using valid votes as a base.

For 1978 the data at the provincial level have been taken from *El País*, May 2 and 3, 1979. DATA is in the process of computerizing at the municipal level the information on the elections of 1933, 1936, 1977, and 1979, and the referenda of 1976, 1978, and 1979.

5. The historical Carlist movement in the nineteenth century was populist, antiliberal, monarchical, and clerical, and had always been prominent on the extreme right. It provided the Franco forces with some of their most devoted and courageous volunteers.

But, under the leadership of the present pretender, Carlos Hugo de Borbón-Parma, it has moved close to the extreme left. (On Carlism today, see Zavala 1976.)

6. Julio Caro Baroja, the leading Spanish social anthropologist, has devoted much of his work to Basque language and culture, and has critically examined the many myths about Basque ethnology and history (Caro Baroja 1958, 1978). On Basque language see Tovar 1950, and sources quoted in Linz 1975. On the languages of Spain, see Díez, Morales, and Sabin 1976; Ninyoles 1977.

7. It should be stressed, however, that language took second place to race in the early ideology of Sabino Arana and the nationalist movement. Arana wrote: "There are many Euskerianos who do not know Euskera. This is bad. There are some Maketos [a derogatory term for Spaniards] who know it. This is worse. A hundred Maketos who do not know Euskera do a lot of harm to the [Basque] fatherland, but greater is the harm done by one Maketo who knows it." See the outstanding monograph by Corcuera Atienza, 1979, on language (pp. 395-401) and on race (pp. 383-86; also pp. 320-22).

8. Not only were Basques overrepresented in the big business elite, but also in the political elite of the Franco regime. In his governments, 25 percent of 102 cabinet members were Basques, compared with 9 percent in the governments of the Second Republic (de Miguel 1973, p. 119). The presence of Carlists and persons linked with the Basque business elite accounts for this.

9. On class structure and nationalism see the studies by nationalist intellectuals and sociologists (Núñez 1977; Beltza 1976).

10. The growth of nationalism with secularization and the role of clergy that dropped out of the church to engage in nationalist agitation in Quebec has been noted by Tiryakian 1979, p. 17.

11. Several works provide information on the ETA. Ortzi 1977 is basic. He is one of the three deputies of the HB. Beltza 1976 discusses the splits in the ETA and their ideological bases, and quotes internal studies. Agirre 1974 gives a dramatic account of the execution of Carrero Blanco, and includes the August ideological declaration of the "Txikia" Commando. See also Hills 1972.

12. On the process leading up to enactment of the Autonomy Statute, see the chronicle by a team of journalists from *Deia* (Bordegarai and Pastor 1979). It includes the draft of the *estatuto* of Gernika and the final text, as well as statements by party leaders. It is particularly revealing on the process of negotiation between Prime Minister Suárez and Garaikoetxea, the PNV leader. For the position of those opposed to the Autonomy Statute, see Ortzi 1977.

3

COMMUNITY FRICTION IN BELGIUM: 1830–1980

Paul Dabin

There are three conflicts, tensions, or cleavages that can be regarded as fundamental features of the structure of Belgian society: the religious/ideological differences, the social and economic cleavage between different classes and, more difficult to define, the regional cleavage between linguistic communities. Three aspects of these cleavages further complicate the analysis of Belgian society; not only do they overlap to some extent, but they also vary in intensity and, above all, in content at different moments in time.[1]

These cleavages are also to be found in other countries of Western Europe. Even regional cleavage is fairly common, as can be seen in the case of Scotland and England and in the struggle involving Basque nationalists and the Spanish central government. But what distinguishes Belgium from other countries is that it remains a country with two roughly equivalent communities, quite a different case from those countries that have a peripheral minority. The really significant point is that the community that was predominant in the past has now become a minority under the hegemony of the other: what Flanders experienced yesterday is today being experienced by Wallonia.

Analysis of the Walloon and Flemish problems shows that both communities now question, if not generally the unity of Belgium, then at least unitary Belgium. This new convergence of attitudes in Wallonia and Flanders, where the problems have different backgrounds and different causes, raises questions about the balance of power between the two communities and about the form the state should take.

Although the community problem dates from one hundred and fifty years ago, its roots go back to the fifth century, the time of the great migrations in

Europe. With the decline of the power of Rome, a wave of invading Germanic tribes settled an area on the coast of the North Sea where the Celto-Roman population was less dense, down to what is now the border of France. This cut the territory of what we now call Belgium into two parts, with populations speaking different languages. The language frontier was never a political frontier, however; the concepts of Flanders and Wallonia as we now know them are quite modern.

The Belgian state, founded in 1830, was fairly homogeneous, however heterogeneous its population. It was a bourgeois state, where only the rich were enfranchised, In 1830 the electorate consisted of a mere 46,000 out of a population of 4 million. In 1893 only 135,000 out of 6 million could vote.

In Flanders after the fifteenth century, as in Wallonia since time immemorial, the bourgeoisie consistently despised regional dialects and languages. Since 1830, French has been the only language used in public life in a country where 57 percent of the population was Flemish speaking. But, at the beginning of the nineteenth century, the whole of Europe witnessed a new wave of nationalism based largely on language. The essence of Flemish nationalism was expressed as early as 1836: "De taal is gans het Volk" ("Language is the essence of the people"). This new awareness of a Flemish character, proceeding from quite definite social and cultural origins, was to give birth to the Flemish movement.

On the other hand, it is only since about 1960 that there has been a new Walloon awareness, developing from principally socioeconomic factors: the declining population and declining economic strength of Wallonia.

The Flemish and Walloon movements, then, have different backgrounds; they are inspired by different goals and, indeed, by different ideologies. What has aggravated the community problem as such is the fact that developments on both sides have occurred at the same time. This has led to an ever-growing and increasingly radical questioning of the traditional unity of the Belgian state.

THE FLEMISH MOVEMENT: FROM LINGUISTIC DUALISM (1870) TO REGIONAL DUALISM (1970)

The Flemish problem was known for a long time as the "linguistic question." But while the Flemish movement began as a cultural movement concerned, on the face of it, with the right to use a particular language, the underlying phenomenon was a social one. Nineteenth-century Flanders was a rural society suffering from a lower standard of living than that enjoyed by Wallonia, then in the midst of industrialization. The target of the Flemish struggle was not, however, the Walloons; the real target lay within the Flemish community—the French-speaking Flemish bourgeoisie. Nevertheless, the major cleavages overlapped: the language barrier was at the same time a social barrier, and the language struggle was directed against the French language as an instrument of class domination. The barrier was, moreover, a partly religious one, for the generally Catholic Flemish regarded the French language as the vehicle of secular ideologies.

Although the language question was linked with other interests—social and religious, but also political, electoral, and economic—the fact remains that for a century the demands raised by Flanders were centered on the desire to "live in Flemish." The foremost means of emancipating Flanders as its political strength grew was to be the legislation governing the country's use of language—the "language laws."

Developments here proceeded in two stages, arising from two different ideals: the period of the language laws in the strictest sense of the word, with the rebirth of Flemish as a language, was followed by a period of regional legislation aimed at achieving the reconquest of Flemish territory.

We make a clear distinction between the two stages in the development of the community problem in Belgium. In the first the emphasis was on linguistic homogeneity; in the second the question of regional autonomy was paramount.

The linguistic period mainly concerned the Flemish, and was in fact a survival operation against the expansion of the dominant French language. It lasted 100 years: from 1872, when the first law on the use of languages in the public sector was passed, to the "Decree of September 1973," which regulated the use of French and Flemish in the private sector. By this time the long-standing Flemish demand for cultural autonomy had been recognized.

The regionalist period, in the strongest meaning of the term (economic as well as cultural), mainly concerns the Walloons (and the people of Brussels), who saw themselves threatened by Flemish dominance in the unitary Belgian state. Their efforts to defend their interests are embodied in a declaration of principle in the Belgian constitution, as revised in 1970—a declaration they are still trying to implement.

Naturally the two periods shaded into each other. While the linguistic era, in its true sense, is now over, the linguistic nerve is still raw and quivering. And the regionalist ferment started to develop during the linguistic era. Since 1932 it has been reflected in national legislation. At that time, the great historian Henri Pirenne still objected to identifying nationality with language, and stated that "The Flemish question is a linguistic conflict, not a national one" (Pirenne 1932, p. 384).

Today it appears that homogeneity of language in the Flemish region and the 1962 law laying down the linguistic frontier followed naturally, after a generation had passed, the compulsory use of Flemish for teaching in Flanders, as decreed in 1932. Thus, the law of 1962 brought out for the first time the two main components of the state—population and territory—and at the same time gave life to two new entities that previously had no legal existence: Flanders and Wallonia.

For the first time in history, the linguistic frontier became a political frontier. The law of November 8, 1962, was the first law with a clearly federalist implication that had been voted by Parliament.[2] The existing state of affairs was not greatly affected at the time, but acceptance of the principle embodied in it—which was revolutionary insofar as the idea of a unitary state was concerned—immediately changed the political atmosphere.

It is interesting to note that the law of 1962 conformed to wishes voiced in 1958 by the Centre Harmel in its final report. This institution had been established by a law of 1948, which designated it as the "research center for the national solution of social, political and legal problems in the Walloon and Flemish regions."

Once it had been launched, the Flemish movement was to reconquer for Flanders the grandeur of its past, as it developed its political strength during the two stages—linguistic survival and regional "new existence"—which have already been defined. During the first phase of the linguistic era, however, the idea of regional autonomy was excluded. The emphasis was on the right to "live in Flemish," and this involved what might be called a program of linguistic rehabilitation.

Legislation regarding language developed gradually, according to tendencies dominant at different times. There was a period when bilingualism was stressed, then monolingualism within particular areas, and finally linguistic homogeneity within designated regions.

The first proposed solution, bilingualism, was inaugurated in the Flemish-speaking region during the last quarter of the nineteenth century. And it was a Parliament in which the majority was still French-speaking that voted for the progressive recognition of Flemish in public life. These laws were badly observed, however; and, furthermore, could not promote the flamandization of daily life.

Another solution, monolingualism within particular cultural areas, prevailed in the 1930s. Its slogan in Flanders was "In Vlaanderen, Vlaams!" ("In Flemish land, Flemish language!") But it was demanded also by a majority of the Walloons. Bilingualism remained only in the Brussels conurbation and in other urban areas where either linguistic minority constituted 30 percent of the population. These laws, although they affected private as well as public life, could not stop a gradual increase in the use of French in areas close to the linguistic frontier.

Thus, a more radical solution was attempted in the 1960s. Flanders finally achieved full territorial and linguistic unity and integrity with the abolition of language censuses and the definitive fixing of the linguistic frontier. In addition, there were diverse laws regarding the use of language even in private enterprises. Meanwhile, the policy of monolingualism in the schools had produced a new Flemish generation that was purely "vlaamsvoelend" ("that which feels itself Flemish").

Up to these last years, the Catholic majority within the Flemish movement placed its hopes in a form of cultural independence, and rejected federalism and separatism. The idea was that the Flemish would come to real power in Belgium simply through their more rapid demographic growth and also through the gradual shift of the country's economic center of gravity from Wallonia to Flanders. A radical affirmation of the national Flemish awakening was therefore, in their view, quite compatible with the existing unitary structure of the Belgian state.

And yet, despite this progress from success to success, there were some in Flanders who felt that the past 100 years had represented an "era of linguistic expedients," and did not get to "the bottom of the problem" (Ruys 1973, pp. 157, 176). In 1962 the Flemish right wing and the Walloon left wing announced agreement on a fundamental solution, the principle of which was proclaimed by the prime minister in 1970, when he said that "The Unitary State of 1830 has been superseded." This principle was embodied in the constitution when it was revised in 1970-71.

Thus, the era of linguistic laws that concerned Flanders first of all but, as a result of simple symmetry, also affected Wallonia—and for a long time were called the "Flemish laws"—was succeeded by an era of regional legislation that concerned the whole country and was as much economic as cultural in character.

Between the two world wars, when the Flemish movement called for cultural autonomy, economic autonomy was not on the agenda, all the more since the shift of the economic center of gravity from Wallonia to Flanders was occurring despite centralism. It was only at the end of the 1950s that structural unemployment in the Flemish region gave rise to a regional economic demand. The cultural slogan "In Vlaanderen, Vlaams!" was supplemented by the economic slogan "Werk in eigen Streek!" ("Work for each in his own region!"). The laws regarding regional economic development, passed in 1959, benefited Flanders the most. The economic breakthrough of Flanders had occurred by 1970-71, when the constitution was revised. Therefore, it was understandable that this revision should differ regarding the main claims of the Flemish and Walloon regions. The new constitution gave linguistic communities definite autonomy as far as culture was concerned. But it recognized only in theory the grant of regional powers in socioeconomic matters.

Certainly the moderates as well as the radicals understand that the existence of the Flemish community depends on much more than the regulation of cultural matters. Moreover, despite its auspicious beginning, cultural autonomy as hitherto conceived has proved disappointing; two or three decrees have sufficed to exhaust the program set up in 1971. A compromise concluded in May 1977 by the parties that negotiated the formation of a governmental coalition in that year, known as the Pact of Egmont, advocated that the concept of culture be extended to include all matters relating to individuals (health care, for example).

BIRTH OF THE WALLOON MOVEMENT

The Walloon movement was truly born, at least in terms of collective action, in the 1960s, when there was a late awakening to the Walloon decline along with the economic development of Flanders, and a change in the ratio of power, to the benefit of Flanders. Its wellsprings are clearly different from those of the Flemish movement: it is not primarily a question of affirming and commanding respect for a nationality, but of preventing the Walloon future from becoming dependent on the Flemish majority in the country.

Previously there had existed an extreme Walloon current ("Wallingant") that arose in response to the militantly pro-Flemish laws. In effect, the Walloon defensive movement was born—in Brussels—as a result of the initial linguistic legislation that led to compulsory bilingual treatment of affairs in Flanders. Walloon functionaries and magistrates, finding themselves losing jobs and responsibilities in Flanders, refused to be deprived of their opportunities by these "exaggerations of *enflamandement*." A little later other Walloons came to feel that Wallonia should not have to subsidize rural, underdeveloped Flanders. They said "no" to Walloon money for the *flamendiants* (a pun based on the French words for "Flemish" and "beggars").

Soon a feeling of being a political minority was added to these frustrations. "We are conquered people governed against our way of thinking," declared the Walloon socialist Jules Destrée in 1911. Thus it is that the cry "Up with administrative separation!," launched in 1910 by a minister of state, was first heard in Wallonia. The Flemish conception of the *groote doorsnee* (the "great rupture") was soon to correspond to it. Since 1912 a "national" Walloon congress has debated the federalist "expedient" as a way to "affirming Wallonia."

In the period between the wars, the Walloon movement was generally hot air, and it remained rather heated in 1945. But in that year new organs were created that could serve both sides as vehicles of expression and instruments of pressure, especially in the economic field. There were also expressions showing a sentimental attachment to France, and calls for new ties with the "natural native land." These corresponded to the earlier Flemish cry, "Los van Frankrijk," which referred to the demand that Belgium terminate its 1920 military agreement with France.

It is not a question of a mass movement, as in Flanders. Adherence to the Walloon movement will be on an individual basis for a long time, and the political party apparatus will always be either reticent or hostile. But a turning point came at the extraordinary Walloon congress of March 26, 1950, two weeks after a popular consultation on the question of the monarchy had separated the majority in Flanders from that in Wallonia. At that time the Walloon movement was given a popular basis by the adherence of the principal leader of the Walloon trade unions, André Renard.

The outcome of a strike at the end of 1960, called in connection with a draft bill on economic austerity, manifested profound differences between Flanders and Wallonia. This had also been the case with previous confrontations regarding the monarchy, the education question, and other matters. At the beginning of 1961, after the failure of the strike—attribted to the Flemish—Renard relaunched the Walloon movement by associating reforms in the political structure (federalism) with economic reforms (socialism).

From that time on, the majority current in the Walloon movement advocated federalism, with an eye to remedying the Walloon decline by making changes in economic and social structures that could not be realized in the framework of a unitary state dominated by Flemish conservative tendencies. But the unions, spearheads of the movement, were not able to rally majority opinion to such a large-scale program. The traditional left even gradually lost its majority in Wallonia, declining from 53.38 percent of the popular vote in 1961 to 43.16 percent in 1978.

But the Walloon movement experienced a "conversion on the road to Damascus" at the same time as the awakening to the Walloon decline and the affirmation of Flemish economic and political dominance. It was caused, first of all, by the trauma of seeing the coal industry collapse in the 1960s. From that time on, the Walloon thesis has been that since Walloon coal "created the prosperity of Belgium, Wallonia has the indisputable right to the solicitude of the entire nation" (Cornez 1959, p. 13). In brief, the objectives of the Walloon struggle became notably "that Flanders not industrialize to the detriment of Wallonia," especially with respect to maritime metallurgy and other industries that traditionally were located in the French-speaking region. But it has been a long time since the industrial zones moved from the Liège-Charleroi-Brussels axis to the Ghent-Anwers-Brussels axis. Moreover, in the 1970s the crisis in iron metallurgy was added to the general decline of the Walloon industrial fabric, more than 70 percent of which was composed of older industries of low profitability.

Furthermore, a social framework built around the community, which has existed in Flanders for a long time, has never been characteristic of the Walloons. The Flemish constitute a "community" in the ethnological sense, and now also in the constitutional sense of the word: one does things *à la flamande*, among the Flemish and for the Flemish. But the Walloons are accustomed to seeing more broadly or more narrowly. More narrowly, since attachments to Walloon soil would never turn a citizen of Mons into an inhabitant of the nearby Borinage district; more broadly, since the Walloons moved about at ease in the Belgian framework, before it split from top to bottom, using their own language.

The trend toward a Walloon identity has thus moved in the opposite way from the corresponding trend in Flanders. While a tendency toward expansion characterizes the latter, the Walloon community is created by contraction.

Moreover, the movement remains less comprehensive than the one in Flanders. Not a few Walloons (and other French-speaking people) feel themselves to be Belgian at the same time, thinking and living as Belgians. It is also true that, by force of circumstances, the intermediate zones that formerly bridged the Flemish and Walloon communities are tending to grow narrower. The young Flemish generations have grown up in an exclusively Flemish milieu; and why should the Walloons learn *néerlandais* if it's now only the language of a foreign region!

PRESENT STATE OF THE FLEMISH AND WALLOON PROBLEMS

Legislative and communal elections have revealed the following political tendencies in Belgium:

The left traditionally dominates in Wallonia;
The right dominates in Flanders and in some rural Walloon communities, above all in the Belgian province of Luxembourg;
The Brussels agglomeration is the most unstable electorally, while the "emerald belt" of Flemish communities beyond the Brussels suburbs votes rather determinedly for the right.

Moreover, the majority of Walloons, often bolstered by the Brussels majority, have almost always been in opposition to the majority of the Flemish in most national debates. Since World War II one can cite the question of punishing Belgian collaborators of the Third Reich, the question of the monarchy, the education question, the Benelux question, and the strikes of 1960–61. And not to be forgotten are rivalries in economic matters, which highlight divisions of opinion about national economic priorities.

But we should also recall that the Belgian political system has long shown its capacity for containing these disputes, as well as others. Until recently the traditional political parties, because of their unitary structure, played a very important moderating role. Other intermediary groups, and not a few individual Belgians, above all those from Brussels, tended to sneer at this conflict of nationalities before they were forced to become involved. Recently, however, there has been a general activation of the antagonisms between the two communities, whose present positions we shall now try to present.

The Flemish problem resulted from the rise of popular government, although this process was more one of bourgeoisization than of democratization. The "common man" still accepts being spoken to in patois; but his son, who has completed his studies and has become a curate, doctor, or manager, no longer tolerates it. The phenomenon of the mass vote explains the alloys of which the

Flemish movement is made: linguistic elements blend with religious, social, and economic elements, each of which reinforces the Flemish way of thinking.

The Flemish movement has gone through successive stages: from the *moedertaal* to the *landstaal*, from *vlaams spreken* to *vlaams voelen* and finally to the *volksgemeenschap* (from the "mother tongue" to the "language of the country," from "speaking Flemish" to "feeling Flemish" and finally to the "community of the people").

At present it would seem that the stage of "problem" has been passed and that the Flemish reality has come into being. As Manu Ruys writes at the end of his book *Les Flamands*, "The Flemish movement nears its end." But, as the subtitle of his book indicates, the Flemish are still "a people on the move, a nation in being" (Ruys 1973). In fact, at present Flemish objectives appear ambiguous, ranging from achieving hegemony in a unitary state through the use of existing centralism, up to autonomy in a more or less federal structure, and including even more equivocal mixtures of the two tendencies.

Moreover, it must be recognized that if the Flemish are bound by a greater solidarity and a greater consensus than the Walloons, they are far from presenting a united front. The very dynamics of the Flemish movement and its successes cause a certain ambivalence. On the one hand, Flemish radicalization, the fruit of popular action movements, leads logically to an institutional affirmation of the Flemish community that extends beyond the expression of cultural values and takes a political form. But, on the other hand, Flemish preponderance in the machinery of the state, a result of the fact that Flemings have become a majority and are conscious of their power, leads logically to more unitary attitudes in the decisive domains, except for the cultural.

This new spirit—this neounitarianism—is characteristic of the new Flemish bourgeoisie, which has assured itself of a growing share of economic and political power within the Belgian state. Moreover, it is the product of purely rational calculation, directed by the economic and political interests of the Flemish people. In that perspective the tendencies toward federalism—and even separatist impulses—seem to be an important part of the Flemish "establishment" like an unpleasant reminder of the past—a repetition of slogans that have been outmoded by changing historical circumstances. But the new Flemish bourgeoisie, like the ousted Flemish bourgeoisie that shared a French culture, is perhaps able only to slow down the centrifugal forces, which are essentially popular and over which the bourgeoisie has relatively little hold.

Having awakened to its political power and the economic possibilities open to it, the Flemish community more easily accepts the Belgian state. Flemings are no longer the poor relations of Belgium; they have become full-fledged citizens. Being the majority in the country, they no longer let Belgium bully, restrain, or even bother Flanders. Nevertheless, some Flemish expressions continue to be based on a complex about being a minority rather than a true state of being a minority.

French-speaking enclaves in Flanders have been eliminated. Louvain was "reconquered" in 1968, with the exclusion of the French-speaking university, which emigrated to the new town of Louvain-la-Neuve. Then the peripheral communes of Brussels were absorbed in 1970, in spite of their French-speaking majorities. From the Flemish point of view, there remain only the thorny problems of the Brussels conurbation and of the communities with limited bilingualism (known as bilingual "facilities"). The dream of reconquering these will continue to haunt the Flemish movement for a long time.

Where will the Flemish movement eventually go? One can say only that since its beginning, this movement has responded to complex motivations, as though it were tormented by some subterranean force—which one could call nationalism or even romanticism—and that it is impossible to predict where its evolution will lead it.

Turning now to the present state of the "Walloon problem," we have seen that for a long time the Walloons and most of the people of Brussels could identify themselves with the unitary French-speaking state; they had no reason to question structures that worked to their advantage.

The growing awareness of a Walloon problem—and also of a partly distinct Brussels problem (and the birth of a specific party)—dates only from about 1960. While a Walloon consciousness and a Brussels consciousness are coming into being, an observable consensus as in Flanders, whose consciousness has had ample time to mature, is still a long way off.

And yet, overnight, the Walloon and Brussels problems have become as serious and as menacing as the Flemish problem has ever been, chiefly because of the simultaneous appearance of two phenomena: the decline of Wallonia and Brussels, and the renewed power of the Flemish.

The Walloon (and Brussels) population is one of the world's most elderly. The industrial potential that made Wallonia so prosperous for a century is also aging, and is its main handicap today. Wallonia's share of the national wealth is shrinking in proportion to the decline in the importance of the Walloon community in relation to its Flemish competitor.

One key factor in the new Walloon consciousness is that since 1955 per capita income in Wallonia has been below that in Flanders. By 1974 the per capita share of the GNP in Flanders was 13.37 percent higher than in Wallonia (*Institut national de statistiques* 1976). The first signs of the crisis appeared earlier, although many Walloons did not take warning: from 1947 to 1961, the number of persons of working age fell by 150,000 in Wallonia, while it grew by 100,000 in Flanders.

From 1960 to 1975, Wallonia experienced what might be called a free fall. In 15 years it lost nearly 15 percent of its relative weight in the national economy, while Flanders gained 25 percent. This is shown in Table 3.1.

Flanders' economic blossoming dates, in effect, from the 1960s. It was founded on numerous young enterprises—many of them American—that were

TABLE 3.1

Percentage Shares of the Regions
in the Belgian GNP, 1960 and 1975

Year	Wallonia	Flanders	Brussels
1960	32.1	45.1	22.8
1975	27.6	56.9	15.5

more attracted by an industrially virgin maritime region than by an old industrial base dating from the first industrial revolution. Industrialization of Flanders was encouraged also by the first measure for regional economic expansion, enacted by the central government in 1959, and by the birth of the Common Market, which attracted foreign industrial establishments.

Flanders had ended up acquiring a preponderant voice in political life also, not only through the weight of its population in a one man, one vote democracy, but also through the gradual turning of the economic tables. So it is easy enough to understand that while the skeleton of a Walloon movement already existed before 1960, the flesh was put on the bones only as a result of this growing awareness of the decline of Wallonia in parallel with the new economic strength and political preponderance of Flanders.

Because of the origins of the Walloon problem, which are more socioeconomic than sociocultural, Walloon aims were clearly different from those of the Flemish movement. It was not, first of all, a question of affirming and commanding respect for a nationality, but of protecting oneself against a minority status that would be prejudicial to a Walloon economic renaissance.

This having been said, it should be observed that the Walloons and Bruxellois (85 percent of whom are French speaking) are far from sharing a consensus on a single large-scale program. Their recent solidarity is more fragile than that of the Flemish, and their interests remain more divergent.

A majority of those in the Walloon movement, whose spearhead has been provided since 1960 by the unions, preach federalism with a view to remedying the Walloon decline by "structural reforms." These reforms cannot, however, be accomplished within the framework of a unitary state where Flemish conservative tendencies dominate. (The majority of the French-speaking population of Brussels also calls itself federalist, although for other reasons.)[3] Furthermore, the fact that federalist theses or slogans are voiced does not mean that there is such a thing as a well-considered federalist program. The idea of federalism should be taken in its general meaning.

At least until very recently, the French-speaking economic world in Belgium, a majority of the notables, and perhaps the silent majority in Wallonia doubted that the region's renewal could come about if it simply fell back on its own resources. But the Walloon groups and parties that advocate federalism emphasize that such is not their objective. Instead, a "Walloon power" would exist within a Europe of regions. (The expression "Europe of regions" risks creating confusion, however. Is it not illusory to think that Europe may in the foreseeable future become anything but an association of sovereign states?)

The Walloon federalists look to the Belgian state for assistance in accomplishing the renewal of Wallonia, but at the same time refuse to rely on the country's Flemish majority where Wallonia's vital interests are concerned. In answer to the Flemish contention that strict equality among Belgians means "one man, one vote," the Walloons admit that this principle clearly applies in a unitary state. But, they continue, since Belgium will henceforth comprise two communities—at the demand of the Flemish—one can no longer decide vital questions by counting individuals. Instead, one must count by the community and by the region, leaving to each of these the power to determine its own development.

Nor are French-speaking Belgians above playing games with numbers. On the one hand, they have obtained a certain equality at the level of executive power: the same number of ministers and principal civil servants. On the other hand, Wallonia receives a larger share of the allocations from the central treasury, which form the major portion of the budgetary resources of the regions, than its population would entitle it to. This is because the allocations are adjusted according to "objective criteria," which include the area of each region and the proportion of taxes paid by individuals within it. The relationship of population to regional subsidies, since passage in 1974 of the law on provisional regionalization, is shown in Table 3.2.

Aid should not be distributed according to strictly "objective" criteria, the Walloons contend, because of the "objective" needs of the region. Flanders has reached the takeoff point, while Wallonia must catch its breath. It is more difficult and more costly to renovate an industrial establishment than to build

TABLE 3.2

Percent of Belgian Population in Each Major Region and Percent of National Subsidy to Each Region

Region	Population	Subsidy
Flanders	56.38	51.14
Wallonia	32.72	39.27
Brussels	10.90	9.59

from nothing, and aid to Wallonia from the central government that was too strictly proportional—and a fortiori aid that fell below that level—would accelerate the region's decline even more.

THE SITUATION OF BRUSSELS

The Brussels problem is not just a simple appendix to Belgian dualism. The "Brussels fact" has without question long been the principal stumbling block in the path of institutional remodeling.

Brussels is an enclave, in which some 80 or 85 percent of the inhabitants are French speaking, within Flemish territory and about 10 kilometers from the linguistic frontier. Its prodigious development is due to its status as capital of a unitary state. The Bruxellois have been seen as the principal beneficiaries of this centralized system, and were thought to display a certain "pretentious" arrogance toward the provincials. Residents of the capital were much envied by the Liègeois and even by the Anversois, who are still the rather spoiled children of all Belgium. In reality, however, Brussels has lost its haughtiness, and all economic indicators now show its decline. This process of decline, and especially of depopulation, has been fed by the excessive growth of the governmental function, to the detriment of the residential and industrial functions. But a specific cause of the wasting away of the Brussels agglomeration lies in its reduction to 19 urban communes cut off from their natural hinterland.

Brussels has, in fact, been hemmed in for 15 years by an "iron collar," which was imposed on it to stop the "oil spill" (*olievlek*) of "Frenchification" on the neighboring populations of Flemish origin. The "oil spill" has continued to spread, despite the linguistic laws, but the "iron collar" did have the effect of inhibiting economic development by cutting the urban center off from its periphery, and even from its suburbs. This dissociation between the field of political power and the field of natural economic expansion exists for no other great metropolis, except possibly West Berlin.

Meanwhile, the 80,000 to 100,000 French-speaking inhabitants of the Flemish environs of Brussels feel bullied, despite a regime of limited bilingualism (the "facilities"). And the Flemish residents of these environs are resentful of the invasion by citizens who often have a superior socioeconomic status.

The French-speaking majority of the Brussels agglomeration wants to throw off the "iron collar" that cuts it off from its periphery and, at the very least, to be "master of its own house"—that is, to manage local and regional life as it pleases. This is seen notably in the fact that in 1976, of 115 mayors and deputy mayors in the 19 communes, there were only seven Flemish-speaking deputy mayors and one Flemish-speaking mayor.

For most of the Flemish, Brussels must at the very least be a completely bilingual capital of a binary country. An active minority will not rest until there has been a "reconquest of Brussels." However, it should be noted that a great part of the French-speaking Bruxellois are of Walloon ancestry. They are in a different position from the "fransquillons," those people who are Flemish by origin but who use French and can be accused of being *volksvervreemd* (alienated from their people).

A unanimous Flemish demand is that there be an organic connection—an umbilical cord—between Flanders and the Flemish of Brussels, and that there be a national agreement to assure them of guarantees beyond those arising from the simple law of numbers in the agglomeration. These guarantees for the benefit of the Flemish minority of Brussels would deal not only with cultural life and matters related to the lives of individuals, but also with the administration of the Brussels region.

At the present stage of the process of regionalization, some Belgians consider the Brussels agglomeration to be a region entirely apart, while others see it as a city that has some aspects of a region. In the last analysis the Flemish are really obliged to accord Brussels a relative autonomy, but they balk at allowing it to have instruments and prerogatives equivalent to those of Flanders and Wallonia. Thus, there results a federalism of "two-and-a-half"—which the Bruxellois resent as a humiliation.

TOWARD A FEDERALISM LEADING TO UNION OR TO DISLOCATION?

Interaction between socioeconomic phenomena and cultural identification in the two communities has led to the "Flemish problem" and the "Walloon problem." They represent the two poles of the "community problem" in Belgium.

But there is a major difference between them. It has been said:

> In Wallonia the socialist ideology of class struggle has led the militants involved in the Walloon action to a movement for "national liberation"; while in Flanders a late 19th century type nationalism has led to a true class struggle which, however, refuses to see itself in terms of marxist ideology, since its driving forces are above all christian democratic. (Meynaud, Ladrière, and Perin 1965, p. 142)

Flemish nationalism originally developed among the masses in the context of a struggle with the traditional "Frenchified" bourgeoisie of Flanders and the bourgeoisie of Brussels. Nowadays the CVP (Christian People's Party), which represents the majority of the Flemish community, is basically Flemish in outlook because its grass roots are working-class and agricultural.

The evolution within the two communities has led both of them to propose a break with centralism. This had had an effect both at the level of the organization of parties and groups and on the structure of the state. At the organizational level one can see internal separation occurring (with organizations splitting into two distinct wings), leading to the rupture of most private institutions, including the parties and the unions. At the national level the conjunction of the Flemish and Walloon movements has manifested itself in a ten-year search for a large-scale and definitive solution to the community problem in the context of constitutional reform. The objective was defined by a (Social-Christian and Socialist) government in 1961: "The government desires to strengthen national unity by regulating in a lasting way the relations between the Flemish and the Walloons." Ten years later Prime Minister Gaston Eyskens, quoting the "Working Group of 28," officially recognized that "The unitary state, insofar as its structure and functioning are governed by laws, has been outrun by the facts."

As of about 1970, a new phase with uncertain horizons, which might be called "postunitary" or "prefederal," was inaugurated. Four "linguistic areas" are recognized (including the bilingual area of Brussels), and three "cultural communities" (including the small German-speaking areas in the eastern part of the country). There are also three "regions," including the Brussels region. But these regions are not further described by the revised constitution. Are they political divisions, or do they have principally an economic significance, in that in a large sense their social, administrative, and political aspects are tied to economic problems?

Actually, a law that decentralized certain socioeconomic functions to the regions was passed slightly before the constitutional revision of 1970–71 that provided for the autonomy of the cultural communities. It was understood that legislative jurisdiction must remain the prerogative of the national state, although shared by the two linguistic groups, and that decentralization would affect only regulatory and advisory functions.

Nevertheless, a fast-moving trend toward the development of two communities can be seen, even if a large-scale community settlement has remained in abeyance. Thus, since the Egmont Pact cultural autonomy has been strengthened and made more clearly federal in nature. The range of matters coming under the jurisdiction of the communities would be supplemented by matters called "personal" (dispensing of health care, protection of the youth, and so on). And the community councils would be granted their own executives responsible to them, which—in fact if not in law—already exist.

As for regional autonomy, it is in the process of building itself on the same model. Regional councils, freed from the tutelage of national authorities, are called upon to pass ordinances that modify existing laws for their region.

Until recently, interaction among the various cleavages of Belgian society tended to produce a balance, even if it also engendered a certain immobility.

The political parties, in particular, while based on divisions of opinion on some matters, nevertheless cut across the lines dividing opinion on other matters. But today the traditional parties are separated into linguistic wings, even though ideological positions remain as their potential bases. And there are new parties that are based specifically in a particular community or region. This recent evolution explains why the party system, which previously was able to contain most disputes in Belgium within acceptable limits (see Lorwin 1966b) has more difficulty in dealing effectively with the community problem. The parties have lost their traditional moderating, and even integrating, role (see Ladrière 1970).

Meanwhile, other groups continue to play an integrating role. These are the economic interest groups, which, even if declining in significance, are not less important than the parties, the mass media, and the sociocultural groups. King Leopold I noted the "extraordinarily embracing quality of material interests" among the Belgian people (quoted in Van Kalken 1930, p. 57), and there continues to be plenty of room for "undivided and indivisible policies" (Wigny 1969, p. 141). Indeed, the structures of the Belgian economic system constitute the cement with which contemporary Belgium is held together. These structures are still relatively centralized, despite the development of groups of industries that are tied to a particular community and the arrival of multinational industries for which Belgian space is simply a part of European space.

We must also recognize that behind the forward flight toward federalization there is a certain *je ne sais quoi* that restrains the Belgian majority from "carrying the idea of decentralization to its paroxysm, all the way to the heart of the federal state" (Dabin 1939, p. 322). If both communities demand autonomy—the Flemish a cultural autonomy and the Walloons a greater economic autonomy—the two sides agree on the necessity for a national authority that would not be "the weakest link in the chain." This appears to be a sign that in reality the Belgian state is not just a superstructure. It is the political expression, since 1830, of a reality that is no mere historical phenomenon but is still a very present reality (in economic terms among others).

A part of the reason that further decentralization has not occurred is the difficulty of finding a formula that will satisfy all parties. The constitutional revision of 1970–71 recognized cultural autonomy but only the principle of regionalization, and thus sanctioned Flemish demands; but it did not satisfy Walloon desires and it shrank from the Chinese puzzle that Brussels represents. It left it to the legislative branch to modulate the degree of actual regionalization, depending on the relative strength of the decentralized and federalist tendencies.

A law of August 1, 1974 on *provisional* regionalization, completed by a law of July 5, 1979, created regional institutions "preparatory to the application of art. 107 (#4) of the revised constitution." But the two last governments fell in October 1978 and on April 8, 1980 because both the unitary principle and the federal principle have proved till now impracticable in actuality.

Carl J. Friedrich considers, furthermore, that "a dual regime compounded of two hostile groups . . . provides no foundation for the operation of the difficult system that is federalism" (Friedrich 1968, p. 128). He also thinks that "federalism may be a separating factor in countries like Canada and Belgium" where regional feeling is of greater importance than national feeling. But, on the other hand, "it strengthens national unity in countries such as the U.S.A. and the German Federal Republic where a greater degree of national integration has already been achieved." However, he adds, paradoxically, that in Belgium, the federalist element leading to separation is "possibly the only way of avoiding total dismemberment" (Friedrich 1970, p. 964).

But, since 1970, no definitive formula has been found, essentially because the Flemish do not accept the idea that the Brussels region—Flemish territory with a French-speaking majority—should become a separate political entity in its own right. It must be added, moreover, that the Walloons are not keen on a three-sided federal structure if this would give Brussels the power to act as arbiter and to block matters relating to Walloon economic development. The era of institutional reform is thus far from over; difficult solutions to complex problems require time to mature. As Jean Monnet wrote in his *Mémoires*: "What one does prematurely, time will take care to undo."

The present situation in Belgium is, therefore, that of a growing de facto federalism: the two communities are moving apart, even separating, and the number of fields in which this is happening is increasing all the time. The trend appears to be the opposite of that in the United States, where unity was forged out of a far greater ethnic diversity.[4] Nonetheless, if the Flemish and Walloon problems have called into question a unitary Belgium, they have not (yet) destroyed Belgian unity. As many citizens have become more Flemish or more Walloon, they have still adhered to the "Belgian fact." And so there are two opposite but simultaneous trends.

It should be noted, moreover, that community frictions in Belgium have always resembled a domestic quarrel more closely than a civil war. All in all there are many Belgians who prefer an arranged marriage to a judicial separation. In spite of a certain feverish micro-nationalism and pressures from extremist tendencies, it appears that Belgium, a country between two civilizations, Latin and Germanic, will forever be the country of the *Houdt middelmaete* (seek the "happy mean"), as in the song of Jan Houwaert, the fifteenth century Brabant poet.

A NOTE ON PUBLIC OPINION RESEARCH

It has been said that "Knowledge of public opinion in a democratic system, more than in all other systems, is clearly a preferred basis for political action"

(Charlot 1975, p. 136). Under the circumstances in Belgium, however, one cannot imply that opinion surveys are of great importance in the solution of the community conflict. This is because patterns of decision making and patterns of compromise between regions and communities do not depend on them.

We thus make a fundamental distinction between surveys as instruments for becoming acquainted with popular attitudes, on the one hand, and as instruments of government or of political action, on the other. In the case of Belgium, they can illustrate political and economic differences among communities, as well as different social values, even if they cannot be said to have appreciable political significance.

One of the rare surveys of regional attitudes in Belgium was that made for the European Communities in 1967.[5] In the Walloon region 54 percent said that they had the impression of being a region clearly "on the decline," as against only 27 percent who saw their region as "on the upgrade." The corresponding percentages for Flanders were 36 and 45. The Walloon percentage expressing pessimism was the highest for any region, except northern Sweden, in the nine countries surveyed.

According to an opinion poll on policy problems and solutions in Belgium, carried out in mid-year 1975, the most salient problems at that time were economic, viz. inflation and employment. The Belgian community problem was not perceived as the most pressing by a majority of the population, but 77 percent of the sample gave the following opinions about the maintenance of the political unity of Belgium (see Table 3.3):

TABLE 3.3

Opinions About the Unity of Belgium (percent)

Absolutely necessary	45.6
Necessary	33.7
Not very important	12.5
Unnecessary	5.3
Improper question	2.9

Source: Institut belge de Science politique, *Res Publica* 1975, pp. 510, 559. Based on a representative sample of 1521 Belgian citizens (502 in Wallonia, 522 in Brussels and 497 in Flanders).

As regards possible solutions to the intercommunity problems, survey results show that a small majority of the Belgian population favors both a three-party federalism and the restriction of the Brussels region to the town of Brussels itself, i.e., not including its suburbs.

Another opinion poll, conducted in December 1978, asked the following question among others: "Do you desire a revision of the constitution intended to: let Belgium remain united or one that would lead to the creation of 2 or 3 separated states?" (see Table 3.4).

TABLE 3.4

Preferences Regarding Constitutional Revision (percent)

	United	2 or 3 states	No answer
Wallonia	74.7	22.0	3.3
Flanders	53.4	44.1	2.5
Brussels	68.4	28.6	3.0
Entire Country	63.4	33.8	2.8

Source: Extract from an inquiry by I.C.S.O.P., Brussels, December 8-12, 1978 (representative sample of 1204 Belgian citizens; results weighted in accordance with the population of the three regions). See Centre d'études des réformes politiques, "Explication des votes de l'électeur belge le 17 décembre 1978" (mimeo) 1978, p. 47.

To take another example, Table 3.5 shows the different perceptions of Flanders, Wallonia, and Brussels in regard to the economic situation as of 1978. As one would expect, the economic crisis was seen as most serious by respondents in Brussels, as less serious in Wallonia, and as even less serious in Flanders. At the same time, paradoxically, there was a substantial measure of consensus among the three regions, suggesting that the community problem is secondary, in that it does not obscure awareness of a national economic crisis.

To take a third example, we find that a poll conducted in 1975, confirming the result of polls of 1974 and 1966, indicates that large majorities of Belgians favor bilingualism as a solution to the community problem. This solution is rated as of either first or second importance by more than two-thirds of respondents in all three regions (see Table 3.6). The same is true, although it is not shown in the table, for inhabitants of cities and villages of all sizes and for men and women of all backgrounds, ages, and professions.

The poll seems to suggest political action, that is, to require the study of both Flemish and French throughout Belgium; but in the light of events, it is clear that an even more significant tendency is not reflected by the survey. This

TABLE 3.5

Perceptions of the Gravity of the Economic Crisis in Belgium, 1978 (percent)

	Brussels	Wallonia	Flanders
1 (crisis extremely grave)	56.1	46.7	36.3
2	6.6	8.3	7.6
3 (middle position)	21.8	32.8	34.2
4	4.4	2.9	5.4
5 (crisis not serious)	11.1	9.3	16.4
	100	100	99.9

Note: Sample of 2,433 persons representative of the Belgian population 18 years and over.

Source: Extract from a study by the Interuniversity Institute for Public Opinion Polls (Brussels, August 1978). The data and comments on the study were provided by Nicole Delruelle.

TABLE 3.6

Rank Order of the Importance of the Solution "Require the Study of Flemish and French in the Whole Country, so That Everybody Can at Least Understand These Two Languages" (percent)

Order of Importance	Brussels (N = 250)	Wallonia (N = 653)	Flanders (N = 1,098)
1 (very important)	46.1	40.6	50.5
2	22.3	25.9	20.8
3	5.8	14.1	11.3
4	6.0	6.9	6.0
5 (not very important)	13.2	10.2	9.7
No reply	6.6	2.3	1.7

Source: Extract from a survey conducted by Makrotest in December 1975. For a commentary see Verdoodt 1976.

is the worry among linguistic groups that they will lose their dominance in particular regions—their regional identity. While there is sentiment in favor of bilingualism, the sentiment in favor of protecting one's language is even stronger.

Nevertheless, the poll results do provide another illustration of a factor to which we have referred several times: the preference of Belgians for avoiding extremes and for seeking compromise solutions. The survey thus provides insight into the popular mood, even though it may not serve as an adequate guide for political action.

One cannot generalize on the basis of a single case. But the situation in Belgium would suggest the hypothesis that the usefulness of polls for indicating solutions to community or nationality problems may vary according to the degree to which mechanisms for social and political expression are already developed. In countries where there are few channels for the expression of popular attitudes, public opinion research may point to new ways of reconciling differences and avoiding conflict. In countries where most of the people, through the cultural and political organizations to which they belong, already have access to decision makers and to the mass media, polls are unlikely to suggest new alternatives. They can, however, help to reveal the mood of various elements of the population. Perhaps they can give courage to voices of moderation.

NOTES

1. For an overall approach to the realities of Belgium, and especially the social and political systems, see P. Dabin et al., "Belgique," in *Encyclopedia Universalis* (Paris: 1968-73), vol. III, pp. 80-135.

2. An ephemeral administrative separation had been enacted by the Germans in 1917, and a law on the same subject was proposed by Flemish nationalist groups in 1931. Their slogan was "Voor 't Belgikske nikske" ("No compromise with little Belgium"). The proposal was quickly buried.

3. The Front Démocratique des Francophones Bruxellois (FDF) was created in 1964, with a program that was initially limited to opposing the linguistic laws of 1963. It opted for federalism in 1970.

4. The contrast can also be seen on a continental scale. The United States, a "nation of immigrants," appears to have achieved a "plural unity" to which Europe still aspires.

5. "Comment les régions voient-elles l'Europe?" release of the European Communities, July 1968.

4

PUBLIC OPINION AND THE SLIPPERY ROAD TO PEACE IN NORTHERN IRELAND

Gerald A. Fitzgerald

In the United Kingdom's 1918 general election, the Irish political party Sinn Fein, whose candidates had vowed never to sit in a British Parliament, captured 73 of Ireland's 105 parliamentary constituencies (Lyons 1973). This nationalist victory margin was then—and still is—roughly equivalent to the Catholic proportion of the island's population. The vote thus represented a clear expression of Catholic opinion on the question of self-determination for the country, even though the British government did not intend the election to be a plebiscite in this sense.

Another unscheduled plebiscite, also stimulated by the imminent prospect of home rule for Ireland, had taken place six years earlier among the Protestant population. That poll, conducted largely in the church halls and vestibules of the part of Ireland historically called Ulster, had produced an equally decisive result: of the 500,000 Protestant adults then living in the region, 447,000 signed a solemn covenant pledging themselves to resist home rule by any means (Rose 1971). Nearly 100,000 also enrolled in an irregular but well-equipped militia called the Ulster Volunteer Force (UVF). When World War I broke out in 1914, a heavily armed and resolute band of Protestants stood ready to fight even the British army to keep the North within the United Kingdom (Stewart 1977).

If the names and issues sound familiar, it is because many of the principals in today's Ulster conflict claim descent, a mandate, or both from events or organized groups of the period 1912-18. The border dividing the two parts of contemporary Ireland also reflects, roughly, the divisions revealed by those expressions of popular feeling in 1912 and 1918. Linked to Britain in the nineteenth century by a direct legislative union, the 26 counties with large nationalist (Catholic) majorities became, after 1920, first a self-governing dominion within the British Empire and then a politically independent republic. The six northeastern counties of modern Ulster, with their overall 65 percent Protestant majority, meanwhile emerged as the United Kingdom province of Northern Ireland. A separate, largely autonomous parliament at Stormont, near Belfast, which the Protestant Unionist party ruled without serious challenge, governed the province between 1921 and 1972.

THE LEGACY OF PARTITION

Partition changed the terms of the dispute in at least three ways. One was that the two directly contending parties no longer occupied roughly distinct territories; while Protestant and Catholic communities are remarkably self-contained in Northern Ireland, they are scattered randomly throughout the province. Also changed was the number of groups involved: instead of three—Irish Protestants, Irish Catholics, and the residents of the British mainland—there were four: Ulster Protestants, Ulster Catholics, the British across the water, and the Irish across the border. The Irish Republic has provided northern Catholics not only an independent voice in world councils, but also support of many other kinds, including some early government assistance to the Irish Republican Army (Rose 1976).

A third outcome of partition was the freedom it gave majorities in both sections of Ireland to pursue their separate national and religious destinies. In the absence of effective counterpressures, among the first things to go were the carefully secular constitutional arrangements implemented with repeal of the direct British legislative union. A northern example was the capitulation by the Stormont parliament in 1930 to the Protestant clergy's demand for a dominant role in managing the legally secular state school system (Akenson 1973). The best-known southern instance was the constitution adopted in 1937, which formally acknowledged the "special position" of the Catholic Church and gave constitutional force to the Catholic church's prohibition of divorce.

Partition formally segmented one Irish society into two, and the part that was homogeneous became united, while the other, containing a one-third Catholic minority, remained divided. In the South the years embracing World War I, the Easter Rising of 1916, the Anglo-Irish War, and the birth of the Irish Free State had witnessed a halving of the Protestant community's relative size. Protestants, who in 1911 made up about 10 percent of the population in the counties later to become the Republic of Ireland, were reduced to a 5 percent minority by 1926. In the end they had no choice but to assimilate politically (White 1975).

Among northern Catholics, on the other hand, fear of permanent political and economic disabilities evoked strong resistance, first to partition itself and then to rule by the Protestant majority. The loss of more than 500 lives in the partition-era troubles weakened support for armed struggle, but not for eventual union with the South; thereafter Catholics refused to fly the Union Jack over their schools or churches, or to treat the six-county Unionist government as other than the product of an illegitimate gerrymander. When a school survey in 1969 asked children in Londonderry to name the capital of their country, more than half of the Catholic youngsters placed it in Dublin. Adults thought and acted no differently. According to one authority, only a single issue during the entire life of the Stormont parliament was trivial enough to create a cross-party alignment—the Wild Bird Act of 1931 (Darby 1976).

The bitterest disputes involved security. A particularly sore spot with Catholics was the retention—long after the disappearance of any credible internal threat from the IRA—of emergency powers allowing warrantless searches, detention without trial, and internment. Especially resented was the use of those powers, exclusively against anti-partitionists, for political advantage as well as for security reasons. Another source of bitterness was the Ulster Special Constabulary, or B-Specials, an exclusively Protestant force originally recruited from the ranks of the UVF, which acquired a reputation for poor leadership and excessive violence. With the judiciary an almost totally Protestant preserve, the courts became identified with anti-Catholic bias, as did many of the local councils. Protestants elected council majorities even in heavily Catholic areas like Londonderry. This was achieved through two strategies. The first was blatant gerrymandering; the second was to restrict the vote to householders and, at the same time, to restrict Catholic access to public housing, which had the effect of keeping down the number of Catholic voters.

Although both sides practiced discrimination in public employment wherever possible, the opportunities for doing so were infinitely greater for Unionists. The extent to which the same pattern existed in private employment is not precisely known, but favoritism toward Protestant workers was encouraged publicly by Unionist party leaders. Perhaps the most remarkable thing about discrimination against Catholics was the degree to which community loyalties seemed to determine attitudes about it. For example, surveys conducted in 1968 and 1969, when civil rights activists were taking the issue into the streets, revealed astonishing differences of opinion on the question: three-quarters of Catholics said discrimination existed; three-quarters of Protestants said it did not (Rose 1971, p. 272).

The problem was that in every public way possible—by the ways they voted, by their ritual observances of past tribal victories, and, in particular, by their refusal to intermarry or to educate their children together—each side persuaded the other that there could be no middle ground. The situation has sometimes been compared with segregation in the American South. Yet there is the essential difference that the segregation practiced in Northern Ireland is largely voluntary, and this fact makes the comparison weak to the point of uselessness.

Though demographically a majority, northern Protestants are in some ways less sure of themselves than northern Catholics. One indication that this is so comes from a question asked in 1969 about national identity. The Catholic response was unambiguous: three-quarters said they were Irish. Protestants agreed much less on how far they constituted a distinct community: the largest fraction (39 percent) sais they were British, but many others identified first with Ulster (32 percent) or with Ireland (20 percent) (Rose 1971, p. 208). Recent events may have caused a shift in identification among Protestants. In a 1978 survey nearly three-quarters of Northern Ireland's Catholics (69 percent) still identified themselves as Irish, but almost as many Protestants (67 percent) identified themselves as British (Rose 1979).

However, on another question, regarding partition, Protestant intensity compensated for this seeming disunity: 52 percent endorsed the use of "any measures" necessary to keep Ulster Protestant, whereas only 13 percent of Catholics were willing to go to the same lengths to end partition (Rose 1971, pp. 192–93).

RECENT SOCIOECONOMIC CHANGES

Ulster changed in the 1960s, in part because the world outside Ulster changed. Among the important sources of change were the education reforms of Britain's postwar Labour government and transformations in the Ulster economy. The education reforms—the first to bring secondary and higher education within reach of the British working classes—ensured a major expansion of the Catholic middle class in Northern Ireland. At the same time diversification, expansion, and modernization of the Ulster industrial base sharply increased the demand for skilled workers and competent professionals; and because many of the new enterprises were divisions or branches of English or overseas firms, with no interest in endorsing the existing community split, their presence posed a threat to the economic supports of the system. Two major established industries, linen manufacturing and shipbuilding, the latter long a preserve of Belfast Protestant workers, meanwhile went into a steep decline.

No part of Ulster society was unaffected by these changes. In the Catholic community the new, better-educated, middle-class recruits resented the sectarian walls impeding their economic and social mobility. They also quickly tired of the Nationalist party's obsession with partition, its lack of formal organization, and its "conventions," often called by the local parish priest, to select candidates for elective office. Most important, however, was the fact that newer members of the Catholic middle class were willing to acknowledge the de facto authority of the state. They still wanted the two parts of Ireland united, but much more immediate was their wish to participate directly in Ulster's government and to end discrimination. And so a civil rights movement was born (McAllister 1977).

The Protestant community has always been, by its very nature, more fragmented than the Catholic, and it has become progressively more so as a result of postwar changes in Ulster society. New educational opportunities, which in Catholic areas produced new leadership, had politically negative consequences in working-class Protestant areas; what the reforms did there was to drain off much of the natural leadership, leaving the Protestant working class and middle class isolated from one another to a degree unknown among Catholics. For example, a recent study comparing two adjacent Belfast Protestant neighborhoods, one working-class and the other middle-class, found them to be as ignorant of each other's ways as a pair of adjacent working-class Protestant and Catholic enclaves in another section of the city (Boal 1972). The nature of the political problem thus aggravated has been described by anthropologist Rosemary Harris in a comment on tensions between Protestants of different classes in the countryside: "The basic political problem of the poorer Protestant was that to secure his independence from the Irish Republic he had to support politically those whom he neither liked nor trusted" (Harris 1972, p. 187).

Extreme Unionism finds its readiest adherents among Protestant workers and small farmers, and the reasons for this are both economic and religious. While Catholics are more deprived than Protestants, Northern Ireland is noticeably more deprived than the rest of the United Kingdom. In 1978 unemployment reached a level not seen in Ulster since 1940; at nearly 12 percent, it represented almost twice the overall United Kingdom rate (*Fortnight*, June 9, 1978, p. 11). But there is also a religious component: the more extreme Unionists have been socialized in a form of evangelical Protestantism that opposes cooperation not only with Catholics but also with Protestants who are willing to deal politically with Catholics. They fear being sold out. These worries—being overwhelmed by Rome, by the island's chronic economic backwardness, or by both— greatly undermined Unionist solidarity in the late 1960s, when the party's increasingly liberal middle- and upper-class leaders began to recognize the legitimacy of Catholic grievances.

One way of viewing the situation as it existed just before the onset of widespread community violence is that in Ulster society there were three main groups: moderate Protestants, intransigent Protestants, and moderate Catholics. Two majorities were possible: a coalition of the first and second groups—the Unionist alignment since partition—or of the first and third. Unionist Prime Minister Terence O'Neill's aim seemed to be to make a transition from the first kind of coalition to the second when events overtook him in 1969 (Whyte et al. 1972).

RENEWED "TROUBLES"

Whether such a transition could have occurred peacefully in Northern Ireland will never be known. Around that time, control over events in the civil rights movement shifted from middle-class Catholic professionals to a generation too young to appreciate the depth and intensity of Ulster's divisions. Street protests favored by the movement's university-based activists triggered intensive rioting, and in August 1969 the British army intervened. This, in turn, led to an IRA military campaign to expel the British troops and reunite the island politically. The IRA campaign sharply increased the level of violence. The Unionist regime responded to IRA physical force with traditional security measures, applied massively, indiscriminately, and exclusively against the Catholic population. In March 1972, with no other choices left, the British government abolished the Stormont parliament and instituted direct rule from Westminster.

The other side's turn came in January 1974, when moderate Unionists, a new, nonsectarian Alliance party, and the new Catholic Social Democratic and Labor party (SDLP) formed a coalition government, the first in Ulster history to give Catholics significant governmental responsibility.

Considering some of its internal contradictions—for example, the SDLP's participation in a government whose police it didn't recognize—many people were surprised that the coalition lasted as long as it did. However, the straw that appears, now, to have broken this government's back was a provision creating a Council of Ireland, a relatively powerless body designed to forge a symbolic link among Dublin, Belfast, and Westminster governments. Four days after the government took office, the Unionist party rejected the arrangement, forcing its leader, Brian Faulkner, to choose between his party and his position as chief minister in the new "power-sharing" executive. Faulkner chose the latter, but in doing so became a victim rather than a controller of events. A national election on February 28, 1974, to elect a new Parliament at Westminster gave 51 percent of the Northern Ireland vote and 11 of Ulster's 12 parliamentary seats to Unionist candidates opposed to power-sharing and the Council of Ireland; Faulkner Unionists picked up 13 percent of the vote and the Catholic SDLP, 22 percent. This provoked the charge that Britain was attempting to govern Ulster through an executive that excluded a majority of the Protestant community. In mid-May the executive fell, the result of a 14-day general strike organized by the Protestant Ulster Workers Council. The strike, enforced by various Protestant paramilitary groups, paralyzed the province by gradually cutting off electrical power (Rose 1976, pp. 30–31).

PUBLIC ATTITUDES TOWARD POLITICAL ALTERNATIVES

Power-sharing—that is, government under a constitution guaranteeing the Catholic minority a share of political power—remains a key provision of the settlement demanded by Ulster Catholic politicians and the governments at Dublin and Westminster. Public opinion surveys reveal very strong support for the concept among Catholics, ranging from a high of 98 percent in June 1974, immediately after the interim power-sharing government collapsed, to a low of 83 percent early in 1978. Protestant opposition to the idea was highest in June 1974, when 34 percent said they would "never" accept it, and lowest in May 1973, shortly before the idea was tried, when 17 percent opposed it. As Richard Rose points out, however, the average Protestant is highly ambivalent: "The median respondent 'does not mind' power-sharing or 'just approves' it or supports it 'sometime in the future' but not now" (Rose 1978, p. 14).

As in 1974, an important sticking point has been the insistence of both sides that any settlement also contain guarantees of their long-term aspirations (*Irish Times*, January 23, 1979). Catholic demands for recognition of an "Irish dimension," an idea promoted by the British and Irish governments, has met a Protestant counterdemand for retention of the majority-rule governing principle. In effect, both sides are saying "Whatever else we do, we don't intend to assimilate."

By mid-January 1978 about 40 percent of the people in Ulster had been exposed to political violence, many of them at first hand (Rose 1978, pp. 9–10). The difficulty since the early 1970s has, thus, been how to restore self-government to Northern Ireland without intensifying violence and disorder. It is not that most people do not want peace. The emergence of an indigenous mass peace movement in 1976 proved that they do. Likewise, public opinion surveys going back into the 1960s reveal the existence of very substantial majorities—Protestant and Catholic—in favor of educating children from the two communities together. However, it is also true that the only people who can make peace are the people who are making war—the IRA, the British army, and the armed Protestant loyalists. Of course, the IRA and the armed loyalists constitute only a very small percentage of the Ulster population; but in a political culture like Ireland's, great problems can be created by very small groups. This fact is illustrated in a poll cited by Conor Cruise O'Brien, a cabinet minister in a recent Dublin government:

> A survey carried out in the Republic (in 1977) showed that while only a tiny minority—two percent—actually approved of the Provisional IRA campaign of violence, a large minority—35 percent—attributed idealistic motives to the Provisionals, and expressed respect for these. True, a majority of the whole sample condemned the Provisionals absolutely, but it was the smallest majority possible: 51 percent. (O'Brien 1978)

Another survey, carried out in 1978 by the Dublin Economic and Social Research Institute, underscored O'Brien's point. In that poll a series of attitude questions yielded an aggregate 60.5 percent majority opposed to the IRA, but only 33.5

percent who were prepared to condemn the motives attributed to the terror campaign (*Irish Times*, October 16, 1979).

An indirect, but still indicative, expression of distaste for violence in Northern Ireland on the part of Irish south of the border was obtained by a survey carried out in Dublin slightly earlier. More than half of a sample of 2,300 agreed with the statement "Northerners on all sides tend to be extreme and unreasonable" (MacGreil 1977).

In Ulster in 1980, devolved government has gone, and with it all forms of old-style state discrimination against Catholics. Among reforms imposed on the province since 1969 have been a law eliminating voting abnormalities, another creating a central executive to allocate housing fairly, and one establishing the Fair Employment Agency. Dispassionate and careful studies of the Catholic employment disadvantage—studies in some cases sponsored by the government—have begun to appear. The studies show that Catholics are 2.5 times more likely to be unemployed than Protestants, and that Catholics are disproportionately clustered in manual, unskilled, and low-status occupations or industries (Osborne 1977). But they do not say how much of the pattern reflects discrimination and how much is a legacy of past disadvantage caused by geography or other impersonal factors. Towns and cities in the rural west and along the border, traditional centers of Ulster's Catholic population, have for decades suffered unemployment several times greater than the industrial east. And, of course, the studies do not say how great is Westminster's commitment to identify and attack actual discrimination in private industry, or to tailor investment policy to meet the needs of people living outside the east. Westminster's response to both challenges will affect Catholic perceptions of fairness.

Direct rule is not anyone's first choice as a solution to the problem of governing Ulster. In the province itself the arrangement achieves a high degree of acceptability if people are asked, not what form of government they prefer, but what they find most expedient under the circumstances: by March 1976 such questions had attracted 72 percent Protestant and 79 percent Catholic majorities to direct rule, and the further reduction of violence since that time probably has not diminished the acceptability of such an alternative. What is, therefore, disturbing is that there is little public support in Britain for the arrangement: the same March 1976 survey showed the alternative most acceptable in Britain to be withdrawal of troops from the province (51 percent); next in popularity was to encourage the north and south of Ireland to unite into one country (49 percent); next was to impose power-sharing in some form (40 percent).

A survey carried out in Northern Ireland by the Belfast Opinion Research Centre on behalf of British Independent Television News in the summer of 1979 indicates that direct rule from England is still the alternative that has the most support among both Protestants and Catholics. But the survey did not ask about this directly. Question wording in Northern Ireland is a subject on which diplomatic as well as polling expertise is called into play. Respondents were asked: "In the long term, which one of these [referring to a set of alternatives] do you think is a solution to the difficulties in Northern Ireland? " Note that the question does not ask for preferences, but for an opinion about "a solution to the difficulties."

The alternative "Being much more closely tied in with Great Britain" was chosen by 67 percent of the Protestants and 26 percent of the Catholics. The alternative "Being much more clearly tied in with the Republic of Ireland" found more Catholic support (37 percent), but was chosen by very few (2 percent) of the Protestants. A third alternative, "A more independent form of government within the United Kingdom," was preferred by about 20 percent of both the Catholic and the Protestant respondents (Rose 1979).

In Northern Ireland it is not unusual to hear even hard-line Protestants concede the inevitability of political union with the Irish Republic. What is conveniently left unspecified is when this union will come about and what form it might take in order to be broadly acceptable both to the mainland British taxpayer and to the various factions throughout Ireland. Until recently there has been little serious public debate about the possibility, partly because it has been assumed that at least one party to the dispute—the hard-line northern Protestant— would not go along, and partly because no party has been willing to risk the much wider conflict that predictably would follow any attempt to impose a solution of that kind unilaterally.

Within the province no London government has yet found an alternative to direct rule, and none is likely to emerge until Ulster's two main confessional groups—Protestants and Catholics—can achieve an acceptable minimum of material and ideological satisfaction. An assumption has been that if a workable governing mechanism could be put in place, these benefits eventually would follow. Upsetting such calculations, however, have been the continuing guerrilla war and the fact that, to many Protestants, the mechanisms proposed have represented a step on the road to a united Ireland. Looking south, they have seen a nation with a social and political life built around a community of alien religious custom; the resulting fears have proved insurmountable.

With the exception of this group, however, public opinion throughout the British Isles seems to accept a political association of the two Irelands as the ultimate solution. The example of Newfoundland, Britain's first overseas colony, was cited in this connection by a political columnist in a Belfast newspaper. Newfoundland, too, wanted initially to retain its direct link with Britain, but was gradually pushed toward a self-governing federation with another country, Canada, by means of a series of carefully worded referenda (*Belfast Telegraph*, June 14, 1979). When, or if, such an objective becomes Britain's declared policy in Northern Ireland, there will again be work for the pollsters, for then the referendum draftsmen will need all the information they can lay their hands on, in order to avoid repeating the mistakes that policy makers have made since 1969.

5

THE CONFRONTATION OF HARIJANS WITH INDIAN SOCIETY

Eric P. W. da Costa

The Indian constitution provides for self-determination, in terms of association and organization, if members of a community seek such organization to protect their rights and to restore their dignity. It is more difficult for a community that cannot be described as a nationality to escape from economic, social, and political disabilities. The scheduled castes and tribes of India, often described ambiguously as either Harijans or Untouchables, constitute such a community.

There are, however, parallels between the confrontation of Harijans with Hindu society and the efforts of minority nationalities to assert themselves within a state dominated by some other national group. The disabilities of the scheduled castes and tribes are prescribed at birth; there is no way that a person born into such a community can change his or her identification. In seeking to achieve their rights and dignity, Harijans encounter the resistance of groups that enjoy a more privileged position in society. Political self-assertiveness within a community that for centuries has been powerless to rebel or to change its future has sometimes led to violence. There is no doubt about the increasing threat to peace.

This longitudinal study of change, supported by two surveys made five years apart, suggests that a way to peace can be found not so much by self-determination as by mobilizing a national consensus that is being wrought by the spirit of our times. It is not the mass attitudes of the Harijan minority that can work the miracle so much as the fortunate convergence of growing economic prosperity and a situation in which the pursuit of social justice becomes a condition of survival for political leaders.

Our starting point has to be the caste system. First, a look at its hoary past as a prelude to what it is today in a democratic, egalitarian society. Amaury de Riencourt (1960) has argued that Buddhism, for all its idealistic philosophy that envisages freedom of knowledge, popular education, belief in the equality of man, and a democratic outlook, never really made a dent in the basic polytheism, religious emotionalism, and acceptance of social inequality that prevailed among the vast majority of the Indian population. The decline of Buddhism was linked to the collapse of India's Universal State after the reign of Ashoka (who died about 230 B.C.), and was intimately connected with the rise of the caste system, which became established in the framework provided by Manu, the Hindu lawgiver. The surprising development is not the sudden arrival of the caste system, but its amazing persistence over the centuries. Thus was developed the bedrock of custom, the hard crust of which never seems to break, more powerful than law or administrative intervention.

The rapid correction of disabilities of all Harijans, as well as of the scheduled tribes, has been hindered by the fact that they do not constitute a single community. There are many hundreds of different castes "scheduled" under the Presidential Order of 1956 that led to their classification. The distribution of Harijans is such that they command no majority in any state, although the scheduled castes in total constitute 14.9 percent, and scheduled tribes nearly 7 percent, of the Indian population, making nearly 22 percent in all. With this percentage they could sway the results of close elections, and they hold in addition the reserved constituencies in which the candidate must be a member of a scheduled caste or tribe. These reserved seats in the Lok Sabha (parliament) constitute about 20 percent of all seats; therefore, if they were united, the scheduled castes should have, by force of numbers, substantial political weight. However, their division among the many states of the Indian union, whose objectives at the capital are often divergent, has meant that there is no concentration of political power such as could be obtained with a united vote of 22 percent of the population.

In the larger states the proportion of the population belonging to scheduled castes varies from 6.2 percent in Assam to 24.7 percent in the Punjab. For scheduled tribes, if we ignore the high concentrations in Manipur, Meghalaya, and Nagaland, the figure varies from 4.1 percent in Himachal Pradesh to 20 percent in Madhya Pradesh. Because of the many divisions in their own ranks, even in states where their numbers are high, there is no concentrated pressure on state governments or on district administrators to correct the injustices inflicted on the scheduled castes and tribes in contravention of the constitution. The political force of large numbers is blunted both by dispersion and by disunity.

FOUR STATES SHOWING RAPID PROGRESS

One should not, however, overlook encouraging solutions that, in particular areas, have operated with remarkable vigor. Consider the case of the Punjab, for example, with a population of which 24.75 percent is scheduled castes and tribes. It has been shown in research studies, which the Indian Institute of Public Opinion has developed on the basis of information from the National Sample Consumer Expenditure Survey, 1973-74, that the "deprivation" in this state, whether of food, clothing, or fuel, is minimal. There are no food shortages experienced by destitutes in the first three degrees of destitution, and there is a minimal shortage, equivalent at cost to 58.5 million rupees in the fourth degree, for "food," which would bring "destitute" consumption up to the poverty line. There is also a shortage of 38.3 million rupees among "poor but not destitute" people. But, in effect, the scheduled castes and tribes of the Punjab, as well as other poor in that state, are relatively well fed and not inadequately clothed. Thus, almost all deprivation, in some mysterious manner, has been abolished in the Punjab for Harijans no less than for other destitutes. This points the way to the possibility that in a vigorously active, growth-oriented society, even one that has the largest percentage of scheduled castes, economic disabilities can be rapidly removed.

In one sense, however, the Punjab environment is unique because of the dominance of Sikhism. This is not the case even in the neighboring state of Haryana, which is more Hindu than Sikh. The Sikh scheduled castes, which are recognized under the constitution, have no inequalities inherent in their cultural lines, since Sikhism is, by and large, a religion in which caste differences have been abolished.

A dent of equal significance in the deprivation of Harijans has been struck by a massive milk cooperative enterprise, known as the Amul Prototype. This consists of six district cooperative milk producers' unions in the state of Gujarat, where a "white revolution" has occurred with incredible speed, and a tremendous liberation of Hindu society has taken place. The reason is that members of the cooperative societies enjoy completely equal treatment and facilities. The Amul Prototype is a marriage of the cooperative network with modern scientific technologies, in which Kaira District pioneered. It has now become a new development model, the vehicle of economic and social change throughout Gujarat.

The cooperatively applied technology has helped to bridge the gap in resources between large farmers, marginal farmers, and landless agricultural laborers. The scheduled castes have a large stake in landless labor, about a third of their number being agricultural workers. The milk contributed is jointly collected; some cooperative societies have Harijan chairmen; and village community welfare is administered by each society: all schools, wells, and playing fields are open to all. The barrier between Harijans and higher castes has been broken down in village after village. For the poorest—landless laborers and marginal farmers—new avenues have been opened, leading not only to economic benefits but also to much greater social equality.

The integration of a village society in which all communities are able to work together without differences caused by religion, community, or class may well prove to be a greater ingredient of rural development than the solely economic benefits that other forms of farming or animal husbandry could conceivably generate by application of increased resources and sound economic principles. In the last analysis the significance of this move toward an equal society should be judged not in money resources but in the capability of individuals to work as members of a village community in the application of a technology.

Integration of village society holds out enormous possibilities for the solution of the problem of confrontation between the scheduled castes and other castes in other states. In fact, the experience in the Punjab, with the largest percentage of scheduled castes, and in Gujarat, with the extraordinary equality among the members of its cooperative societies, can well be the most important elements in the total revolution of attitudes that is the only final answer to the current confrontation.

One can look upon the Punjab's high growth rate as exceptional, and Gujarat's Amul Prototype can be seen as a particular case—both exceptions to a fairly sterile record of intercaste stagnation. Nevertheless, if one regards the elimination of deprivation as a partial solution to the problem of Harijan economic disabilities, one can find encouraging evidence in the National Sample Survey for 1973-74 that a more general force is also at work. In Bihar, where perhaps the confrontation between castes is greatest as of the late 1970s and where more violence has been invoked, the Harijans are acquiring much more confidence. In many villages they now insist on their rights, and caste Hindus are fighting back, sometimes in a last-ditch stand to maintain their privileges. The evidence is that the Harijans have moved from a passive to a militant posture, with the result that the ineffectiveness of change in the first 25 years of independence has, in an extraordinary manner, been diminished. There is support for the view that the socioeconomic forces operating in Punjab since 1960 and in Gujarat since 1964 are gathering strength in other states, such as Bihar and Karnataka, where as recently as the 1970s no change seemed likely.

The distribution of land and of house sites to Harijans in Karnataka has been remarkably effective under a chief minister who has had the vision to place his government's future in the hands of the backward classes. He won a critical election in February 1978 on this strong base, and has extended his power since then. This is a political formula that has worked.

Taken together, these four states—Punjab, Gujarat, Bihar, and Karnataka—present a picture of what is possible even without a major commitment to change by all the instruments of state government and a new determination from New Delhi. In other words, there is progress to report although, quite frequently, it is hidden from the observer by the fact that the magnitude of poverty continues to be demoralizing. There is now evidence that the areas of success are growing slowly but surely. Even while the confrontation grows more acute in some areas, it is abating in others, on a scale suggesting that the 1970s, through a variety of channels—economic growth, imaginative cooperative institutions, political foresight, and struggles in the field—have revealed a vision of the manner in which this oppressed community is slowly gaining ground, disproving the dismal prophets and establishing the fact that the winds of change are affecting fortresses built for centuries on barren rocks of privilege.

DISCRIMINATION AS SEEN BY HARIJANS

It is not possible to express a final view on the total range of progress, or on the likely duration of the long journey to total victory, on the basis of observations that give cause for optimism in four states. It is, therefore, important to consider how the people most concerned see these operations, particularly in the last decade, and to measure what might be described as their subjective valuation against the impression of slow but steady progress that seems to be developing. It is only the toad beneath the harrow that knows how deep each tooth point goes, and in the end the quality of any great economic and social experiment must be decided at the level of those who are the least equipped to imbibe its lessons or to rescue themselves through the benefits accruing from new developments.

It was for this purpose that the Indian Institute of Public Opinion designed its first all-India survey of Harijans in January 1974, in order to assess the feelings of India's downtrodden, particularly of the scheduled castes and tribes. How deep was their misery? How angry their mood? How readily would they join in violence if reform were indefinitely delayed? Answers to these questions were sought in a survey confined to Harijans in 1974. The sample was, it is true, not very large, just 1,500 respondents; and two states, Assam and Kashmir, were excluded because of difficulties in reaching the remote areas at a reasonable cost.

The results of this survey, published in the Institute's *Monthly Public Opinion Surveys* for March and May 1974, presented the first picture of a section of India's downtrodden as seen by themselves. A bench mark was obtained, revealing the extent to which they and their children were subjected to indignities such as social segregation, and the extent to which their frustration with economic and social disabilities was driving them to violent rebellion. Some of the results of this survey are shown in Table 5.1.

What was truly surprising at that time was the abundant patience of the overwhelming majority. Militancy was shown by only 5 percent of the total sample, which might mean 6 million of the Harijan population. But the survey also provided a warning. Anger and violence are not always of the revolutionary kind. While the survey showed no powerful revolutionary component, it would

TABLE 5.1

Conditions Experienced by Harijans, 1979 and 1974 (percent)

	January 1979		January 1974	
	Urban (N = 284)	Rural (N = 552)	Urban (N = 434)	Rural (N = 1,006)
When you visit a house of a different caste, do you have to stand, sit on the ground, or are you offered a seat at a distance or on the cot/chair the family uses?				
Have to stand	19	23	18	24
Have to sit on the ground	27	41	18	33
Offered a seat but asked to sit at a distance	19	21	22	20
Offered a seat on the cot/chair the family uses	35	15	42	23
Are you allowed to fetch water from the common well used by the other castes?				
Yes	40	34	61	43
No	15	23	14	33
Never tried, as we have our own well	45	43	25	24
In schools, are your children segregated in seating arrangements?				
Yes	9	6	12	14
No	89	92	88	86
Can't say	2	2	—	—

TABLE 5.1 (continued)

Have you ever been physically beaten by a member of a dominant caste?				
Yes	4	3	3	4
No	96	97	97	96
Has there been any case in your village/town where other castes combined to use mass violence on your community?				
Yes	25	18	13	14
No	74	81	87	86
Can't say	1	1	–	–
Would you prefer your community to organize itself to fight against injustice committed by other castes? If yes, would you resort to violence in self-defense?				
Yes	34	18	21	19
No	16	20	12	21
Can't say	–	1	–	–
Not applicable (answered "no" to first question)	50	61	67	60

Source: Indian Institute of Public Opinion.

TABLE 5.2

Reflections of Untouchability, by Education and Income, January 1974 (percent)

	Number of Respondents	Allowed to Fetch Water From Common Well		Never Tried; Have Own Well	Children Segregated in Schools		Children Allowed to Play with Children of Dominant Castes		Allowed to Visit Temples	
		Yes	No		Yes	No	Yes	No	Yes	No
Education										
Illiterate	832	47	30	23	14	86	84	16	56	44
Primary or less	380	47	27	26	14	86	84	16	61	39
Middle or less	164	51	24	25	13	87	87	13	63	37
Under matric	60	61	19	20	8	92	93	7	75	25
Higher secondary completed	49	63	21	16	8	92	96	4	70	30
Other	15	47	20	33	7	93	93	7	73	27
Total	1,500	48	28	24	13	87	85	15	60	40
Income (rupees/month)										
Up to 150	572	44	31	25	17	83	82	18	55	45
151–300	688	51	27	22	12	88	86	14	61	39
301–500	184	49	24	27	10	90	87	13	66	34
Above 500	53	60	13	27	7	93	91	9	68	32
No answer	3	66	—	34	—	100	100	—	66	34
Total	1,500	48	28	24	13	87	85	15	60	40

Source: Indian Institute of Public Opinion; Harijan Survey, 1974.

have been a mistake, as seen from experience in 1978, to underestimate the passion of the 5 percent who were truly "aroused." The militancy of the Black Panthers—as they were then described in Maharashtra—might in time have injected anger into large numbers.

Subsequent surveys in March 1977 and in January 1979 show that "violence in self-defense" is now much more widely supported. Almost 50 percent of respondents would like to see the community organize itself to fight injustice inflicted on it by other communities.

The 1974 survey also revealed the weaknesses of the Indian legal system when it came to enforcing the laws on the abolition of untouchability. In Table 5.2 four areas of untouchability, set out in a form to show variations related to education and income, are highlighted. Of all respondents, 28 percent were not allowed to fetch water from the common well, 13 percent of Harijan children were segregated in schools, 15 percent of these children were not allowed to play with the children of dominant castes, and 40 percent were not allowed to enter temples. However, as the table shows, the figures decline significantly with rises in education and income. As education advances, the opposition is partly overcome. Of those with higher secondary education completed, 63 percent were allowed to fetch water from the community well and their children were segregated only about half as often as the children of those who were illiterate, and so on.

A reduction in discrimination also accompanied increasing income. Those with incomes of over 500 rupees per month obviously had more confidence and enjoyed privileges similar to those with higher secondary education completed. The figures, shameful as they might appear in terms of the existence of untouchability in the year 1974, provide at least some encouragement, since education and income are bound to increase. One can therefore expect a softening of the incidence of crude untouchability.

Most of these questions were asked again in January 1979 of a smaller sample of 836. The proportions of those allowed to fetch water or whose children could sit with others at school were larger, although not markedly so in the latter case, and the differences were not statistically significant in either. However, a major form of self-reliance with respect to drinking water appeared on the scene: 45 percent of those questioned did not need to take water from the caste Hindus' well because they had their own wells. The disability is thus reduced not so much because permission is granted by others as because self-reliance has been developed in the community.

If one goes further, and looks at both surveys together, one might record the fact that there had been very little progress on some points. About the same percentage of Harijans were being beaten by members of a dominant caste, and there was no less violence toward the Harijan community than in 1974. In 1979, 50 percent in urban communities would prefer the Harijan community "to organize itself to fight against injustice," whereas in 1974 the

TABLE 5.3

Harijan Experience with Distribution of Land and Housing Sites, and with Voting, January 1979 (percent)

	Urban (N = 284)	Rural (N = 552)
Have you received any land for cultivation during the Emergency?		
Yes	5	2
No	95	98
If "No," has any other person in your community been allotted land for cultivation?		
Yes	10	17
No	83	80
Can't say	2	1
Not applicable	5	2
Have the people of your community been allotted housing sites?		
Yes	23	30
No	77	69
Can't say	—	1
Has there been an attempt by dominant castes in your or nearby villages to eject the Harijans from the housing sites allotted to them?		
Yes	16	13
No	78	80
Can't say	6	7
Has there been an attempt by dominant castes to snatch away the plots of land belonging to Harijans?		
Yes	20	14
No	75	78
Can't say	5	8
Have the members of your community resisted the attempts to forcibly evict them from their lands or housing sites?		
Yes	12	12
No	82	77
Can't say	6	11
Do you think authorities act in a partial manner whenever Harijans are subjected to high-handedness by dominant castes?		
Yes	29	33
No	66	71
Can't say	35	6

TABLE 5.3 (continued)

	Urban	Rural
Are you allowed to vote freely by dominant castes in *panchayat* or civic elections?		
Yes	96	95
No	4	5
Can't say	—	—
Are you allowed to vote freely in Assembly and Lok Sabha elections, or do the dominant castes intimidate you to vote for a particular candidate?		
Allowed to vote freely	96	95
Intimidate to vote for a particular candidate	4	5
Have any members of your community been prevented from voting through use of threats or force?		
Yes, in town	9	—
Yes, in villages	14	12
No, in town	69	1
No, in villages	7	85
Can't say	1	2
Would your community people exercise their right to vote freely if there were adequate arrangements, like protection by police and separate booths for your community?		
Yes, in town	11	—
Yes, in villages	8	15
Yes, in towns and villages	14	11
No	51	49
No opinion	16	25
If "No," why do you think so?		
Fear of reprisals by dominant castes supporting a particular candidate	—	1
My community does not care which party wins the election	—	2
It is undemocratic; will sow the seeds of communalism	12	12
No need; my community exercises the right freely	39	34
Not applicable	49	51

Source: Indian Institute of Public Opinion: All Indian Survey of Harijans, 1979.

corresponding figure was only 33 percent. Of those who would like to see the community organize itself, 34 percent would resort to "violence in self-defense." The result of these attitudes was shown in the late 1970s, when the upper castes and the Harijans often joined battle on fairly equal terms. The rural landlords may still have the advantage in manpower and weapons, but now that the police are being forced by directives of central and state governments to take action to protect Harijans, the boot may soon be on the other foot.

It is now possible to review the answers to new questions asked in January 1979 in regard to the particular areas of controversy that were not covered in the January 1974 survey (see Table 5.3). An intermediate survey showed that very few respondents had received any significant amount of land, but in January 1979 it was found that a large number of housing sites had been effectively transferred. In urban areas 23 percent reported such a transfer, and 30 percent in rural areas. While there have been significant attempts to eject the Harijans from these sites and plots of land, as revealed by the survey, these attempts have been resisted in more than half the cases. The question "Do you think the authorities act in a partial manner whenever the Harijans are subjected to highhandedness by dominant castes?" showed that there is still distrust of the administrative apparatus. However, it is also to be noted that 66 percent in urban areas and 71 percent in rural areas did not complain that authorities are "partial."

In *panchayat* (Communal Council) and civil elections, according to 95 percent or more in both urban and rural areas, Harijans are allowed to vote freely by the dominant castes. In other words, a considerably more liberal atmosphere prevails, at least in local elections. This also applies to Lok Sabha elections, though there have been cases of interference—suggesting that some additional confidence must be created by ensuring adequate police protection and by providing separate polling booths in areas of Harijan predominance. The broad picture is one of a fairly democratic structure in which opportunities for dominant castes to deny the Harijan community the right to vote seem to have disappeared. All these are straws in a favorable wind. No one need describe these signs of betterment as adequate, since untouchability still exists and violence erupts to prevent peaceful cultivation of arable land or enjoyment of house sites that have been distributed. However, the course of change is not in doubt. The Harijans—or at least two-thirds of them—see government machinery as being on their side.

THE IMPORTANCE OF FAMILY PLANNING AND EDUCATION

Are there any areas where the Harijans can take thought for themselves and by their actions save future generations from suffering the same impediments to welfare? There are two areas, which have been the subject of opinion surveys by the Indian Institute of Public Opinion, in which much more self-reliance can be exercised: family planning and education.

Family planning was the subject of one of the Institute's large surveys in 1971, with a sample covering the whole population and including 6,300 interviews in both rural and urban areas. More than half the respondents were either illiterate or had "primary or less" education, and this sector included a very large number of Harijans. Two questions were particularly important: the "ideal" number of children desired, and the current size of the respondent's family. Among the illiterates and less educated, it was observed that the "ideal" family of three children was favored by a very small proportion. Whereas 44 percent of literates considered three as the ideal number of children, 28 percent of illiterates would not favor a limit and 16 percent would not be satisfied with fewer than five or more children. A similar pattern was observed among those with monthly incomes of less than 150 rupees. It is largely the better-off and the better-educated who have accepted the family planning message, although in many cases acceptance of change does not necessarily involve prompt corresponding action. There is no doubt, however, that the response to Indian family planning is well expressed in Table 5.4, which relates education to the number of children respondents actually had.

Most of the Harijans were included among those with large families. It will be seen that 35 percent of illiterates have five or more children, against less than half that proportion of those with secondary education completed or some university education. Families also are large among those with primary education or less. It is arguable, on the basis of these survey results, that as most Harijans advance to the secondary level of education, their approach to family planning will be as positive as among those who now have higher educational qualifications. It is not impossible that within a generation an acceptance of a limit of three will be the dominant position.

The other way that Harijans can take action to improve the prospects of their next generation is by sending children to school. This was part of the January 1974 survey, and the character of the educational block in larger families is well revealed in Table 5.5, which shows how many children in families of various sizes have been admitted to school. It is clear that with the current economic pressure, the larger the number of children in a family, the greater the likelihood that many of them will not attend school. If one reads the table diagonally, it can be seen that the percentage of families that send all their children to school falls from 75 percent for one-child families to 38 percent for families with more than five children. Twenty percent of the respondents had no children admitted to school. Thus, more than half of all Harijan children in 1974 were not attending school. Obviously, poor families would rather make their children earners at an early age than allow them to "waste" time in school. This is a major obstacle to the escape from frustration, deprivation, and despair.

TABLE 5.4

Number of Children per Family, by Level of Education, 1971 (percent)

Level of Education	No. of Respondents	Number of Children								
		One	Two	Three	Four	Five or More	Unspecified	None	Unmarried	Total
Illiterate	1,011	9	14	16	15	35	1	6	4	100
Primary or less	2,531	10	13	15	16	32	1	7	6	100
Some secondary	1,317	9	14	17	15	25	1	8	11	100
Secondary completed or some university	995	13	13	14	12	16	1	9	22	100
University degree or more	446	11	13	13	10	17	1	10	25	100
Total	6,300	10	13	15	15	28	1	8	10	100

Source: Indian Institute of Public Opinion: All India Family Planning Survey, 1971.

TABLE 5.5

School Attendance, by Size of Family, 1974 (percent)

Number of School-Age Children in Family	No. of Respondents	Number of Children Admitted to School						
		One	Two	Three	Four	Five	Above Five	None	Total
One	330	75	–	–	–	–	–	25	100
Two	335	17	61	–	–	–	–	22	100
Three	207	19	19	47	–	–	–	15	100
Four	113	13	18	7	46	–	–	16	100
Five	37	8	14	16	5	46	–	11	100
More than five	35	6	9	11	12	14	38	10	100
Total	1,057	35	26	11	5	2	1	20	100

Source: Indian Institute of Public Opinion: All India Survey of Harijans, 1974.

A "virtuous circle" may be created as education improves among Harijans. This is because practicing family planning is more acceptable to those with higher secondary education than to illiterates. Thus, the number of children per family in these families could drop in a generation from five to three and, as a consequence, the number of children likely to complete secondary school will rise irrespective of government incentives like free meals, scholarships, or other inducements. With higher educational qualifications and, in time, better health, Harijans will inevitably make better use of job reservation policies and more speedily ascend the ladders of promotion in both government and private employment. It is the failures in family planning and school attendance that have been responsible for much of the stagnation since independence.

PERCEPTIONS OF PROGRESS

Notwithstanding the overall depressing picture, do the Harijans perceive any change in their lot compared with what their forefathers had to suffer? Table 5.6 presents data on this question. The broad picture that emerges is one of islands of relative improvement in a sea of economic stagnation. Roughly 30 percent thought they were better off than their fathers: they lived better, were clothed better, and traveled more frequently. Thus, there is little doubt that some benefits have trickled down to a minority of Harijans. The vast majority, however, still wage a losing battle to make ends meet. Nevertheless, the younger people were more optimistic; it may be that longer memories are a disadvantage, since older people will remember more incidents reflecting indignities.

In the same survey another question was "Has the attitude of the dominant castes become better or worse over the last ten years?" Here a contrary result appears: 72 percent said that the attitude of the dominant castes had become "better"; only 3 percent said that it had become "worse." The breakdown of this particular answer by state shows that except for Andhra Pradesh and Orissa, there was a majority in every state indicating that things were better; in Punjab this was unanimous (see Table 5.7).

When the same question was asked in January 1979, for a single year instead of the last ten, the answer was equally encouraging: 31 percent thought that the attitude of dominant castes had changed for the better; 8 percent thought that the change was for the worse. However, 21 percent thought that the change for the better was only outward behavior and that there had been no change of heart, as against only 6 percent in 1974 (Table 5.8). Thus, both optimism and distrust are shown in the most recent studies. The question is how to interpret these subjective assessments of experiences during a period of change in which the memories of the old are not yet anchored in the present and in which both young and old, remembering the past, can misread the future.

TABLE 5.6

Quality of Life, by Age and Education, 1974: Do You Think You Live Better Than Your Grandfather or Father? (percent)

	No. of Respondents	Live Better Than Grandfather				Live Better Than Father			
		Better	Worse	Same	Don't Know	Better	Worse	Same	Don't Know
Age									
21–35	652	32	43	13	12	33	44	21	2
36–50	519	25	42	23	10	30	43	25	2
More than 50	329	28	41	18	13	29	43	22	6
Total	1,500	29	42	17	12	31	43	23	3
Education									
Illiterate	832	20	46	21	13	23	49	24	4
Primary or less	380	36	37	15	12	35	39	24	3
Middle or less	164	41	40	11	8	46	38	15	1
Under matric	60	40	47	8	5	46	37	17	—
Higher secondary completed	49	58	28	8	6	62	22	16	—
Other	15	60	20	7	13	67	20	13	—
Total	1,500	29	42	17	12	31	43	23	3

Source: Indian Institute of Public Opinion: All India Survey of Harijans, 1974.

TABLE 5.7

Perceived Attitude Change of Dominant Castes,
1974 Compared with 1964 (percent)

	No. of Respondents	Better	Worse	No Change	Qualified: Apparent Change for Better Is Only Outward; Inwardly No Change	Total
State						
Andhra Pradesh	75	39	1	45	15	100
Bihar	100	78	1	19	2	100
Delhi	50	66	2	30	2	100
Gujarat	120	59	8	30	3	100
Haryana	30	94	—	3	3	100
Madhya Pradesh	120	83	5	9	3	100
Maharashtra	150	72	5	17	6	100
Karnataka	75	80	—	15	5	100
Orissa	50	4	16	72	8	100
Punjab	60	100	—	—	—	100
Rajasthan	75	84	1	8	7	100
Tamil Nadu	150	94	—	4	2	100
Uttar Pradesh	300	70	2	17	11	100
West Bengal	145	72	3	21	4	100
Total	1,500	72	3	19	6	100
Area						
Urban	434	80	4	13	3	100
Rural	1,066	69	3	21	7	100
Total	1,500	72	3	19	6	100

Source: Indian Institute of Public Opinion: All India Survey of Harijans, 1974.

TABLE 5.8

Perceived Attitude Change of Dominant Castes, 1979 Compared with 1978 (percent)

	No. of Respondents	Better	Worse	No Change	Qualified: Apparent Change for Better Is Only Outward; Inwardly No Change	Total
Area						
Urban	248	30	9	45	16	100
Rural	552	32	7	37	24	100
Total	800	31	8	40	21	100
Age						
21–35	426	27	9	44	20	100
36–50	260	37	5	34	24	100
More than 50	114	36	7	38	19	100
Total	800	31	8	40	21	100

Source: Indian Institute of Public Opinion.

TABLE 5.9

Personal Ratings on the Hadley Cantril Self-Anchoring Striving Scales, 1963 to 1974

	1974			1972			1970		1968		1967		1963	
	Primary or Less	All Groups	Illiterates	Primary or Less	All Groups	Illiterates	Primary or Less	All Groups	Primary or Less	All Groups	Primary or Less	All Groups	Illiterates	All Groups
Past	4.4	5.5	2.8	3.5	3.6	3.0	3.6	3.7	3.7	4.3	4.8	4.6	3.0	3.5
Present	3.8	5.1	3.0	2.7	4.0	2.9	3.3	3.5	3.4	4.5	4.1	4.7	3.2	3.9
Future	3.3	4.8	3.3	4.5	5.2	3.2	3.5	3.8	3.0	5.0	4.7	6.0	4.1	5.4
Past to present	-0.6	-0.4	0.2	0.2	0.4	-0.1	-0.3	-0.2	-0.3	0.2	-0.7	0.1	0.2	0.4
Present to future	-0.5	-0.3	0.3	0.8	1.2	0.3	0.2	0.3	-0.4	0.5	0.6	1.3	0.9	1.5
Overall change	-1.1	-0.7	0.5	1.0	1.6	0.2	-0.1	0.1	-0.7	0.7	-0.1	1.4	1.1	1.9

Source: Indian Institute of Public Opinion.

It is well established in studies of the scheduled castes in India that they can both deplore their hopeless situation and yet entertain high expectations for the future. An interesting study of scheduled-caste students in eastern Uttar Pradesh showed that most of them were expecting to earn the baccalaureate, and some sought doctorates and technical qualifications superior to the B.A. degree at a time when their parents, many illiterate, were greatly deprived.

The Hadley Cantril striving scale (Cantril 1965) has been administered to samples in India in several studies starting in 1963. The self-anchoring striving scale usually involves the use of a picture of a ladder, with rungs numbered from 1 to 10. Respondents are asked to imagine the first rung as representing the worst possible life they could experience and the tenth rung as the best possible life. They are then requested to indicate the rung that corresponds to their condition at present, the one corresponding to their condition in the past (for instance, five years ago), and the one that represents the life they expect in the future (for instance, five years from now). From the answers the researcher can derive numerical scores showing whether, subjectively, conditions of life are improving or deteriorating, and whether expectations for the future are optimistic or pessimistic. (See Table 5.9.)

The results clearly indicate some pessimism in the less educated groups, measured against all groups together. In 1963, when the personal rating for all groups for the future was 5.4, that of illiterates was only 4.1. There was not much change in 1974, when the rating for the future for all groups was 4.8 and that for illiterates was 3.3. However, "future" ratings by all groups and by those with primary or less education were, except in 1974, higher than "present" ratings. It would seem, with only one exception, that the expectations for the future of less-educated respondents are higher than judgments about the past or the present. This must apply to Harijans in particular, since they constitute a very significant fraction of those illiterate or with primary or less education.

THE UTILITY OF ATTITUDE SURVEYS

Can the evidence provided by public opinion research help to bring about peaceful change for the better in the case of the Harijans? Our answer must be equivocal—perhaps it can, although the relationship between what we know, or can find out, and what actually happens may always remain obscure.

As to the probability of increased violence, we can document that a small but increasing proportion of Harijans will not take things lying down—but only if the provocation is extreme. About 40 percent would opt for organizing the community if and when they are faced with mass violence from dominant castes; 20 percent would not hesitate to use violence in self-defense. But 60 percent could not think of violent action even if physically assaulted on a large scale.

The existence of tension, even growing tension occasionally erupting in violence, is not necessarily proof either that no improvement is taking place or that only militancy and force can succeed. Although the truly Gandhian frame of mind has largely disappeared, it did not, by and large, belong to the privileged classes. It was a mass instrument significantly blunted by the arrival of independence, when much Indian passion for equality lost its motive force. Nevertheless, the great majority of the Indian people deprecate accentuation of violence; and the state administrations, whether Congress or Janata, are sensitive to criticism of inaction or of favoring the well-to-do. Thus the environment has improved.

The inevitability of some gradualness does not imply that there is no will to accelerate the process. Indeed, the evidence, both objective and subjective, as revealed in the January 1979 survey, is that much improvement occurred during the year 1978, in which violence, whether in Bihar or Maharashtra, was at its worst. Monitoring public opinion by surveys may help to spread the assurance that improvements are taking place, although it is unlikely to arrest violent incidents that are sporadic, often occur in inaccessible areas, and are totally unpredictable. Provided no complacency grows, this conflict can be patiently resolved.

Nor does the downtrodden condition of the Harijans seem to have inspired substantial political radicalism. Great injustices can persist, and even increase, partly because they are old and partly because organization of the very poor and uneducated for a show of force is almost impossible. It is often said, with truth, that an old tax is no tax; shoulders, even when weak, quickly broaden to bear the burden. In much the same sense great indignities borne for centuries are accepted passively as part of human destiny or Hindu karma. Indeed, the Hindu view of rebirth fits neatly into a pattern of mental escape in which one awaits another life for correction of the miseries one suffers now. All in all, Indian revolutionary fervor has always been dampened by acquiescence rooted partly in religious fatalism and partly in inertia due to ill health and ignorance. It is only with a major transformation of mental attitudes, largely proceeding from education and economic well-being, that decisive turns in political and economic behavior have been born.

Survey evidence thus brings into sharp relief the point that deprivation in India does not breed extreme radicalism. All the studies carried out by the Indian Institute of Public Opinion show that the Harijans by and large shun both the extreme rightist and leftist parties. A large minority has always supported the Congress, a party of the center. They are, indeed, painfully aware of their condition, and fervently desire a better life. But the downtrodden are not willing to take uncalculated risks; they sense their helplessness. A majority of them partake of a reality world in which there are no personal ascent ladders in the future. The past, the present, and the future are merged in a static and unfriendly world of endurance and want.

As for the optimism that some Harijans express, it is possible that among the truly poor, small additions to welfare, or little acts of consideration, are wildly exaggerated. It has been a major finding in recent opinion research in Western Europe that even when major benefits in housing and unemployment relief are obtained, the working-class mind records no increasing level of satisfaction. The revolution of rising expectations has the effect of neutralizing in the mind the physical additions to consumption or comfort, once a certain critical level is reached. But below that level, and below the poverty line, the opposite effect may well be generated. The mind has a curiously contrary, perhaps a compensating, response to small, unexpected improvements. It requires very little to give good cheer to the desperately poor: the silver lining is thus, by act of God, more frequently seen. Once basic wants are satisfied, the dissatisfactions, sometimes hidden in every person, assert themselves. For Harijans that time has not yet arrived. For the great majority the search for basic necessities and for equal dignity must go on, perhaps for a generation. Nevertheless, at present the great majority see the future with optimism.

Monitoring mass attitudes can certainly help to prevent the belief that the current situation is more critical than it really is, although the study of opinions can do little to provide assurances that tendencies toward violence and radicalism will not increase in the future. But a more positive role is also possible. In the case of the Harijans, surveys enable the identification of the areas where improvements are seen as taking place and of programs that appear to be effective. Especially it is possible to suggest actions that can be taken by the people themselves to improve their condition—for example, with respect to family planning and education.

Nevertheless, insofar as a study of public opinion of the poverty-stricken in India permits one to generalize, it is to the effect that the degree of degradation and helplessness of the most afflicted groups will not permit them to raise themselves by their own bootstraps. Until education and skills are acquired by those of them who are better-equipped, physically and mentally, leadership will be very slow to arrive. In any case, financial (and often physical) resources must come from outside. It is not the opinion of the underprivileged that will be decisive, but the quality and imagination, no less than the compassion, of the privileged in each community where deprivation has separated millions of human beings from dignified communication with their kind.

It is the force of public opinion created by economic and political winds of change that is likely to accelerate the processes of achieving more equal citizenship all over the world. There are no minorities anywhere who have suffered so long and so patiently as the Harijans of India. It would be a triumph for the last quarter of the twentieth century if it witnesses, as I believe it will, a massive liberation of their creative energies directed not only to self-reliance but also to equal partnership—which is the nation's greatest need. In the last quarter of the century, India must give new meaning to the grand preamble in its constitution. Only in that way will India's truly democratic commitment become an enduring monument to the Gandhian age.

(*Editor's Note*: Because of difficulties experienced in communicating by mail and cable between Princeton and New Delhi, it was not possible for the author to check the final version of the above article. [One letter from New Delhi arrived after three months with a postmark indicating that it had come by way of the Teheran post office. A manuscript sent from Princeton was delivered in New Delhi after a delay of four months.] The editors apologize to the author and to the reader in the event that any errors have been introduced in the process of preparing the article for publication.)

6

ARAB MINORITIES IN ISRAEL

Don Peretz

The Arab minorities of Israel differ from other groups discussed in this volume in that in 1948 they were transformed from a majority of their country's population into the minority they have remained since the establishment of the state of Israel in that year. They have not recovered from the trauma created by the events of 1947-48: the partition of Palestine, the departure of Great Britain, the subsequent Arab-Israeli war, and the mass flight of most Arabs from areas that became part of the Jewish state. The war between Israel and the Arab states in 1947-48, and the three wars that followed in 1956, 1967, and 1973, have strongly colored perceptions and attitudes of Jews and Arabs toward each other, and have formed the basis of official policies by Arab and Jewish governments in the Middle East toward their respective Jewish and Arab minorities.

On the eve of the establishment of Israel, Arabs constituted some two-thirds of the approximately 2 million inhabitants in the British Mandate of Palestine; Palestinian Jews and newly arrived Jewish immigrants from Europe made up the other third. Over 700,000 Arabs fled from the areas acquired by Israel during the 1948 war, leaving an Arab minority of approximately 156,000 living among 700,000 Jews.

Since 1948, Israel's Arab population has risen by more than 300 percent, mostly as a result of natural increase, although a few thousand other Arabs were incorporated into Israel after it acquired several former Jordanian villages under the 1949 armistice agreements and through Israeli annexation of Arab East Jerusalem from Jordan after the 1967 war. The Jewish population has more than quadrupled during that period, but nearly half the increase is due to immigration. Despite the growth of the Arab population from 156,000 to 580,000 by the end of 1977, the minority has never recovered fully from the political, the economic, or the social effects of the 1948 war and the subsequent decades of Arab-Jewish hostilities.

During the era of transition from Palestine to Israel (1946-48), most Arab politicians and intellectuals, a large proportion of the professional and middle class, and many religious and national leaders left the country and have been unable to return. The loss of Arab leaders was accompanied by the disappearance of credible institutions with which to identify. The Arab political and social structure that had characterized Palestine vanished. Political groups and social organizations established during the mandate were not reconstituted in Israel. As the Arab community became subservient to and dependent upon the Jewish majority for the essentials of life, new economic relationships and a completely revamped class structure replaced those that had emerged during the 30 years of British hegemony. These changes in conditions and in status have taken place against a background of continuing war between Israel and the Arab states and imposition of special security measures by the Jewish majority on the Arab minority.

Furthermore, as Israel has increasingly emphasized its unique ties with world Jewry and the Zionist movement, the status and roles of non-Jews in a Jewish state have remained ambivalent. In the minds of many, both Jews and Arabs, Israeli identity and the mission of the state are decidedly Jewish. Even if the uncertainties, the mutual mistrust, and deep apprehensions caused by the continuing state of war between Israel and most of its Arab neighbors were to be terminated, the status of Arab minorities would continue to be ambivalent.

For example, peace between Israel and the Arab states would remove one major area of tension in relations between majority and minority, but would do little to ease strains caused by Arab dependence on the Jewish economic system. Since 1948 the country's Arab economy has been transformed from a self-sufficient, albeit subsistence, one to an integral and dependent sector of the Jewish economy, in which primary consideration is given to the pursuit of Zionist goals, such as increasing Jewish productivity and expanding Jewish settlement in all regions of the country. Development of the Arab economy receives secondary consideration and is encouraged to the extent that it is consistent with Jewish development.

During the first generation, from 1948 to 1967-68, the trauma of Israel's Arab minorities had few political repercussions. Indeed, as of 1966 there was so little political or social unrest in the Arab community that security authorities agreed to terminate the military government responsible for supervision of the Arab sector. Control of the Arab community had been relatively easy. There were no major disturbances, strikes, or eruptions of antigovernment sentiment. Few Israeli Arabs were involved in espionage or sabotage activities against the state. No antigovernment political movements of consequence had developed, nor had any charismatic personalities that could threaten Israel emerged.

To the extent that there was an Arab opposition, it was concentrated in the Israel Communist movement, which never obtained more than 5 percent of the country's total vote in the six national parliamentary elections held before 1966. Some analysts attribute absence of political upheaval in the Arab community to astute and efficient methods of control through the military government. Others, less critical of the authorities, assert that benefits derived by the minority from policies of a benevolent government were responsible for the lack of problems. Actually, a balance was maintained between policies of control and rewards for cooperation. Furthermore, conditions for an effective opposition did not emerge until the late 1960s. The generation of 1948-68 was still suffering from the shocks of the 1948 defeat and the disruptions it created in the Arab community.

During this 20-year period Israel's Arabs were isolated from the majority of Palestinians living beyond its borders. Only a small number of Christians were permitted to cross the armistice lines once or twice a year for celebration of religious holidays. Political contact with the wider Arab world was banned by both sides. It was considered treasonable by the Israeli authorities, and was not permitted by the Arab countries. Israeli Arabs were regarded with suspicion by Israel's Jews because they were Arabs, and by the Arabs because they were Israelis.

By 1967 the situation began to change, and has continued to change. Events within the minority community and in the surrounding countries converged to spark more militant attitudes and to create a more effective Israeli Arab opposition to government policies and programs. This activism continued and spread until, by 1979, it had aroused apprehension in many government circles, which became aware of the wide range of problems that had existed in relations between Israeli Arabs and Jews for the previous 30 years. As of the end of January 1979, hardly a day passed without discussion of these problems in the press.

CHANGING SOCIAL STRUCTURE

The most profound change in Israel's Arab community and in the larger Palestinian Arab community since 1948 has been the process of "depeasantization." I use this term instead of "proletarianization" because few Palestinians became proletarianized in both the economic and social senses. While they were removed from the land, and agriculture ceased to be the major source of their subsistence, and while many moved to cities and larger towns, the traditional rural social structure, with its manners and customs, still exists in more instances than not. In Israel, although most Arabs work in nonagricultural urban pursuits, such as construction, industry, trade, and services, a majority of Arabs continues to live in villages from which they commute to urban employment. Although economically they are situated at the bottom of the urban working class, many aspects of their lives follow traditional rural patterns formed by the village environment in which they continue to reside.

At the end of the mandate, in 1947–48, about two-thirds of the Palestinian Arabs were employed in the rural sector, mainly in agriculture. By 1963 the proportion of Arab agricultural workers in Israel had declined to 38 percent of employed Arabs. The decline continued, so that by the end of 1977, 16.7 percent worked in agriculture, forestry, and fishing; 22.9 percent in construction; 22 percent in public, community, and personal services; 12.2 percent in commerce, restaurants, and hotels; 17.7 percent in industry (mining and manufacturing); 6 percent in transport, storage, and communications; 2 percent in finance and business services; and 9.1 percent in scientific, academic, and other professional pursuits. (These figures total over 100 percent because of overlap between certain categories.) Employees constituted 70.8 percent of the Arab work force in 1977; 24 percent were employers, self-employed, or members of cooperatives; and 5.2 percent were unpaid family workers. A little over half, 52.3 percent, worked in the locality of their residence, and 47.7 percent worked elsewhere (Israel Government *Statistical Abstract of Israel*–1978, no. 29, Jerusalem: Central Bureau of Statistics, 1979; p. 351).

Despite a decline in the proportion employed in agriculture, the percentage of Arab agricultural workers was more than three times the percentage of Jews in agriculture (5.2 percent of the total number of Jews were so employed); four times the percent of Jews in construction (5.7); less than the percent of Jews in services (36); the same as the percent of Jews in commerce, hotels, and restaurants; less than the percent of Jews in industry (24.8); roughly the same percent of Jews in transport, storage, and communications (7.1); and only about a quarter of the percent of Jews in finance and business services (7.8) (*Statistical Abstract of Israel* 1978, pp. 354–55).

Decline of the autonomous Arab rural economy, and increasing economic dependence of the minority on the Jewish sector, was caused by several factors. Most important was the transfer of land in Israel from Arab to Jewish ownership. Estimates vary as to the amounts of Israeli Arab land expropriated, but a reasonable figure is that about 1.25 million of the minority's more than 2 million *dunams* was taken by the government for diverse reasons. Arab land has been sequestered for military purposes, for national development, and for establishment of new Jewish settlements. The result is that by 1977 non-Jewish farms in Israel constituted a little over 20 percent of the cultivated area, totaling 4.3 million *dunams*. This total figure is somewhat deceptive, for it does not distinguish between irrigated and unirrigated farms, the latter producing much less than irrigated ones. Whereas more than half the Jewish farm area was irrigated, less than 10 percent of Arab farms were irrigated (*Statistical Abstract of Israel* 1978, p. 400).

The relative lack of development in the Arab agricultural sector was also due to the diminishing importance of crops that had been its mainstay. Tobacco and olive cultivation was reduced, and competition from mechanized and irrigated Jewish farms often made continuation of agriculture unprofitable. Even with the decline of the Arab sector, however, its relative importance in the country's total economy remained significant. About a quarter of total agricultural production still comes from Arab farms. In some areas, such as production of vegetables, melons, and grapes, the Arab contribution is major.

THE LAND QUESTION

The land question is one of the most crucial in relationships between the majority and minority. It is symptomatic of major differences in the national goals of Jews and Arabs, and is a focal point on which many minority grievances have centered. To the Arabs, expropriation of land was perceived as not only, or even primarily, an economic loss. It was regarded as loss of national status and relegation of the minority to a subservient position. The centrality of the land question antedated establishment of the state. During the mandate era it was a major issue in the dispute between Arab and Jewish nationalist movements. From the beginnings of Zionist settlement in Palestine, opposition to Jewish acquisition of Arab-owned land was a source of protracted conflict.

On the other hand, acquisition of land, its settlement, and its development have been central concerns of the Zionist movement since it was founded. Although establishment of the Jewish state removed most obstacles to realization of these goals in Israel, land acquisition and settlement continue to be major concerns of the state and of the world Zionist movement with which the Israeli government is so closely associated. A major objective of every government since 1948 has been to expand Jewish settlement of areas within the pre-June 1967 borders where Jews are relatively few in number. Most vital is the Galilee area, where about two-thirds of the Arab minority lives. At present, Arabs own about 10 percent of the land in that area. The remainder is state land or is controlled by Jewish settlements. The area also occupies a vital location along the country's frontiers with Jordan and Syria. Despite its strategic importance and predominant Jewish control of the land, the Jewish population in the region is being overtaken by the Arab inhabitants. Estimates as of 1979 were that the population was 52 percent Jewish and 48 percent Arab. In 1965 the ratio was 57 percent Jewish to 43 percent Arab (*Jerusalem Post*, January 1979).

Despite their lack of success, Israeli government policies aimed at reversing this trend through intensification of Jewish settlement in Galilee have become a major source of controversy between majority and minority. The government and the Zionist organization have periodically called attention to the problem of

Jewish settlement in Galilee. In January 1979 the Jewish Agency, executive arm of the world Zionist movement, announced that it would set up 28 "lookout points" on 300,000 *dunams* of state-owned Galilee lands. These points are to serve as forerunners of permanent Jewish settlements.

An announcement in October 1975 by the Ministry of Agriculture emphasized the "special problem of the Galilee region." Not only was the Jewish population outnumbered, but

> The lack of center of attraction in Galilee is borne out by the 1960–70 estimate of Jewish population growth there—the lowest in Israel. ... It is necessary to change the existing situation ... by means of implementing a long-range development program. ... Since the fundamental task of the proposal—making Galilee into a region with a Jewish majority—cannot apparently be implemented in the immediate future, the plan was devised to comprise an early part extending to 1980 and a later stage from 1980 to 1990. (*MERIP Reports*, no. 47, May 1976, p. 4)

Confrontation between Jews and Arabs over these policies came to a head in March 1976, when Arabs organized protest demonstrations and a general strike against a cabinet decision to confiscate several thousand *dunams* of Arab-owned lands in Galilee and the southern Negev region. Arab leadership was divided over response to the proposed strike. More than half of the 68 Arab local council heads voted against it. But a meeting of Arab mayors and several council leaders led by the popular Communist mayor of Nazareth, Tewfik Zayad, organized the National Committee to Continue the Defense of Lands, which assumed leadership of the strike movement. After Israeli authorities attempted to persuade many of the traditional leaders affiliated with the government to oppose the movement, the issue became the source of politicization in the Arab community. Advocates of cooperation with the government opposed the strike, while those in the opposition, centered mostly in Rakah, the New Communist party of Israel, supported it.

Attempts by Israeli security authorities to impose a curfew on several villages in the Galilee region during Land Day, as the March 30 event was named, culminated in violence, with three Arab deaths, several injured soldiers and policemen, and scores of arrests. The militancy of the Land Day protests was probably strengthened by similar unrest that broke out simultaneously in several West Bank towns occupied by the Israeli army after the 1967 war. There, too, protests were organized against Jewish settlement of Arab lands and as manifestations of solidarity with the Israeli Arab Land Day demonstrations.

Israeli authorities blamed Rakah for the eruption of violence, although many observers noted that Rakah leaders had cooperated with the government in preventing clashes following Land Day. After Land Day, Rakah's prestige increased substantially and the party was regarded by a larger number of Arabs as spokesman for the minority.

Land Day and the subsequent political manifestations associated with it illustrated the growth of the minority's dissatisfaction with its relative economic deprivation and its relegation to a dependent status. Disagreements over land increasingly became the focal point of Arab dissatisfaction. Although the economic position of Israeli Arabs has improved greatly since 1948 in relation to conditions during the British mandate and in comparison with the status of Arabs elsewhere in the Middle East, their basis of comparison is neither conditions before the establishment of Israel nor the status of workers and peasants in other Arab countries. Their reference group is the Israeli Jewish population among whom they live, or workers in Western countries.

GROWING RADICALISM

From the time Israel was established, until 1965, the Israel Communist party (Maki) was the most vocal and politically influential spokesman on issues of concern to the minority. It did not hesitate to take up the land issue and other questions related to minority rights, such as military government, citizenship requirements, and the social and economic gap. It was the only party in the Knesset (parliament) that was non-Zionist and consistently took positions on both domestic and foreign policy questions that were likely to find favor with the minority.

In 1965, Maki split into two factions over several nationalist issues. Rakah, the New Communist List, broke away from the parent group to become the larger of the two. Most, but not all, of its members and leaders were Arab. After 1967 its positions became antagonistic to the Jewish consensus and much more in accord with prevailing Arab opinion. For example, it supported Soviet policy in the Middle East, especially demands for Israel's withdrawal from all territory occupied in the 1967 war.

The original Communist party, Maki, soon lost its support, fragmenting into several factions until it disappeared completely. Increasing Arab support for Rakah has been demonstrated by Knesset elections since the 1965 split. It received 23.6 percent of the Arab vote in 1965, 29.6 percent in 1969, 37 percent in 1973, and nearly 50 percent in 1977. It galvanized support beyond the small circle of party faithful—doctrinaire Communists or intellectuals—by broadening its activities in several groups established to deal with issues of vital concern to the minority. These groups included organizations such as the National Committee for Defense of Arab Lands; trade unions with large numbers of Arabs, such as the construction and agricultural workers; the organization of Arab mayors and local council leaders; and groups of women, students, and intellectuals.

In an attempt to broaden its constituency during the 1977 elections, Rakah expanded its Knesset slate to attract non-Arab and non-Communist support. Its Democratic Front for Peace and Equality included an Oriental Jewish leader of the Israel Black Panthers and the non-Communist chairman of an organization of Arab local council heads. Although the new list increased its Knesset representation from four to five of the 120 seats, some analysts had predicted that the Communists would win as many as six or seven places because of rising Arab unrest. However, some Arab dissatisfaction with government policies was indicated through abstention rather than an increase in votes for Rakah. Fewer Arabs participated than in several previous elections. Opposition to Rakah's relatively moderate positions was expressed in attacks on it by a new militantly nationalist group, *'Abna' al-Balad* (Sons of the Village).

This movement began in 1969, organized by an Arab law student at Tel Aviv University. Its original objective was to prevent government attempts to divide the Arab community through stirring up old feuds among families represented in the local and municipal councils. Established to deal with community problems, Sons of the Village participated in the 1976 Land Day demonstrations and in the Committee for Defense of the Land. As international issues and the Palestine question increasingly involved Israeli Arabs, the organization adopted a militant nationalist position close to that of the Rejectionist Front in the Palestine Liberation Organization (PLO). Sons of the Village asserts that the PLO is the only true and legitimate representative of the Palestinian people, and that all Palestinians, including the Arabs of Israel, share one identity. Because the Israeli Arabs are an integral part of the Palestinian people, according to this reasoning, any solution of the Palestinian problem must include official recognition and international guarantees for the national identity of Israeli Palestinian residents. These guarantees would include the right to remain in their homeland, the return of confiscated lands and property, and full implementation of cultural, social, civil, and political rights.

Sons of the Village is a spiritual descendant of *al-'Ard* (The Land), a group outlawed by the Israeli authorities because of its militant nationalist stance. It, too, considered Israeli Arabs an integral part of the Palestinian Arab nation. It called on Israel to recognize the Arab national movement, and to retreat to the 1947 partition borders. The Israel Supreme Court ruled that *al-'Ard* sought to negate Israel's territorial integrity and the Zionist mission, and it was made illegal in 1965. While acknowledging their spiritual debt to *al-'Ard*, Sons of the Village has criticized it for dissipating political energies on non-Palestinian causes, such as support for Egyptian President Gamal Abdel Nasser and Arab unity—an association made by most Palestinian Arabs until 1967. Since then, Palestinian nationalism has focused more on its own national redemption and on Palestinian identity. *Al-'Ard* was also criticized as a band of intellectuals who formed their organization from the top down, whereas Sons of the Village aspires to be identified as a "mass movement" working at the village level on day-to-day problems, as well as on larger national issues and issues of cultural identity.

The growth of nationalist sentiment among Israeli Arabs after 1973 was marked by the emergence of several groups identified with programs similar to those of the Sons of the Village. In the village of Taibah a similar group was formed. It was called *Nahada* (Resurrection or Rebirth). In Arara the name was *al-Bayadir* (Threshing Floor or Place of Harvest). Sons of the Village adherents, or those with a similar orientation, became council members on nationalist platforms, or formed united fronts with groups like Rakah.

PALESTINIAN IDENTITY

Manifestations of intensified political activity among Israeli Arabs have increased, with more open support for the PLO and its programs. At a conference in Nazareth during January 1979, the Committee for Defense of Arab Land and more than half the chairmen of Arab local councils approved a resolution supporting the PLO-led struggle against Israeli occupation being made by West Bank and Gaza Arabs. The meeting also called for a halt to "torture of Arab prisoners in Israel jails and collective punishment as a revenge." Other positions taken by the conference included opposition to appointment of a special minister to deal with Arab affairs, and to a proposal that would require Arabs to perform national service in lieu of military duty. Imposing such a duty on the shoulders "of our people is tantamount to depriving them of their rights," the conference stated (*Jerusalem Post*, January 2, 1979).

The growth of Palestinian Arab nationalism among Israel's approximately 1,800 Arab university students is another indication of the challenge to existing conditions. The national Arab student organization and Arab student organizations at each of the country's six universities have become divided between supporters of Rakah, considered "moderate" in the present political context, and those identifying with Sons of the Village or more militant groups.

When the fourteenth session of the PLO National Council convened in Damascus, Syria, during January 1979, a new group of Arab students with the name Progressive National Movement (PNM) emerged in Israel. Although few people could account for its origins or give an accurate estimate of its membership, it aroused a flurry of front-page press coverage and widespread editorial comment when six of its members allegedly sent a cable of support to the Damascus meeting. At a press conference in Jerusalem, spokesmen for the group said they supported a Palestinian and regional revolution that would replace Israel with a democratic secular state. Furthermore, the spokesmen refused either to condone or to condemn a recent terrorist attack in the Jewish town of Netanya. They were "aggrieved by bloodshed among both Jews and Arabs," but insisted that "violence begets violence."

Most of the Arab student committees at the six universities were divided between "moderates," who were members of Rakah or close to its positions, and students with more militant positions, close to those of the Sons of the Village or the more radical PNM. Most of these committees and their national organization supported the PLO as "the sole legitimate representative of the Palestinian people," but they also sought a Palestinian state in the West Bank and Gaza, rather than the democratic secular state advocated by the militants who wanted to replace Israel. (Heads of the Arab student committee at the Hebrew University estimated that not more than a quarter of the Arab students backed the PNM.)

Frequently the moderate wing of the Arab student movement cooperated with left-oriented Jewish student groups, as they did in response to expulsion orders for the PNM members. Such orders, issued under the 1945 Defense Emergency Regulations of the British mandate government, were perceived as a danger to Jewish dissidents as well. Rather than support the "rustication" orders, a Jewish student group declared: "It is the university's function to protect the students who are its wards, and to guarantee freedom of speech" (*Jerusalem Post*, January 29, 1979).

These incidents illustrate how the Arab minority was becoming Arabized and Palestinianized, a trend that gathered increasing momentum after 1967. They also signaled the emergence of the PLO as the most visible and outspoken manifestation of Palestinian Arab nationalism. Until 1967 no clear political trends had existed, nor had any minority parties or countrywide leaders emerged. The military government was the most decisive factor in the political life of the minority until 1966. From the establishment of the state until termination of the military government in 1966, the military administration was part of the Ministry of Defense, controlled by the Labor party. Labor's control of the military government facilitated extension of its political influence through the Arab community, as evidenced by the consistently large Arab vote for Labor in all elections until the late 1960s.

Termination of the military government preceded by one year the traumatic Arab defeat in the Six-Day War. The psychological effects of this war were nearly as great on Israel's Arabs as on those in the surrounding countries. There was general disillusionment with all existing Arab leadership, an awakening of self-analysis and introspection, and the start of a reevaluation of institutions, traditions, and customs in the Arab East. These trends were even more sharply articulated among Palestinians. There was particular disillusionment with Egyptian leadership and with the role played by that country's president, Gamal Abdel Nasser. Until the defeat Nasser was perhaps the most charismatic Arab leader. His photograph could be found in homes throughout the Arab world, including those of Israel's minority. Propagation of Arab unity won him many followers, especially among Palestinians. A major theme of Egyptian propaganda was that Palestinian national rights would be redeemed under President Nasser only as an element of Arab unity.

This line was espoused by the small group of militant Israeli Arab nationalists represented by *al-'Ard*. On the other hand, Arabs affiliated with the dominant Labor party and even those in the Communist party avoided too close identification with Nasser and the Egyptian line.

The Arab defeat in the Six-Day War gave birth to a new phase in the Palestinian national movement. Like the Arabs of Israel, the larger Palestinian community was leaderless and in disarray until 1967. The prevailing image of the Palestinian community living outside the country was of a homeless refugee mass. In the international community and in the rest of the Arab world, the term "Palestinian" was generally equated with "refugee." After 1967 the younger generation emphasized their Palestinian rather than their refugee identity.

The reborn national movement found its distinctive Palestinian heroes, poetry, literature, and political themes. Although it continued to associate itself with larger concepts like Arab unity, the primary concerns were redemption of Palestine and revival of Palestinian national and cultural identity.

Nearly all of the several dozen groups that emerged among Palestinians after 1967 emphasized the reconquest of Palestine by forceful destruction of the Jewish state. Zionism became the principal enemy, so that the clash with Israel acquired strong ideological overtones. The more leftist groups also sought to extend the revolution to the whole Arab world, with overthrow of all "reactionary" regimes in the region. Between the mainstream guerrilla groups, of which Yasser Arafat's Fatah was the largest, and the smaller, more Marxist organizations, such as George Habbash's Popular Front for the Liberation of Palestine (PFLP), there was a major ideological debate over the issue of whether Palestine could be redeemed without a total Arab revolution, or whether the Palestinians should concentrate their efforts on their own struggle before participating in the larger revolution.

These ideological and tactical differences were reflected in attitudes among Israeli Arabs. Intensification of Palestinian Arab guerrilla activity inside Israel and the increase of terrorism against the Jewish community began to undermine the relative stability that had prevailed since 1948. By 1969 the number of Israeli Arabs who had been arrested for collaborating with Arab terrorists increased several times, to over 100. Most startling was the growing number of Israeli Arab youths among the educated elite, including high school and university students, who were apprehended. The militant PFLP seemed to be among the more favored organizations.

A survey among a small group of Israeli Arab schoolchildren undertaken in 1968 by Hebrew University sociologists, indicated the shifting attitudes and the intensification of Arab nationalism. According to the survey, "A feeling of uncertainty and marginality is ... one of the 'Leitmotifs' in the identity of the Israeli

TABLE 6.1

Influence of 1967 War on Attitudes of
Arab Schoolchildren Toward Israel (percent)

	Rose	Remained same	Fell	N
Respect	43	17	40	299
Fear	52	34	13	282
Hatred	73	23	4	201

Source: Eisenstadt and Peres 1968, p. 27.

Arab, or in the phrasing of one of our respondents: 'I sometimes think that we are neither real Arabs nor real Israelis because in the Arab countries they call us *traitors* and in Israel—*spies*'" (Eisenstadt and Peres 1968). The influence of the 1967 war on attitudes of Arab schoolchildren toward Israel is shown in Table 6.1.

Trends toward closer identification with nationalist movements in the surrounding countries have gathered momentum since 1967–68. Until 1967 there were few contacts between Israeli Arabs and those outside the country. When opportunities for such contact arose outside the Middle East—for example, among students at universities in America or in Europe—the Israeli Arab was often regarded as a pariah, disloyal to his or her people, and blamed for not becoming a refugee like the other 1.5 million Palestinians. Few publications or periodicals from the Arab world reached the average member of the minority. The principal contact was via the airwaves, dominated by Radio Cairo. The virulent broadcasts from Arab capitals made no concession to the Israeli Arabs, with the result that radio contact tended either to alienate them from other Arab countries or to engender guilt feelings.

After 1967 and removal of barriers to visiting the West Bank and Gaza, contact became more frequent. Initially West Bank Arabs tended to regard those from Israel with suspicion or at least with doubt. But as contacts became more frequent, and Arab intellectuals on the West Bank began to comprehend the situation of their fellow nationals in Israel, suspicions and doubts began to dissipate. Furthermore, by 1977 there were some 80,000 Arabs from the occuped areas employed in Israel, many of them in construction, agriculture, and services, the occupations in which Israeli Arabs were to be found. Closer contact often led to greater mutual understanding. Eventually understanding led to feelings of identity and occasional political collaboration.

International events also strengthened ties between Israeli Arabs and Palestinians outside Israel. These included growing recognition of the PLO by third world countries and the United Nations, several U.N. resolutions recognizing the PLO as the legitimate representative of the Palestinian people, Arafat's appearance before the international organization, and the U.N. General Assembly resolution condemning Zionism as racist. The aura of legitimacy that formed around the Palestinian cause, and legitimization of the PLO, tended to remove obstacles to identification with the Palestinian movement by Israeli Arabs.

POLARIZATION OF ATTITUDES

Despite the small number of students associated with the PNM, its appearance seemed to polarize differences between majority and minority as much as any event since Land Day in 1976. The government, university officials, and several Jewish members of the Knesset rose to the occasion with statements against the radical student organization. Education Minister Zevulun Hammer threatened: "A citizen who does not recognize the state or one who supports a violent struggle against it, cannot enjoy its good services" (*Jerusalem Post*, February 8, 1974). In response to posters put out on the campus by the PNM supporting the PLO and equating Zionism with racism, the president of the Hebrew University in Jerusalem expressed shock, and threatened disciplinary action against any students who proved to be involved. Six members of the Arab student organization were subsequently confined to their villages, under provisions of the 1945 Mandatory Emergency Defense Regulations, for posting the offending notices in university dormitories; and the military command of the West Bank issued an expulsion order against a young militant in the occupied region who was suspected of inciting the Israeli minority students.

The cable and poster incidents, and the political ferment generated by the March 1976 Land Day, dramatized several cleavages within Israeli society related to the minority question. Most significant was the widening political rift between the Arab and Jewish communities. But each community was also divided in its reactions.

The consensus of the Jewish community was that the PLO was a terrorist group beyond the pale of civilized society. Acknowledgment by Fatah and other PLO-affiliated groups of responsibility for frequent terrorist attacks on Israeli civilians and civil institutions, the refusal of the PLO to amend its national covenant, which denied legitimacy of the Jewish state, and the general demeanor and rhetoric of the Palestinian guerrilla leaders discredited them in the eyes of most Jews. Institutions or individuals that accepted PLO claims to legitimacy were denounced as enemies of Israel. Failure to denounce the PLO and its affiliates was considered by most Jews as tantamount to sympathy for the "saboteurs," as they were identified in most of the Hebrew-language media.

When Arab student groups refused to publicly denounce statements by small militant student groups, such as the PNM, in support of the PLO, there was a public campaign to elicit Arab loyalty to Israel. Presidents of the country's six universities met with the minister of education, and a special ministerial committee was recommended to decide on steps against Arab students who publicly identified with the PLO, advocated violence against the state of Israel, or were otherwise disloyal to the country. The university executives could not agree among themselves whether they or the government should be responsible for taking action against Arab students whose statements or actions might be seditious or treasonable. Some felt that such action should be the sole responsibility of the police and government, on campus as elsewhere. The education minister expressed the view that Arab students who were loyal were fully entitled to all opportunities and privileges enjoyed by other students. He asserted, however, that such rights should not be extended to those who deny Israel's right to exist as a Jewish state, "since our universities are Zionist institutions . . ." (*Jerusalem Post*, February 8, 1979).

Reacting to pressure for loyalty statements, the Committee for the Defense of Arab Lands and the conference of heads of Arab local councils in Israel jointly declared their support for the PLO. Arab students at the Hebrew University published a request in their Jerusalem magazine, *el-Tahdi* (The Challenge), that Israeli Arabs in Galilee and the Triangle,[1] where 80 percent of the minority live, be given the right to join the autonomy plan for the West Bank and Gaza. The proposal was considered so radical that even Rakah rejected the suggestion that it join in support.

Arab students found little public sympathy for the right to support the PLO. The consensus called for expulsion of PLO supporters from Israeli universities, and the head of the Hebrew University Student Union asked the minister of education for legislation prohibiting enrollment of PLO supporters. At Tel Aviv University a group of Jewish students demanded expulsion of the Arab students who signed the alleged cable to Damascus.

Foreign Minister Moshe Dayan issued a general warning to the country's Arabs not to be "carried away by the mood of fanatical Islam" sweeping the region. Referring to student support for the PLO, Dayan urged the country's Arabs to "remember 1948," when they threw away the chance for peace and a separate Palestinian state, with the result that many ended up as refugees in Lebanon. "We . . . are not foreign rulers here," and if Israeli or West Bank and Gaza Arabs propose the replacement of Israel, ". . . they will have to pay very dearly for it" (*Jerusalem Post*, January 24, 1979).

Jewish settlers in Galilee expressed their anxiety by calling for increased government support for Jewish settlement in the region, and met to discuss the Arab resolutions expressing support for the PLO. The *Jerusalem Post* labeled the Arab conference resolutions "a provocation that cannot serve the interests of the Arab minority in Israel." Although it warned against the "marked rise in vociferous verbal radicalization of a pro-PLO type," it also "recognized that many other Israeli Arabs do not share these feelings, and to the contrary, try as best they can to juggle their complex identity as loyal Arab citizens of Israel while being horrified at the barbarities committed in the name of Palestinian nationalism." Israeli Jews, the paper argued, had taken many years to overcome their suspicions of the Arab minority as potential fifth columnists and subversives. "But they have done so." The recent declarations in support of the PLO "can only undo, in record time, the rudiments of mutual trust that have been built up so painstakingly over the years by men and women of goodwill in both communities." The silent majority in the Arab community was being silenced "by a vociferous, and at times a violent, minority of young hotheads" (*Jerusalem Post*, January 23, 1979).

More nationalist declarations of this type, continued the *Jerusalem Post*,

> ... and it will become nearly impossible to resist demands for the expulsion of PLO-lining Arab students from the universities. For how is it possible to explain to Jewish students, most of them war veterans and some war invalids, that they must continue to sit with fellow students sworn to the destruction of all they hold dear? More cries of 'We will liberate Arab Galilee with our blood,' to quote a leading Arab politician ... and well-intentioned Jews will cease to respond to legitimate demands for more schools, roads, houses and electricity for the Arab population. (*Jerusalem Post*, January 23, 1979)

The newspaper called, on the one hand, for the government "to finally begin paying attention to the problems of the Israeli Arab minority," as had been urged in vain by its own Arab affairs advisers, and, on the other hand, for "a return to sanity" by the upcoming generation of Israeli Arab leadership. "Providing leadership for a minority in the delicate situation in which Israeli Arabs have found themselves for three decades, requires more maturity and greater wisdom than that shown either by the student hotheads or by their elders in Nazareth."

In a later editorial the *Jerusalem Post* argued that those who advocated replacement of Israel with a "democratic secular state" were exploiting the institutions of freedom and civil liberty in Israel "in order to destroy it" (*Jerusalem Post*, January 31, 1979). When such subversive forces also acted to support an external foe engaged in active hostilities against the state, such as the PLO, then the state "has not only a right, but a responsibility to protect itself." Disagreeable as was the use of the Emergency Regulations, "it is too facile ... simply to condemn" their use. Because the PNM "has crossed the line of what can be accepted even in our free-wheeling democracy.... The laws of the state should therefore be wielded to protect freedom in Israel from itself."

Some Israeli intellectuals and left-oriented political figures regarded mainstream reactions in the Jewish community as retrogressive and a threat to the entire peace process. At a meeting of nearly 1,000 Arab and Jewish students in Jerusalem, several prominent Jewish professors and academicians charged that public reaction to Arab nationalist manifestations could lead to reintroduction of military rule over Israel's Arabs. Professor Yeshayahu Leibowitz believed that the government's counteractions would be an obstacle to normal relations between Israel and its neighbors. Normal relations "will remain impossible to achieve so long as one people rules as a conqueror over another," he asserted. Several Jewish academicians at the rally condemned the Hebrew University's silence regarding confinement of six Arab students, and demanded that the university "protect the freedom of expression of its students" (*Jerusalem Post*, February 8, 1979).

The Israel Council for Israel-Palestine Peace, an organization supported by several former members of the establishment who advocated recognition of Palestinian national rights and negotiations with the PLO under certain conditions, expressed "its grave concern over the campaign of fierce incitement being waged against Israel's Arab population." It accused journalists, retired army personnel, and public figures of banding together to create "an ugly, unprecedented atmosphere—an atmosphere laden with, among others, threats and plans for the expulsion of Arab citizens from the country." It charged the media with legitimization of "racial incitement" against the country's Arabs, and believed that the academic freedom of Arab students was being violated. The confinement orders and intervention of the army "constitute a flagrant violation of human freedoms, and are deserving of every condemnation." The Council appealed to "the enlightened sector of the population to condemn the wave of racism and incitement engendered against the Arabs of Israel, and to lend its hand in their integration into all walks of civilian life as equal and proud citizens of their country" (*Jerusalem Post*, February 8, 1979).

RECENT SURVEY DATA ON ATTITUDES OF AND TOWARD ISRAELI ARABS

Intensification of Palestinian national sentiment, with consequent polarization of differences between Israeli Arabs and Jews, was revealed in several opinion surveys made after 1967. In a study conducted by Mark A. Tessler among 348 Arabs of northern Israel during 1974–75, respondents were asked whether, in their judgment, the term "Israeli" described them very well, fairly well, a little, or not at all (Tessler 1977, pp. 317, 318). They also were asked to make the same judgment for the term "Palestinian" (see Table 6.2). Most respondents could not identify themselves as both Israeli and Palestinian, but accepted one designation or the other.

Tessler's survey found a strong correlation between the quality of a respondent's relations with Jews and the acceptance of Israel's legitimacy.

TABLE 6.2

Self-Identification as Israeli or Palestinian (percent)

The Term "Israeli" Describes Respondent				
Very Well	Fairly Well	A Little	Not at All	N
14	39	23	24	342
The Term "Palestinian" Describes Respondent				
63	22	10	5	342

Source: Tessler 1977, pp. 317, 318.

Those who had good relations with Jews were likely to accept Israel's legitimacy, whereas Israel's right to exist was questioned or rejected by those who had poor relations with Jews (see Table 6.3).

TABLE 6.3

Relations with Jews and Attitudes About Israel's Right to Exist (percent)

Respondent Considers His Relations with Jews					
Excellent	Good	Fair	Minimal	Bad	N
15	24	37	11	13	328

Respondent Believes Israel Has the Right to Exist			
Yes	Yes with Reservations	No	N
40	35	25	328

Source: Tessler 1977.

The growing impact of the Palestinian question and its influence on Israeli Arab consciousness was indicated by answers to a series of Tessler's questions suggesting that

> ... political identity is not associated in any important way with views about the desirability of various solutions to the Palestinian problem. Especially between those who identify as Palestinian and those who consider themselves Israeli, levels of support for making Israel a "secular" state and for establishing a Palestinian state alongside Israel do not differ. This suggests that Arabs in Israel, regardless of their subjective political identification, view the Palestinian problem in similar terms. Even Arabs who believe the problem does not involve them directly, those who consider themselves Israeli and not Palestinian, have the same views about justice for the Palestianian people as do other Arabs. (Tessler 1977, p. 326)

The influence of Palestinian political leaders on those identifying with the national movement was also demonstrated, whereas those who identified as Israelis rarely considered themselves to be represented by Arab leaders outside the country. "The clear implication ... is that identification as Palestinian among Arabs in Israel is not purely a cultural or abstractly national phenomenon. It has direct political connotations and implies ideological solidarity with a greater Palestinian Arab nation" (Tessler 1977, p. 327).

The intensification of Palestinian national identity among Israeli Arabs has run parallel with a deepening of alienation between the majority and minority of the country's population. Closely related to alienation of Israel's Arabs is the tendency of Jews to stereotype Arabs as inferior and dishonest, and to suspect that Arabs hate Jews. This is revealed in a survey of reciprocal attitudes undertaken by Sammy Smooha and John E. Hofman of the Haifa University sociology department in 1976. Their survey also indicated that only a minority of Arabs accepted Israel's legitimacy, felt at home in the country, or had hope for Arab youth in Israel. They concluded that "... most Arabs reject the State of Israel and would prefer a different political arrangement" (Smooha and Hofman 1976, p. 11).

The researchers attributed this deep alienation to persistence of the Arab-Israeli conflict, which they called "an overriding factor in the perceived Jewish distrust of Israeli Arabs," and to the "deep feeling of superiority among Jews as the institutionally dominant group in the Jewish state, as the modern and more socio-economically privileged sector, and probably as European in origin or at least in orientation." Despite the deep chasm separating the two communities, the authors observed that both Arab and Jewish readiness for closer relations far exceeded actual contact. "There is no doubt, a large untapped reservoir of good will underlying the gap between the readiness for and the scarcity of intergroup social relations" (Smooha and Hofman 1976, p. 11).

They also observed that the asymmetry between Jewish and Arab attitudes was diminishing. Arab eagerness for contact with Jews, although declining between 1971 and 1975, was still greater than that of Jews for contacts with Arabs.

TABLE 6.4

Reciprocal Attitudes of Jewish and Arab Israelis, 1967–76

	Percent Yes
Identity	
Jewish attitudes	
"Being an Israeli" important or very important (1974)	91
"Jewishness" important or very important (1974)	74
Arab attitudes	
"Israeli" describes me fairly well or very well (1974)	53
"Palestinian" describes me fairly well or very well (1974)	85
Alienation	
Jewish attitudes	
It would be better if there were fewer Arabs in Israel (1967)	91
Every Arab hates Jews (1967)	76
Arabs will not reach the level of progress of Jews (1967)	64
The police ought to keep an eye on Arab activities (1976)	92
Would accept an Arab as a minister in government (1976)	69
Would accept an Arab as a top official in government (1976)	76
Would accept an Arab as a top official in a mixed city (1976)	85
Believe Arab youths have future in Israel (1971)	67
Believe Arab youths have future in Israel (1975)	61
Arab attitudes	
Israel has the right to exist (1967)	31
Israel has the right to exist (1974)	40
Feel more at home in Israel (1967)	31
Feel more at home in Israel (1974)	25
Favor an Arab state in the entire territory of Palestine (1967)	19
Favor a secular state where Jews and Arabs will have equal rights (1974)	48
Believe Arab youths have future in Israel (1971)	54
Believe Arab youths have future in Israel (1975)	61

(continued)

TABLE 6.4 (continued)

	Percent Yes		Percent
Social distance	1971	1975	Difference
Jewish attitudes			
Have opportunity for contact	21	16	− 5
Consider contact essential	11	23	+12
Try to foster contact	19	25	+ 6
Do not avoid contact	77	77	0
Consider contact possible	78	77	− 1
Consider contact desirable	97	88	− 9
Arab attitudes			
Have opportunity for contact	74	34	−40
Consider contact essential	80	61	−19
Try to foster contact	88	60	−28
Do not avoid contact	94	83	−11
Consider contact possible	75	66	− 9
Consider contact desirable	86	73	−13

Source: Smooha and Hofman 1976, p. 12.

The decline in Arab readiness "signifies a decrease in their unilateral dependence on Jews as well as a rise in their self-esteem and self-assertiveness following the war of 1973" (Smooha and Hofman 1976, p. 13).

A third observation was that a significant discrepancy existed between personal and political attitudes of Arabs. "While a majority of Arabs consider relations with Jews as possible, desirable, or even fair to excellent, the majority have reservations about or reject Israel's right to exist." This "inconsistency" was more marked among urban, educated, and younger Arabs. "It appears that with advancing modernization Arabs come closer to Jews on the personal level but get further away on the national level" (Smooha and Hofman 1976, p. 13). Some of the data on which Smooha and Hofman base their conclusions are shown in Table 6.4.

Many of these observations were reinforced by still unpublished results of an extensive survey of Arab opinion undertaken by Smooha for the Ford Foundation in 1976 (Smooha n.d.). The growing identification of Israeli Arabs with the term "Palestinian" was clear. Nearly 60 percent believed that "Palestinian" described them best, and results of the questionnaire indicated that some 70 percent felt that there was a distinct contradiction between being identified as Israeli and as Palestinian. It was clear, however, that the Arab community was deeply divided about its identity and that many preferred to keep their Israeli identification. But even those who identified themselves as Israeli had ambivalent feelings about Israel's role as a Jewish and/or Zionist state. Nearly two-thirds of the respondents, including a substantial number who preferred to be classified as Israeli rather than Palestinian, perceived the Zionist movement as racist, focusing their dissatisfaction on repeal of the Law of Return, which gives automatic immigration rights to Jews and facilitates their acquisition of citizenship. Some 70 percent felt that Arabs could not be equal to Jews in Israel. Their perceptions were shared by 51 percent in a cross survey of a small Jewish sample.

Only a third of the Arab sample were willing to accommodate themselves to the status quo; the rest preferred some type of change leading to a democratic secular state or establishment of a Palestinian state alongside Israel.

It was quite evident, however, that Israeli Arabs perceived the term "democratic secular state" much differently than did the country's Jewish population. Whereas most Israeli Jews and their friends abroad have come to associate "democratic secular state" with liquidation of Israel, Arabs are more divided. A few, such as those affiliated with the PNM, agree with the PLO rejectionists that a democratic secular state would mean the end of Israel. Less militant pro-Palestinians, and even many Israeli Arabs who prefer to be identified as Israeli, associate the term with more equitable opportunities for the minority in the Jewish state. Opposing perceptions of this emotion-laden term have served to embitter relations between Jews and Arabs, and are a major source of misunderstanding. It has tended to ideologize differences, so that sympathizers with one view tend to become anathema to their ideological opponents. The differences became obvious in the January 1979 dispute over statements of Arab students at Israeli universities. The widespread student sympathy for the PLO and aspiration to achieve an equal status in Israel through "de-Judaizing" it were incomprehensible and totally unacceptable to the overwhelming Jewish majority. Israeli Arab aspirations to terminate or even to diminish the Jewish and/or Zionist character of Israel are perceived by most Jews as efforts to destroy the country.

Conflicting orientations in the Arab-Israel conflict are inevitable, given the differing backgrounds of Israel's Arab and Jewish citizens. Yet it is often reactions to issues in the conflict that the average Israeli uses as criteria for loyalty to the Jewish state and sympathy for Jewish aspirations. Among the most controversial questions has been the right of Arab refugees to return to Israel. In the Arab world repatriation of the Palestine refugees has been a sine qua non for a settlement with Israel. But refugees' return to their homes has for years been regarded by Israelis as tantamount to destruction of the state. Although the Israeli government was willing to negotiate the issue for the first few years after it was established, the subject has disappeared from its peace agenda.

The return of refugees is more than a political or security question for the average Israeli Arab. Most members of the minority have close relatives who became refugees in 1948 and now live in one of the surrounding "enemy" countries. Thus refugee return means reestablishment of severed family links.

In Smooha's 1976 survey the wide gap in perceptions of the issues in the conflict was all too evident. Some 85 percent of the Israeli Arabs polled favored the right of the refugees to return, in contrast with only 5 percent of the Jews. An even larger proportion of the Arabs, 87 percent, favored recognition of the Palestinians as a nation—in contrast with only 9 percent of the Jews.

Differing perceptions of the road to peace were also significant. Whereas 92 percent of the Jews desired a settlement based on the post-1967 borders, with minor modifications at most, only 13 percent of the Arabs considered these borders acceptable. A small minority of 8 percent of the Jews was willing to return to the pre-1967 frontiers, but no significant number would accept the borders of the 1947 U.N. partition plan. In contrast, 59 percent of the Israeli Arabs believed that the U.N. partition borders were acceptable, and 28 percent would accept the pre-1967 borders. The Arab response should be considered in light of the fact that more than three-quarters of them live in the Triangle and Galilee, which Israel acquired after the 1947 partition plan had been advanced.

The political implications of these results were revealed in questions related to leadership of the Israeli Arab community. The answers in Smooha's survey corresponded to election returns in which Rakah has steadily increased its strength among the minority. They also confirm the observation that Israeli Arabs identify more with political groups than with charismatic individuals. The loss of credibility among local leaders was underscored by the fact that they received support from only 14 percent of the sample. On the other hand, Rakah received about the same amount of Arab backing in the survey as it obtained in the 1977 election—49 percent. Other groups that obtained significant support were also prominent in the news. They included the conference of Arab local council heads and the National Committee for the Defense of Arab Lands. The conference was seen by 48 percent as a reputable spokesman for Israeli Arabs, and the Committee for Defense of Lands received more backing than any other group—80 percent—underscoring the central significance of the land question for Arabs. The growing nexus of political identity between Israeli Arabs and other Palestinians was evident by the 31 percent who supported Palestinian personalities in the West Bank and the 48 percent who regarded Palestinian personalities abroad as their spokesmen.

The increase of political activism was demonstrated in Smooha's survey in growing support for more militant ways of achieving political objectives. Nearly two-thirds advocated the use of more radical tactics, such as protest demonstrations, strikes, and boycotts. The overwhelming majority still wanted to keep within the law; only 19 percent supported extralegal tactics involving the use of force.

Israel's Arab minority is more exposed to the media than is the Arab population in most of the surrounding countries. According to Smooha's survey, nearly 50 percent read a newspaper and over 80 percent listen to the radio or watch television frequently. Exposure covers diverse sources, both Israeli and from the Arab world. A high percentage of Israeli Arabs know Hebrew and are familiar with both the Hebrew and the Arabic press, radio, and television. Among the literate population there is wide exposure to both government and opposition views through the Arabic-language press of each side. After 1967 the newspaper fare was broadened considerably as Israeli Arabs obtained free access to Arabic-language publications from East Jerusalem, which usually follow a line critical of the Israeli government and frequently favorable to the PLO. The Israeli Communist biweekly newspaper *al-Ittahal*, published by Rakah, is read by at least a third of those who habitually keep up with the press. Thus, it is not difficult for the themes of Arab nationalism, Palestinian revival, and the PLO ideology to reach Israeli Arabs. They are no longer isolated from the most recent political developments in the Arab world or from the mainstream of Palestinian sentiment, as was the case before 1967, when they were cut off from direct contact with other Arabs.

Simultaneous exposure to Western material culture through Israel, and to their Arab cultural roots in the Middle East, has stimulated a diverse range of aspirations among Israeli Arabs. As mentioned above, there are strong desires for upward mobility through improving material conditions, educational opportunity, and social conditions. Although national and cultural aspirations since 1967 have been strongly identified with movements in the surrounding countries, Smooha's survey indicated strong reorientation of values toward Westernization. There has been a sharp decline in support for such traditional Middle Eastern values as blood vengeance or the right of males to marry their cousins. Three-quarters would abolish blood vengeance, and two-thirds would do away with the marital rights of cousins. On the other hand, protective feelings toward women are still strong. Only 3 percent favored abolishing customs concerned with women's honor, as traditionally conceived, and 83 percent would leave these customs untouched. Smooha concluded that the potential for social change was much greater than the change that is actually taking place. Many traditional social structures in the villages are still intact, not because there is a lack of changing values but because there has been little opportunity to alter many aspects of the rural life-style.

SUMMARY AND CONCLUSIONS

This brings us full circle to the land question. Controversies between Israeli Jews and Arabs over land are both the substance of conflict and symptomatic of a wide range of related issues that have created deep political, economic, social, national, ideological, and psychological tensions between the two communities. Related to disputes over land are questions of economic opportunity, equal rights for all citizens in the country, and expression of national sentiment.

"The land" has far more than economic significance to both Jews and Arabs in Israel. It has great symbolic significance for both peoples. It is associated with national values and ancestral memories as much as with economic survival. Attempts to "rationalize" land use, or to devise plans to reorient the Arab economy toward nonagricultural occupations, have often become the source of tension. They are seldom appreciated by the minority, even by groups that cooperate with the authorities.

The government closely associates its security policies with those related to land and to Jewish settlement. Plans to "Judaize" the Galilee are not solely, or even primarily, based on economic considerations. They are interwoven with Zionist aspirations to establish a Jewish presence in the region, and are influenced by the large Arab population in an area that is of major security and strategic concern.

Implementation of these land policies has resulted in "depeasantization" of the Arab minority and transformation of the economic basis of its existence. The major sources of its economic life are now in the Jewish urban sector; thus the minority has become economically dependent on the Jewish community. Although this minority is economically integrated into the Jewish economy, and has experienced many improvements in its material conditions and life-style since the creation of Israel, Arab economic opportunity is still limited.

Arab labor in the Jewish urban sector is concentrated in certain occupations at the lower end of the wage scale. Upward mobility is limited for various reasons. Approximately a quarter of the economy is devoted to defense-related occupations, which exclude most non-Jews. In many occupations preference is given to absorption of Jewish immigrants in fulfillment of Israel's mission to facilitate Jewish settlement. Some sectors of the economy prefer to employ those with special skills or training to which the Arab minority has not yet had access. In an economy dominated by one ethnic group, there will inevitably be discrimination against "outsiders," especially those perceived by the majority as unreliable or unqualified.

The existing circumstances have created an Arab working class that perceives itself as relegated to a dependent position with limited opportunity. In some instances the answer is emigration. A constantly growing trend is radicalization of the Arab worker, indicated by the steady decline in votes for government-supported candidates and substantial increases for the opposition, Rakah.

Along with circumscribed opportunities in the urban sector, changing conditions in the rural residential environment contribute to social unrest. The Arab population has increased by approximately three times since 1948, but the village area where this population is concentrated has remained about the same. In some instances it has decreased because parts of some village lands were absorbed into the Jewish economy. The village life-style has become more compact. Living quarters, although often improved, are more crowded. Innovations, customs, and material assets from the Jewish sector brought to the village by Arab workers often create conflict with traditional village values, and cause deep social unrest.

Concomitant with internal developments has been the tremendous political and social ferment in the surrounding Arab world. Social unrest and political disaffection are growing at the same time as the minority's increased identification with the Palestinian national movement. Radicalization occurs on two fronts—inside Israel on economic and social questions, and outside on the national question.

These developments in the minority community are taking place at a time when the Jewish majority, too, is facing a number of crises. Israel is experiencing serious economic problems derived in part from its ties with the West: low economic productivity, the gap in balance of payments, and spiraling inflation threaten to disrupt the relative economic stability that has given Israel a high standard of living. Under these changing conditions little attention is given to the economic status or the peculiar problems of the Arab minority. Economic instability makes realization of the goals of Zionist ideology more difficult; it is related to such questions as immigration, the balance between agriculture and industry, growing water shortages, and extension of Jewish settlement within the pre-1967 borders. Lack of clear-cut policies and the inability to cope with these problems overshadow the dilemmas of integrating the 15 percent of Israel's population who are not Jews.

The problems of coping with an increasingly radicalized Palestinian minority inside the country have grown with the rise and fall of expectations for peace, and with the emergence of ambiguities regarding autonomy for Palestinians in the occupied areas. Development of a coherent policy acceptable to the minority cannot be postponed indefinitely. An effective minority policy also is closely related to Israel's relations with Western countries, especially with the United States, where it has become an important consideration in evaluating many other policies of the Jewish state. Many countries, rightly or wrongly, perceive Israel through the lenses of the Arab minority. Thus the question is closely related to the process of seeking peace with the Arab world. Even more important becomes the question of whether the minority itself will have the patience to wait for development of a policy that will harmonize its needs and aspirations with those of the Jewish state.

NOTE

1. The "Triangle" is an area in east-central Israel, largely Arab populated, that was acquired from Jordan in the 1949 armistice agreement.

7

THE CASE OF THE MOROS IN THE PHILIPPINES

Linda S. Lichter

The Philippine nation has been called "a remarkably homogeneous society with a strong sense of national unity" (Vreeland 1976, p. 75). A major exception to this unity has been the ongoing confrontation between the Muslim minority and the Christian inhabitants in the southern part of the country. This centuries-old conflict has led to the loss of thousands of lives since the 1960s, an economic drain, and international criticism. At present the conflict shows little evidence of abating. Several attempts have been made to reach an agreement between the Manila government and the Muslim elements, but misunderstandings and bloodshed have continued.

Information about the conflict is difficult to obtain because of censorship and travel restrictions imposed under the martial law that has existed in the Philippines since 1972. Visitors to the United States from the Philippines who were interviewed in 1979, said that much of what they were able to learn about the situation in the southern islands came from word of mouth, and that accounts carried in the Philippine press were sparse.

Muslim Filipinos (commonly called Moros) make up about 5 percent of the total Philippine population. An official publication estimated their numbers as approximately 1.5 million out of a population of 44 million (*The Philippines*, 1977). They are concentrated in the southern islands of Mindanao, Palawan, and the Sulu Archipelago. For over 300 years the Moros have considered themselves a people apart from their Christian countrymen, striving to preserve their cultural autonomy in the face of attempted domination by external authorities—first the Spanish, then an administration controlled by the United States, and now the government of President Ferdinand E. Marcos.

By 1521, when the Spanish colonial period began in the Philippines, Islam had already established itself firmly in parts of the southern islands. As a result of the lack of organized resistance, Spanish authority was rapidly extended over most of the country, but the Spanish did not initially make a serious effort to subjugate the Muslims in the remote South. Thus, the Muslims in that area enjoyed de facto independence. With the strengthening of their navy in the mid-19th century, the Spanish were able to conquer some Muslim regions, but Spanish culture and religion never permeated the entire area (Lightfoot 1973, p. 103). "What cultural diversity and conflict exists in the modern Philippine nation can be attributed in large part to the failure of the Spanish to bring the entire population under control" (Vreeland 1976, p. 80).

The American colonial administration, which existed from 1898 to 1946, succeeded in bringing the Muslim regions under the central government as a result of the "Moro wars" (Corpuz 1965, p. 26). But the smoldering resentment of the population against foreign interference in their way of life was never extinguished.

During the period following World War II, when government resettlement programs designed to ease crowding in densely populated areas were intensified, large numbers of Christians migrated to Mindanao. This triggered conflicts over the ownership of land that had long been under Muslim control. While Moros claimed ownership on the basis of original settlement, the Christians invoked a law that provided for public settlement of all lands in Moro territory that were not occupied by specific individuals. As an official Philippine publication has put it:

> The modern concept of the purchase and sale of land to be registered under permanent and exclusive title guaranteed by the government was not easily understood or accepted by a people who still lived under what was more or less feudal rule, dating back to the times when all land was held in common under the sultans. (Guerrero 1975, pp. 43-44)

In the late 1960s violent confrontations resulted from land disputes in Mindanao. By that time Christians were in control of a large proportion of the agriculture and industry in the area. During 1970 and 1971 over 1,000 lives were lost as Muslims clashed with local security forces.

Rather than concentrating on specific economic grievances, the Moro resistance increasingly shaped itself as a separatist movement. The Mindanao Independence Movement (MIM) was formed in the early 1970s, and in 1972 the Moro National Liberation Front (MNLF), which encompassed more diverse elements of the population, emerged as the dominant revolutionary force in the Muslim fight against the Christian government.

The declaration of martial law on September 23, 1972 (which Marcos claimed was at least partially caused by the Moro conflict),[1] strengthened government control over some of the area, but it only intensified Moro resistance. Since martial law required the rebels to surrender their weapons, compliance with the law would have crippled their quest for autonomy. Thus, martial law actually served to encourage the mobilization of Muslims to fight. Intense conflict broke out in western Mindanao in October 1972, and spread to four other islands. By 1975 warfare in the South had caused the death or displacement of several thousand persons.

In 1973 and 1974 the Marcos government took several steps toward a political settlement. The president publicly recognized the existence of official corruption and the "grasping attitude" of some Christian landowners. The use of Muslim dialects in primary schools was authorized, Muslim customs and traditions were studied for incorporation into Philippine law, and 4,000 scholarships on the college level were offered to qualified Muslim youth (Marcos 1973).

Overtures were also made toward giving the Moros a limited measure of autonomy; but since the government was at the same time arming Christians in the southern islands, these overtures were regarded with suspicion by the Muslims. Manila claimed success for its "policy of attraction," under which insurgents were offered monetary loans or commissions in the Philippine armed forces if they "rejoined the government," but the actual success of this policy is not known (Vreeland 1976, p. 380). In any event, the fighting continued.

This unrelenting conflict brought negative economic consequences for the Moro area and the nation as a whole. Costs of troop maintenance drained the national treasury, and the potentially lucrative fishing, mineral, and agricultural resources in the southern islands remained underdeveloped.

MUSLIM-CHRISTIAN DIFFERENCES

Disputes over land sparked the Muslim-Christian conflict, but behind them are a long history of independence in the southern islands and the unique social and political structures of Islam. Even before the Spanish arrived in the Philippines, the Muslim lands in the South were divided into sultanates, each ruled by a sultan who was the overseer of all secular and religious activities. The economy of the sultanates revolved around trade with other countries in the area. Extension of the Manila government's authority destroyed the power of the sultans, but the *datus* (local leaders whose prestige and wealth enabled them to attract followers) have maintained some of their power. While the *datus* do not hold political offices, Philippine government agencies frequently operate through these local leaders (Vreeland 1976).

Some traditional Muslim practices, especially polygamy and slavery, alienated Christian Filipinos, even though these have been much less in evidence during recent times. In general the state did not interfere with polygamy, which could be afforded only by wealthy men. But slavery raised more problems, since one of the Moros' means of acquiring slaves was to raid Christian settlements. Slavery was outlawed in 1968, and raids to obtain slaves no longer occur; but it is still recognized that debtors may work off their debts without pay. Another divisive factor is the Muslim concept of *maratabat* (translated as "face"), which requires extreme measures to maintain the honor of oneself and one's family. Violation of *maratabat* demands that one kill the violator oneself, rather than seeking justice through the courts.

The evangelizing character of both Islam and Christianity has made the coexistence of the two religions extremely difficult. While the Muslims have not actively attempted to convert the Christians, they have insisted on noninterference with their religion. Even when confronting the immense military power of the United States during its colonial administration, they refused to make peace until they were assured that their religious practices would be respected.

In spite of their devotion to traditional practices, the religious consciousness of the Moros had not always maintained the intensity of its present level. Before World War II there were few mosques in the southern islands, the people were generally unfamiliar with the Koran, and few practiced the prescribed prayer rituals or made the required pilgrimage to Mecca. Their religion was significant mainly as a rallying point around which resistance to foreign interference could be mobilized. After World War II, however, there was a major religious revival, at which time numerous mosques and schools were built. Islamic societies were formed and pilgrimages to Mecca became more frequent.

Differing educational systems also have divided the Muslim and Christian Filipinos. Moro students have accused the Philippine schools of using history books that present a negative image of Muslims, and many Moros will not permit their children to attend the state-run schools.

In an effort to narrow these differences, the nationally mandated curriculum in Philippine schools was revised in the early 1970s to provide for the inclusion of information about the customs and traditions of the nation's cultural minorities, notably the Muslims, in textbooks and other instructional materials. At the same time the administration of Mindanao State University became largely staffed with Muslim academicians, and Muslim culture was given prominence in the Museum of Philippine Traditional Cultures and in other museums.

Whether such measures have had the effect of moderating Muslim suspicions cannot be ascertained on the basis of the available information. It seems doubtful. The Muslim struggle to preserve their way of life in a nation dominated by what appears to them to be a hostile Christian majority is, after all, centuries old.

NEGOTIATIONS FAIL TO PRODUCE A SETTLEMENT

The first formal negotiations between the MNLF and the Philippine government took place in January 1975. A deadlock resulted, with the government refusing to consider the MNLF demand for autonomy in the southern region.

But Marcos could not afford to abandon efforts for a peaceful solution. A military solution did not seem to be possible. Even though 70 percent of the Philippine armed forces were fighting in the South, the Moros showed "a bold persistence that has continued to challenge our forces," as Defense Secretary Juan Ponce Enrile put it (Tasker, December 31, 1976, p. 8). Further, the conflict had attracted international attention, and the 42-nation Islamic Conference had twice called for dialogues between the Moros and the Marcos government. Material aid was coming to the rebels from Libya, where MNLF troops were being trained. The threat of being denied oil from the Islamic countries made it necessary at least to appear to be seeking an accommodation.

In the fall of 1976, Imelda Marcos, the president's wife and roving ambassador, accepted an invitation from Libyan leader Muammar el-Qaddafi to visit his country and discuss a solution to the conflict. The meeting produced little in the way of immediate results, since the Philippine government would not agree to the Libyan request that the MNLF be granted virtual government status in the negotiations. But Mrs. Marcos' trip and a visit to Manila by representatives of the Islamic Conference were enough to lay the groundwork for a second round of talks between the Moros and the central government.

These negotiations began in December 1976, in Libya, under Qaddafi's personal supervision. The MNLF was represented by Nur Misuari, formerly a professor of political science at the University of the Philippines. The Manila government appeared to be demonstrating more flexibility than it had in 1975, since it agreed to negotiate with Misuari, who was still strong in his insistence on autonomy for the South (Tasker, December 31, 1976, p. 9).

But it was doubtful that the government had actually changed its stand. Prior to the conference Marcos announced that a separate Muslim state would not be permitted. He also rejected the possibility of a federal solution, thereby seeming to reject autonomy as well. At the same time an anti-MNLF campaign was launched in the Philippine press. Newspaper articles claimed that many Muslims opposed MNLF demands and that the group did not represent the Muslim Filipino people. The government also released what it claimed were the demands made by the MNLF at the 1975 negotiations. These included Muslim control of the government in the South, which would have jurisdiction over the economy, security forces, the judiciary, and schools up to the secondary level; MNLF participation in the national government; and the establishment of Islamic life and society in the southern region.

There was some reason to doubt that a formal peace agreement would be accepted by all elements of the Muslim population, assuming that it could be concluded. It could be expected that some rebel leaders would be reluctant to relinquish control over lands that they had acquired in the course of the conflict. There were also groups of Muslim guerrillas battling the Marcos forces, and these groups were concerned only with local matters. Further, the question remained as to how the Christians in the South, many of them deeply hostile to the Muslims, would react to the prospect of Muslim autonomy.

Nevertheless, with the strong hand of Qaddafi guiding the MNLF, the December 1976 talks did in fact produce an agreement for a cease-fire. The Philippine government promised to grant autonomy (not defined in detail) to 13 provinces in the South, and amnesty to political prisoners. Further talks to work out details were scheduled for February 5, 1977. After the final settlement Marcos was to appoint a provisional government for the South, which would make preparations for the election of a legislative assembly.

Any hopes that this agreement would put an end to the conflict were crushed before the February meeting convened. In January 1977 a new group, calling itself the Moro Reform Liberation Front, appeared and sent its representatives to Manila to demand that the Philippine government negotiate with them. Although the origins of this group are cloudy, it is believed that at least some of its members are MNLF defectors. Some support is given this hypothesis by the fact that the government-controlled press reported claims by the new group that it had 26,000 members fighting in the South—the exact number of MNLF rebels that the Manila government had previously reported as surrendering to government forces (Tasker, January 28, 1977).

Much to the dismay of the Islamic Conference, which had declared the MNLF to be the only legitimate representative of Muslims in the Philippines, Marcos agreed to meet with the Moro Reform Liberation Front. Further friction between Marcos and the MNLF came as a result of Marcos' proposal that he draw up a code of "personal laws" to regulate legal and personal relations in Muslim areas. The MNLF insisted that such a code could be drafted only by Muslims after the southern region had gained autonomy.

The most significant threat to the cease-fire came when Marcos announced that autonomy for the 13 provinces in the South, negotiated at the December meeting, would not be granted by presidential decree but would be contingent on the outcome of a plebiscite to be held in the southern region. This put the question of autonomy into serious jeopardy, since only five of the provinces in question have a Muslim majority.[2] Marcos claimed that the vote was necessary because the Philippine constitution required that any boundary changes must be approved by a majority in the affected areas.

This announcement infuriated the MNLF. Misuari, the group's leader, was reported as saying that the move would rekindle fighting in the South. But

Marcos stood firm. He said he would agree to postpone the plebiscite until final negotiations with the MNLF were concluded, but not to abandon it.

A further complicating factor was that the two sides interpreted the concept of autonomy differently. Marcos did not want to lose all control over the southern provinces, some of which include rich agricultural regions. Also, oil exploration had recently started off the island of Palawan, one of the disputed areas. The MNLF definition of autonomy would have denied the Manila government dominant economic authority. Indeed, the MNLF favored a virtually independent state, to be known as the Bangsa Moro Islamic Region, with its own flag, seal, language, and capital city.

In this tense situation both sides geared up their armed forces in the South, and the future of the cease-fire became increasingly bleak. Marcos then made a bold diplomatic move. Sidestepping the MNLF, he instituted direct negotiations with Qaddafi, Mrs. Marcos representing the Manila government. The resulting agreement provided for immediate autonomy for the 13 provinces, with the establishment of a provisional government that would include the MNLF. After this a referendum to settle "administrative details" would be held in the region (Tasker, April 8, 1977).

A STATE OF NEITHER WAR NOR PEACE

The referendum that finally was submitted to the residents of the southern islands asked them to approve or reject autonomy. This, in effect, rendered the prior promise of autonomy meaningless. It is not known how many Muslims turned out to vote, since the MNLF urged that the plebiscite be boycotted, but the majority Christian population voted to remain under the central Manila government. In any event, it was reported that autonomy had been rejected by approximately 98 percent of the voters (New York *Times*, April 20, 1977).

Not surprisingly, the referendum settled nothing. Qaddafi objected that it was a misinterpretation of the agreement he and Marcos had reached, claiming that the principle of full autonomy had already been accepted and that the vote could settle only minor administrative arrangements. Marcos, on the other hand, interpreted the vote as meaning that the population in the South had rejected a break with the central government. He was, however, willing to see representation by the MNLF in a provisional government for the South, which in turn would remain under the control of the central government, and invited MNLF leader Misuari to join this provisional government. Misuari did not reply to the invitation.

On May 1, 1977, both the MNLF and the Philippine government officially announced that talks had broken down. The former claimed that the government would not grant meaningful autonomy, while the latter said that the MNLF demands exceeded autonomy and amounted to secession. The Islamic Conference, meeting shortly thereafter, rebuked Marcos for his "irresponsible" actions, but did not impose economic sanctions on the Philippines, as the MNLF had hoped. Instead, it called for continued negotiations and a peaceful solution to the conflict, and also requested that all Islamic countries assist the Philippine insurgents in all ways possible.

Marcos made attempts to mollify the Moros by publicizing plans for the development of neglected Muslim areas. He also attempted to undercut the MNLF in a more subtle manner by announcing that he was abandoning a long-standing Philippine claim to Sabah, an island administered by Malaysia. This move appeared to be an effort to persuade the Malaysians not to allow the continued use of Sabah as a training ground for MNLF forces.

Although the cease-fire was not officially abandoned by either side, violence in the southern region escalated substantially. Thousands of lives were lost, and many more were left homeless.

A Muslim-Christian conference in the summer of 1978 focused on alleged atrocities committed by government soldiers in the South, including retaliation against Muslim women, children, and old persons, the disappearance of Muslims arrested by Philippine soldiers, and the destruction of property. The conference report stated: "It is our observation that the most cruel atrocities . . . are inflicted by misguided military who do not even consider Muslims as Filipinos." It called for educational efforts to change the negative image of Muslims held by their Christian countrymen. The conclusions of the conference are underscored by a survey that found that only 13.5 percent of those sampled in the Philippines considered the Moros to be Filipinos (Tasker, August 18, 1978).

There are some indications that talks between the MNLF and the government may be resumed. But given the complexity of the issues involved, and the firm positions of both sides, it is unlikely that a lasting peace will be soon in coming. Regional elections were held in southern Mindanao in the spring of 1979, but at the same time that President Marcos installed the newly elected Muslim officials in office, he appealed to Muslims who were still resisting to join in the effort toward national unity.

NOTES

1. Marcos later wrote that declaration of martial law had been compelled by "seven grave threats to the existence of the Republic. These were the communist rebellion, the rightist conspiracy, the Muslim secessionist movement, the rampant corruption on all levels of society, the criminal and criminal-political syndicates . . . , the deteriorating economy and the increasing social injustice" (Marcos 1973, p. 127).

2. According to an official Philippine source, only two provinces have a Muslim majority (Guerrero 1975, p. 42).

8

SELF-DETERMINATION IN QUEBEC: LOYALTIES, INCENTIVES, AND CONSTITUTIONAL OPTIONS AMONG FRENCH-SPEAKING QUEBECERS

Maurice Pinard

In any nation there develops a certain degree of national loyalty in the population. Its strength of course varies from nation to nation, and an important source of this variation is the degree to which people feel that their fate is dependent upon, and well taken care of by, the nation and its institutions. But in ethnically segmented societies, precisely because of this segmentation, people tend to develop a system of dual loyalties. Apart from their national loyalty, they develop loyalty to their own ethnic group and its segmental institutions, on which so much of their fate depends. This is particularly the case with an ethnic group that occupies a subordinate position or is in a minority.[1] Moreover, while both forms of loyalty may be of equal strength and/or compatible with one another in a system of dual loyalties, it often happens that ethnic loyalty predominates over loyalty to the nation as a whole and that the two are incompatible.[2]

Such loyalties affect people's positions vis-à-vis the nation's political institutions and, in particular, their preferences for various constitutional options. These preferences, obviously, will tend to differ from those of citizens living in

---

Major portions of this paper were presented at the Canadian Political Science Association and the Israel Political Science Association twinned workshop on "Cultural Cleavages and Their Impact on the Socio-Political System," Sde Boker, Israel, December 12-15, 1978. Some sections draw heavily on a paper originally published in French (Pinard 1975). The support of a Killam Senior Research Scholarship while engaged in this work is gratefully acknowledged, as are the comments of Richard Hamilton on an earlier draft.

nonsegmented societies. To be sure, factors other than loyalties affect people's choice of certain constitutional options. Such factors include the system of segmental institutions, ethnic grievances and aspirations, interethnic competition, patterns of accommodation, and the tempo and patterns of social change, as well as the ratio of positive and negative incentives implied by these options.[3]

We shall examine the loyalties and systems of incentives present in the Canadian situation. While strong ethnic loyalty might lead a person to support any of a number of political options, we find great variations in the support for various options. One of the reasons for this is that the costs and rewards to be derived from the support of each option are quite different.

More specifically, we shall consider the system of dual loyalties prevailing among Francophones in Quebec, how it differs from that of other Canadian groups, and the trends in this regard. We shall also examine the degree of support for three different constitutional options—constitutional revision and greater autonomy for Quebec, complete independence, and sovereignty-association—and again assess the trends in the support for each of them. Finally, we shall assess the impact of loyalty on each of these options, and the role of incentives in accounting for differences in the level of support for each.

But first, let us examine some basic demographic dimensions.

DEMOGRAPHIC DIMENSIONS

Contrary to what one might think, Canada is not a bi-ethnic, but a multi-ethnic society. To the two so-called Charter groups, the French and English Canadians, many waves of immigration, and in particular some important ones after World War II, have added many other ethnic groups which together formed in 1971 no less than 27 percent of a total population of some 21 million people. This is not a minuscule aggregate, when compared to the 29 percent of French ethnic origin and the 45 percent of British ethnic origin. Over the last 100 years, this third group has steadily increased (from 8 to 27 percent), but largely at the expense of the British, whose proportion decreased from 61 to 45 percent. The French, on the other hand, succeeded in maintaining their position (declining only from 31 to 29 percent). This success for the most part has been due to their exceptionally high birth rate, one of the highest ever recorded, which declined only recently (Arès 1975; Lieberson 1970).

But if by ethnic origin Canada has become a multi-ethnic society, it has largely remained a country with two languages only; immigrants have tended to assimilate to the language of the dominant group. Indeed, if instead of peoples' ethnic origin, we consider their mother tongue, the proportion "English" increases from 45 to 60 percent; and in terms of the "language most often spoken at home," it increases further to 67 percent. Correspondingly, by these two criteria, the proportion "French" does not make any net gains. On the contrary, it suffers slight net losses from 29 to 27, and 29 to 26 percent, respectively, with the "Others" experiencing important net losses, from 27 to 13 and 27 to 7 percent, respectively.

These linguistic groups, moreover, are dispersed very unevenly over Canada's ten provinces. Nine of them are predominantly English, with the French highly concentrated in only one province (but the second-largest one), Quebec, which had a population of almost 6 million in 1971. Indeed, 77 percent of all Canadians of French origin reside in Quebec.

Conversely, Quebec is also predominantly French, with about 80 percent of its population belonging to this group (79 percent according to ethnic origin, and 81 percent according to mother tongue or language used at home). That proportion has been relatively stable (78 percent of Quebec was French, for instance, in 1871). But while the rest of the Quebec population was predominantly of British origin in 1871 (20 percent), that same proportion has decreased to 11 percent in 1971. The rest of the population is now made up of other ethnic groups (10 percent). But again, most of these "Others" tend to assimilate to the English, as 81 percent of the Quebec population speak mainly French at home and 15 percent, English.

Overall, the linguistic polarization within Canada is extreme: in terms of language spoken at home Quebec is 81 percent French and 15 percent English; but the rest of Canada is 87 percent English and only 4 percent French. It should be noted, however, that there are strong concentrations of French in areas immediately surrounding Quebec, as in northern New Brunswick and eastern and northern Ontario (Joy 1972, Chap. 3).

Furthermore, the situation in Quebec has been seen by many as possibly changing in the future. The sharp drop in the birth rate of French Canadians over the last 20 years or so means that they can no longer compensate by a higher fertility for the strong tendency of immigrants to join the English-speaking community. This could mean a decline in the proportion of the French-speaking population of Quebec, particularly in its large Montreal metropolitan area, where the immigrants and the English-speaking community are concentrated (Henripin 1972).

THE DUAL LOYALTIES OF FRENCH CANADIANS IN QUEBEC

In some ways the nationalism of French Canadians can be seen as the expression of a strong loyalty to their ethnic group and its institutions, and hence as the desire to defend and promote the latter's interests. Given that about 80 percent of all French Canadians are concentrated in Quebec, where they constitute the overwhelming majority (about 80 percent of the Quebec population)—given, in other words, a high degree of ethnic territorial segregation—it should not be too surprising to find that loyalty to their ethnic group easily takes the form of loyalty to Quebec and its institutions, and, in particular, to the Quebec government as the supreme corporate body of their own group.

Though there are disagreements among historians as to the beginnings of French-Canadian nationalism as an articulated elite ideology (Ouellet 1964), there is no doubt that a system of dual loyalties has existed in Quebec for a very long time. We will consider how widespread it has been in recent years, whether it is different from the pattern of loyalties in other groups or regions of the country, and how it has changed since the 1950s and 1960s.

That there is very strong attachment to Canada among many Quebecers, particularly among many French-speaking Quebecers, is certain. In a 1970 survey of Quebec, we asked respondents: "What degree of attachment do you feel towards Canada: do you feel a very strong attachment, a fairly strong attachment, a rather weak attachment, or a very weak attachment?" About 80 percent of the French Canadians answered that they felt either a very strong (40 percent) or a fairly strong (39 percent) attachment to Canada. Only 9 percent reported a weak or no attachment, with 12 percent reporting a moderate attachment. As one could expect, English Canadians in Quebec were even more strongly attached to Canada: 75 percent reported a very strong attachment, with almost all others (20 percent) reporting a fairly strong attachment. Only 4 percent fell in the other categories.

But if French Canadians in Quebec are strongly attached to Canada, this is counterbalanced by what appears to be an even stronger loyalty to Quebec. A direct measure of attachment to Quebec, similar to that of attachment to Canada, is not available; but this is not too serious a problem, given that there are many measures of relative attachment to the two, and they all clearly indicate an attachment to Quebec that is somewhat or much stronger than attachment to Canada.

In 1970 we asked respondents: "In general do the provincial and federal governments appear to you as being both equally important, or does one seem to you more important than the other? (If one is more important) Which one?" In general, respondents from all groups saw both governments as equally important. Among French Canadians the vast majority (71 percent) held that opinion (see row 1, Table 8.1). This most directly reveals the presence of a system of dual loyalties, which are not incompatible. But those who put the provincial government first were more numerous than those who put the federal government first (15 percent versus 9 percent). For English Canadians in Quebec, as well as for Quebecers of other ethnic groups, this last ratio was clearly inverted (6 percent versus 26 percent, and 4 percent versus 21 percent, respectively) and slightly fewer (62 percent in each group) found both governments to be equally important.

With a question that did not explicitly offer the alternative "equally important," a poll taken in 1969 among French-speaking voters in Quebec

TABLE 8.1

Loyalty to the Provincial or Federal Government Among French-Speaking Quebecers, 1965-79 (percent)

	Provincial	Federal	Both Equally	Other Answers	N
1. Most important government	15	9	71	6	(4,874)
2. Most important government	47	41	–	12	(428)
3. Politics that interest you more	27	7	30	36	(5,524)
4. Government taking best care of your interests	36	9	16	39	(5,524)
5. Government looking after interests and needs best	38	23	25	14	(806)
6. Government coming first to mind when hearing "about your government"	58	24	16	3	(600)
7. Government coming first to mind when hearing "about your government"	53	27	17	3	(806)
8. Government supported in case of conflict	40	15	–	46	(945)
9. Government supported in case of serious conflict	44	27	9	21	(600)

Note: In this and subsequent tables, some rows do not add to 100 percent because of rounding. The N's in rows 1, 3, and 4 are weighted. Figures in row 8 are approximate.

Sources: Row 1, Pinard 1970; row 2, Jenson and Regenstreif 1970; rows 3 and 4, SRG 1965; row 5, MAIQ 1979; row 6, SORECOM 1977; row 7, MAIQ 1979; row 8, Lemieux et al. 1970; row 9, SORECOM 1977. (Pinard 1970, SRG 1965, and SORECOM 1977 are unpublished surveys).

yielded results that differed in some respects. To the question "Lequel de ces deux gouvernements, le gouvernement provincial ou le gouvernement fédéral, vous semble le plus important?" 47 percent answered "provincial government" and 41 percent, "federal government," with 12 percent undecided (see row 2, Table 8.1). Again a slightly higher proportion of French Canadians chose the provincial government. It seems, however, that most respondents' perceptions were not highly polarized, nor too intense, given, on the one hand, their willingness to say "equally important" when explicitly given a chance and, on the other hand, their willingness to side with one government when asked to make a choice. At any rate, these data do not indicate that either government would be strongly preferred over the other by French Canadians in Quebec.

This has been confirmed in a more recent study commissioned by the Quebec government. The following question was put to a Quebec sample: "Some people think one can *easily* be a Canadian and a Quebecer at the same time. Others think that this is difficult. Do you think it is *easy or difficult* to be a Canadian and a Quebecer at the same time?" Among French-speaking Quebecers, 66 percent answered that it was easy, and only 30 percent said it was difficult, the rest being undecided. This clearly suggests that for the majority their dual loyalties are not incompatible (MAIQ 1979).

So far the system of dual loyalties to Canada and Quebec has been measured in terms of attachment to or identification with, and perception of importance of, these two levels of government. While attachment and identification are good measures of loyalty, perception of importance is less appropriate because it is more likely to reflect the assessment of an objective situation, based on the range of powers of each level of government. Only secondarily would it measure one's sentiments toward each government. Fortunately, other data are available.

In the 1965 study of interethnic relations carried out by the Social Research Group on behalf of the Royal Commission on Bilingualism and Biculturalism, members of a representative sample of all Canadians were queried about their relative interest in federal and provincial politics. They were asked: "Which interests you more, federal politics or provincial politics?" Among respondents from Quebec who identified as French Canadians, many more manifested a greater interest in provincial politics (27 percent) than in federal politics (7 percent), though a still larger proportion (30 percent) answered they were equally interested in both. (See the third row of Table 8.1.) A further 26 percent answered "neither one nor the other," and 10 percent were classified as undecided or gave qualified answers. These results suggest that loyalty to Quebec predominates over loyalty to Canada.

This is further supported by answers to a second question from the same survey: "In your opinion, which government takes best care of the interests of people like you: the federal government or the government of your province?" This is an interesting question because it taps what can be considered the prime determinant of loyalty to a group—the feeling that one's fate is dependent upon that group. If people feel their interests are better taken care of by a particular government, they should have greater loyalty to it. Here again, we find that in 1965 the provincial government was greatly favored over the federal government by French Canadians from Quebec (fourth row of Table 8.1): 36 percent of them mentioned the provincial government, while only 9 percent chose the federal government. On the other hand, 16 percent answered "both equally well" and 17 percent "neither one nor the other." The rest—22 percent—were either undecided or gave qualified answers.

In 1979, that is, fifteen years later an almost identical question was asked in the Quebec government study mentioned earlier: "Which government do you think looks after your interests and needs the best: the government of Quebec or the government of Canada?" About the same percentage answered "the government of Quebec" (38 percent), but a larger number than in 1965 answered "the government of Canada" (23 percent) (fifth row of Table 8.1). The latter figure may reflect the stronger identification with the federal government developing among federalist citizens as the referendum approaches: greater polarization among options occurs. (Using the same question, except for a reversal of the order of the options, the Radio Canada/Canadian Broadcasting Corp. (CBC) poll of February 1979 found 39 percent naming the government of Quebec and 30 percent the government of Canada.) Despite this shift, these data reveal that many more people feel a greater loyalty to Quebec than to Ottawa.

An even better measure of loyalty, of a semiprojective nature, has been employed in more recent polls in Quebec. The following question was asked: "When you hear about *your* government, which one comes first to your mind: the government of Canada or the government of Quebec?" In a telephone poll conducted by the Société de recherche en science du comportement (SORECOM) in 1977, a clear majority of Francophone Quebecers (58 percent) answered "the government of Quebec," while less than half as many (24 percent) answered "the government of Canada" (sixth row of Table 8.1).

Moreover, practically identical results (within 1 percent of each other) were obtained by the Centre de recherches sur l'opinion publique (CROP) among Francophones in Quebec in a poll for Radio Canada in April 1976, and in a poll for *Reader's Digest* in August 1977. But as the referendum approaches, a slightly stronger pro-Canada loyalty seems to be developing. The same question, asked in 1979 in the MAIQ survey, yielded the results presented as row 7 of Table 8.1: 53 percent answered "the government of Quebec" and 27 percent "the government of Canada." Even more pro-Canada results were obtained in a Radio Canada/CBC poll of March 1979 with the same question, the respective figures being 48 percent versus 30 percent. Paradoxically, this suggests that the Quebec referendum campaign may strengthen the Canadian loyalty of those who plan to vote against the independence of Quebec. Whether these effects will be lasting or largely transient remains an open question. Again, it should be stressed that all these data reveal an unmistakable preference for Quebec.

Similar results are provided by another excellent measure of relative loyalty, the level of government one would generally favor in case of conflicts between the two. In a March 1970 poll in Quebec, people were asked: "Could you tell us which government you are generally in agreement with when the governments of Quebec and Ottawa are in conflict with one another?" About 40 percent of the French-speaking Quebecers answered "Quebec," while only about 15 percent answered "Ottawa." The rest were in agreement with neither (24 percent), or had no opinion (22 percent).[4] Thus, for almost every person who favors Ottawa in a conflict, there are almost three others who prefer Quebec (see eighth row of Table 8.1).

Similarly, in the SORECOM 1977 poll people were asked: "If within the next four years there develop serious conflicts between the government of Quebec and that of Ottawa, which one will you personally tend to support?" This time, compared with 1970, about the same proportion of Francophone Quebecers (44 percent) chose the government of Quebec, but about twice as many (27 percent) chose the government of Ottawa; 9 percent answered "both" and 1 percent, "neither," with the rest (20 percent) answering "don't know" or refusing to answer (see row 9 of Table 8.1). Again, the change probably reflected increased loyalty to Ottawa on the part of many federalists after the election that brought the Parti Québécois (PQ) to power a few months earlier (November 1976), and the new meaning the expression "serious conflicts" was then taking. The ratio of Quebec to Ottawa mentions was, however, still larger than one (1.6).

Why do we observe variations from question to question?[5] I can venture only an ex post facto interpretation. When faced with the request to make an abstract, "objective" assessment, French Canadians from Quebec tend to divide about equally in the predominance they attach to Ottawa or Quebec, if it is defined by their relative importance (the first two questions). They can easily rank them as equally important if given the opportunity to do so. Indeed, some of the data presented indicate that those dual loyalties are not perceived as incompatible by a large majority of the French-speakers. But when they are asked about their personal, subjective feelings as to which level interests them more, serves them better, first comes to their mind as their government, and would more easily receive their support in case of conflict, the Quebec government is much more likely to be favored, though more recently some reversal seems to be taking place. Why is this so? The reasons are not very complex; but before discussing them, let us briefly consider how French Canadians in Quebec differ in this regard from other ethnic groups in this and other regions of the country. These data are interesting in themselves, but they also reinforce the interpretation to be given.

THE LOYALTIES OF OTHER CANADIAN GROUPS

Loyalty is rarely directed to only one group. As suggested, loyalty to the nation as a whole in segmented societies often goes together with loyalty to one's communal group, be it religious, ethnic, linguistic, or other; but in any type of society it can also be directed to one's local community, to one's region, or even to a unit larger than the country as a whole. When we used the concept of dual loyalties above, it was to stress the importance of ethnic/regional loyalty compared with national loyalty among French-Canadian Quebecers.

But are they the only group in Canada for whom such strongly divided loyalties exist? Of course not. The "limited identities" of Canadians have often been the object of attention (Careless 1959; Schwartz 1974). But these limited Canadian identities are nowhere so pronounced as among French Canadians in Quebec. Thus, in almost all studies from which data are available, we find that in contrast with French Canadians, the minority groups in Quebec are much more likely to exhibit a stronger loyalty to Ottawa than to Quebec. In our own 1970 data, for example, we find that a large proportion among the non-French considers both governments equally important (62 percent), but that there are more who claim the federal government is more important (23 percent) than there are preferring the provincial one (5 percent). Similarly, in the 1965 Social Research Group study for the Bilingualism and Biculturalism Commission, 31 percent of those in Quebec who identified as English Canadians said they were more interested in federal politics, as opposed to 13 percent opting for provincial politics (compared with 7 and 27 percent, respectively, among those identified as French Canadian). These French and English patterns are therefore diametrically opposed. Similar findings appear in Vincent Lemieux et al. (1970, p. 92).

The 1965 study for the Bilingualism and Biculturalism Commission, which is not limited to a Quebec sample, allows us to compare French and other groups in Quebec with similar groups in other regions of the country. We find, first, that with both questions used (degree of interest in, and care received from, each level of government), all groups in all regions of the country are less disproportionately loyal to their provincial government than are French Canadians in Quebec. The latter are clearly the ones with the greatest degree of loyalty to their own province. Second, all groups in Ontario (including French Canadians) are like the non-French in Quebec in being relatively more loyal to the federal government. Third, except for the French in the Maritime Provinces, who are close to their Quebec counterparts, the Maritimers and the westerners from all ethnic groups tend to stand in between. As could be expected, they are more "provincialist" than Ontarians but, despite strong regional pulls, they are less so than French Canadians in Quebec.[6]

In a 1977 paper Lawrence LeDuc reported similar results. He found that affect for Canada (as measured on a thermometer scale) was lowest in Quebec, particularly among Francophone respondents, and highest in Ontario. While Newfoundland was close to Quebec, all other provinces clustered near the average, either slightly below (New Brunswick) or slightly above (all other provinces). Moreover, while affect toward their province was not the highest among Francophone Quebecers, they nevertheless were the only ones, along with Newfoundland respondents, to exhibit higher affect for the province than for Canada.

Finally, in the 1979 Radio Canada/CBC poll reported earlier, in the nine provinces other than Quebec, more respondents mentioned the government of Canada (62 percent) than the government of their province (28 percent) as the one coming first to mind when hearing about "their government" (the corresponding figures being 30 percent and 40 percent for Francophones in Quebec). This pattern held separately in every region of English Canada, though less strongly in the prairie and Maritime provinces than in Ontario and British Columbia.

Clearly, the system of dual loyalties found among French Canadians in Quebec is stronger in its provincial component than that exhibited by almost any other Canadian group. Many other studies have reported similar results (Schwartz 1974, pp. 215-23; Manzer 1974, pp. 138-54; Meisel 1977, pp. 13-23; Johnstone 1969, pp. 1-8, 16-22; Lamy 1975, pp. 263-80; Forbes 1976; Taylor et al. 1972; Blishen 1978, pp. 128-32; Elkins 1979; Elkins and Simeon 1980, ch. 1). It should be noted, however, that when loyalty to Canada is compared with loyalty to Britain and its symbols rather than with loyalty to one's province, French Canadians appear more loyal to Canada than English Canadians, who are strongly loyal to Britain, the royal family, and other British symbols (Schwartz 1967, chs. 4,6; Elkins and Simeon 1980, ch. 1).

DETERMINANTS OF LOYALTY PATTERNS

According to Arthur Stinchcombe (1975), the main determinant of loyalty to a group and its institutions is the degree to which one's fate is dependent upon the group and its institutions. But the degree to which one's fate is dependent upon a group is largely determined by the degree of segmentation of that group, where segmentation is defined as the compartmentalization of groups into a more or less complete set of analogous, parallel, noncomplementary institutions and organizations and more or less distinct cultures or subcultures. (On the definition of segmentation, see van den Berghe 1967, p. 34.) That is, one's fate is determined by one's group to the extent that the group has its own complete set of institutions that serve one's interests well. Loyalty to one's group is also determined by the degree to which the position occupied by the group is a subordinate one.

Given that French Canadians in Quebec constitute the most segmented group in Canada and that in many regards their position is a subordinate one, the finding that they have the greatest degree of loyalty to the provincial government, as the supreme corporate body of their group, is in line with the theory. At the other extreme, people living in Ontario are probably the group that is most strongly dependent on national institutions, and their fate is most clearly determined by the federal government. Hence their greater identification with the latter. This is, incidentally, also the case for the English-speaking minority in Quebec. It is highly segmented, and it perceives its fate as especially dependent on the federal government. We therefore find that these people resemble the Ontario groups with regard to interest in federal politics and are the most pro-Ottawa group of all with regard to the government that serves them better (or that comes first to mind). The latter result is also replicated in LeDuc's data (1977, Table 4). In between is the relative segmentation of the extreme eastern and western regions, and their middle position in terms of loyalty.[7]

TRENDS IN FRENCH CANADIAN LOYALTIES

Unfortunately, due to the lack of longitudinal data, it is impossible to establish quantitatively the changes in the loyalties of French Canadians, if any, that have occurred since 1960 or 1970. The questions we have mentioned have not, to our knowledge, been used, with the same wording, at different points in time. To the extent, however, that loyalty is largely determined by the extent of ethnic segmentation and the degree of subordination, and to the extent that these phenomena are relatively stable over time, one would expect a permanently high level of loyalty to their own group among French Canadians. On the other hand, levels of accommodation and conflict between communal groups could vary over time as a result of other factors. This, in turn, could have a feedback effect on loyalty. One could therefore expect some variations in the levels of loyalty over time, as well as possibly greater variations in its forms, its intensity, and its degree of consciousness.

There is no doubt in our mind that such changes have taken place in Quebec since 1960. In particular, largely as a result of the development of new political options, and as is common during periods of rapid mobilization (Melson and Wolpe 1970; Breton 1972, p. 36), one alteration in the form of French-Canadian loyalty during this period is the redefinition of the group's boundaries. Rather than simply identifying as French Canadians, along with the other members of their group elsewhere in Canada, some French Canadians in Quebec have redefined the relevant boundaries as being those of Quebec, and are now identifying simply as "Québécois." This phenomenon represents a change in identity and form of loyalty. It was undoubtedly limited at first to the more politically conscious and militant, and the more educated, segments of the population. The extent to which this redefinition process was actually taking place was first revealed by the answers to the following question in our 1970 survey: "Comment vous définissez-vous *en tout premier lieu*: comme un Canadien français, un Québécois, un Canadien tout simplement, ou autrement?" A minority of 21 percent of French-speaking respondents identified as "Québécois," while larger proportions identified as French Canadians (44 percent) or even as "Canadien tout simplement" (34 percent). The rest were uncertain or chose some other identity.

In 1977, in a poll conducted by Radio Canada and the Centre de Sondage of the University of Montreal, results were obtained that suggested that the "Québécois" identification was making substantial gains. Though the question used was slightly different ("Présentement, vous définissez-vous surtout comme Canadien, comme Canadien français, comme Canadien d'une autre origine ou comme Québécois?"), it is doubtful that the changed wording alone accounted for the difference. This time, among Francophone respondents the proportion defining themselves as French Canadians remained about the same (45 percent), while the proportion identifying as "Québécois" doubled (41 percent) and the proportion saying "Canadians" decreased to 13 percent. These results indicate an increased level of identification with and loyalty to Quebec. (See also Blishen 1978, pp. 130ff.)

SENTIMENT FAVORING GREATER AUTONOMY FOR QUEBEC

The Union Nationale under Maurice Duplessis dominated Quebec politics from World War II to 1960. The importance of provincial autonomy during that period is well known (see, for instance, Quinn 1963). This was followed by a brief interlude of a few years, but during the 1960s many politicians and other public figures resumed proposing various constitutional arrangements that, short of independence, would have given Quebec more power within Canada, such as a "particular status" or an "associate state" arrangement. Let us examine, first, the degree of support among French Canadians for measures envisaging greater autonomy for Quebec (see Table 8.2).

All studies or polls with which we are familiar indicate a relatively widespread desire for greater autonomy among French Canadians. For instance, in the SORECOM poll carried out for CBC a few months after the 1976 provincial elections, a representative sample of Quebec respondents was asked: "If the constitution were revised, would you be in favor of granting more power to the federal government, more power to the provincial government or leave the division of jurisdiction pretty much as it presently stands?"

The distribution of answers given by French Canadians to that question is presented in the 1977 column of Table 8.2. The single most common answer was that more powers should be given to Quebec, an answer offered by 44 percent. Another 34 percent preferred the status quo. Only 9 percent wanted to increase Ottawa's powers, and 12 percent were undecided. Results along more or less the same lines were obtained by other surveys.

In Table 8.3 we reproduce the results of a poll carried out among French-speaking Quebecers in December 1971 and published in 1972. This poll is particularly interesting because the format of the question permits one to measure both the direction and the intensity of respondents' views on greater autonomy for Quebec, and because it permits a comparison with the more radical option, complete independence, also measured in terms of direction and intensity.

The question was asked by the Montreal Centre de Recherches Contemporaines, and read in part as follows: ". . . Please tell me how you would personally react if the following event were to happen in Quebec within a year or two:

... Greater autonomy for Quebec but within a Canadian framework.... Complete break of Quebec from the rest of Canada, that is, complete and final independence or separation." Sixty-eight percent said they would react with happiness to greater autonomy for Quebec, with half of them being "somewhat happy," and only 10 percent would express any degree of disappointment.

Finally, in some recent Quebec polls respondents have been asked how favorable or unfavorable they were to four constitutional options: the status quo, renewed federalism (defined as more powers to Quebec and to other provinces), sovereignty-association, and independence. The one to which Francophone Quebecers feel most favorable is renewed federalism. In a Radio Canada poll of October 1977, 77 percent said they were very or rather favorable to it. By

TABLE 8.2

Desires for Greater Quebec Autonomy Among French-Speaking Quebecers 1965-79 (percent)

	1965	1967	1973	1977	1979
Federal more control over Quebec ('65); closer Quebec-Canada links ('67); more power to Ottawa ('73, '77, '79)	9	36	3	9	7
Quebec position remain same ('65); Quebec maintains present position ('67); keep federal system as is ('73); leave division of power as is ('77, '79)	33	30	21	34	29
Federal less control over Quebec ('65); special status for Quebec ('67); more power to Quebec ('73, '77, '79)	23	21	39	44	56
Quebec separates ('65); Quebec should separate ('67); Quebec an independent country ('73)	5	7	13	—	—
Other solutions ('65); other options ('77)	2	—	—	2	—
Don't know, etc.	26	6	23	12	8
N	(5,504)	(502)	(861)	(600)	(942)

Notes: 1965 N is weighted; it excludes a few French-speaking Quebecers who identified simply as Canadians.
Dashes in table indicate that the option was not available in the question.
Percentages may not total 100 because of rounding.

Sources: 1965, Social Research Group Study for Bilingualism and Biculturalism Commission; 1967, ORC International's study for the CBC; 1973, Hamilton and Pinard study of 1973 Quebec election; 1977, SORECOM study for CBC; 1979, Radio Canada/CBC study.

TABLE 8.3

Reactions of French-speaking Quebecers to Two Hypothetical Events, Greater Autonomy and Complete Independence, 1971 (percent)

	Greater Autonomy for Quebec	Complete Independence
Extremely happy	21	7
Somewhat happy	33	7
Vaguely happy	14	6
Total happy	68	20
Vaguely disappointed	4	7
Somewhat disappointed	3	19
Extremely disappointed	3	37
Total disappointed	10	63
Indifferent	6	7
Don't know	16	10
N	(514)	(514)

Note: Study carried out among Quebec Francophones only.
Source: Centre de recherches contemporaines, *Synapse* (October 1972). "Ce qu'en pensent les québécois francophones."

comparison, many fewer respondents were (very or rather) favorable to the status quo (46 percent) or to sovereignty-association (51 percent), with even fewer (21 percent) being favorable to independence. From the sum of these proportions (195 percent) it is clear that on the average each respondent was favorable to about two of these options. More recent polls have failed to define renewed federalism as giving more powers to Quebec. Apparently largely as a consequence of this, but possibly for other reasons too, these polls have registered a decline in support for that option. Thus a CROP poll for Radio Canada in February 1980 showed renewed federalism favored by 47 percent of the French in Quebec. The same survey showed 48 percent of them favorable to sovereignty-association; 34 percent, to the status quo; and 26 percent, to independence. The choices are now more exclusive, as the sum of these proportions is only 155 percent.

When asked to specify the option they favored most in a CROP poll for *Reader's Digest* in August 1977, French Canadians answered: status quo, 11

TABLE 8.4

Preferences of French-Speaking Quebecers Regarding the Distribution of Power Between the Federal and Provincial Governments in Ten Areas, 1977 (percent)

	Provincial Only	Federal Only	Shared by Both	Don't Know
Education	57	5	31	7
Cultural affairs	46	7	36	11
Natural resources	40	12	38	10
Radio, television, other communications media	28	13	47	12
Social welfare: old age pension, family allowances	27	17	49	8
Transport	27	16	46	11
Trade, industry	26	13	50	11
Immigration	21	20	49	10
Monetary policy	14	32	39	16
Armed forces	10	47	32	12

Source: 1977 SORECOM Study for the CBC.

percent; renewed federalism, 39 percent; sovereignty-association, 24 percent; independence, 12 percent; don't know, 13 percent. (Many other polls have replicated this result.)

Clearly, support for greater autonomy for Quebec is very widespread. On the other hand, it should also be realized that from about a third to about half of the French Canadians in Quebec would not be dissatisfied with the status quo, a result at odds with almost unanimous elite expectations in Quebec today.

In the 1977 SORECOM study respondents were asked, with regard to a series of ten specific areas, whether "the power should belong only to the provincial government, only to the federal government or be shared by both." The results obtained among the French in Quebec are shown in Table 8.4, where the areas are listed in descending order of "provincial only" preference.

Many interesting observations can be made from these results. First, considering just the "Provincial Only" and "Federal Only" columns, we find that there are many more areas (eight out of ten) where "Provincial Only" dominates over "Federal Only" than vice versa (two out of ten); this clearly reflects the autonomist orientations of the French in Quebec. On the other hand, when we examine those two columns plus "Shared," we find that there are more areas (six out of ten) where the largest proportion thinks that the power should be shared by both than there are areas where the largest proportion thinks that the power should belong only to the provincial government (three areas) or only to the federal government (one area). Moreover, in each of these situations, the largest proportions are substantial, being close to 50 percent. There are, overall, no strong wishes for exclusive jurisdiction, and the preferences of the majority are for arrangements that correspond quite closely to the existing ones.

Thus, joint jurisdictions are preferred in the areas of trade and industry, transport, social welfare, immigration, communications, and monetary policy. This corresponds to existing, if not constitutional, arrangements, except for monetary policy and, possibly, communications. Exclusive provincial jurisdictions are preferred for education, cultural affairs, and natural resources, which again largely corresponds to the actual situation, though the federal government plays some role at least in the first two of these areas. Exclusive federal jurisdiction for the armed forces corresponds to both legal and de facto arrangements.

Thus, while a large proportion of the French (44 percent) in that same study (see Table 8.2) wanted more power to be granted to the provincial government, the results just examined indicate that because of lack of expertise and/or of strong feelings, this demand is not translated into a set of specific requests that differ appreciably from the existing arrangements.

All in all, the data presented in this section support the notion of widespread autonomist attitudes among French Canadians in Quebec, but they fail to corroborate elite opinion that these attitudes are very intense.

TRENDS IN SUPPORT FOR GREATER AUTONOMY

Support for greater autonomy is also an area in which there is a scarcity of longitudinal data. But those that exist, though not exactly comparable, would seem to indicate some increase in the desire for greater autonomy, as can be seen in Table 8.2. In that table a comparison of the 1977 data (just reported) can be made with 1965, 1967, 1973, and 1979 data. In each case the questions were phrased somewhat differently, and therefore the trend that can be inferred must be considered with caution. In 1965 the question was "Which solution do you prefer concerning the political future of the province of Quebec? (1) That Quebec separates from the rest of Canada; (2) That the federal government have more control over Quebec than it has now; (3) That the federal government have less control over Quebec than it has now; (4) That the position of the province of Quebec in Confederation remain the same as it is now; (5) Other solutions." In 1967 the question was "Among the following statements, which one best describes what you think should be the relationship between Quebec and the rest of Canada? Would you say that the province of Quebec should strengthen its links with Canada; that it should maintain its present position within Confederation;

that it should obtain a special status within Confederation; or that it should separate from Confederation?" In 1973 we asked: "Would you say that the present federal system should be kept as it is, that one should give more powers to Ottawa, that one should give more powers to Quebec, or that one should abolish the federal system so as to make Quebec an independent country?" In 1979 the question was "If the Constitution of Canada was revised, whose powers would you increase: would you give more power to the federal government, the provincial governments, or would you leave the powers much the same as now?"

Disregarding the differences in the wording of the questions—and they did certainly affect the answers, particularly in the 1967 study[8]—we find that about equal proportions chose the two categories implying a greater role for Quebec (more powers or independence) in 1965 and 1967, a total of 28 percent each time, but that the corresponding proportions rose to a total of 52 percent in 1973, 44 percent in 1977, and 56 percent in 1979. This represents an increase of close to 30 percent since 1965.[9]

We cannot here examine in detail the factors behind these attitudinal changes, but can only mention that ethnic segmentation, ethnic loyalty, and ethnic grievances cannot alone account for such changes. We would like to suggest that the intensification of the conflict between Quebec and the rest of the country, which was itself brought about by processes inherent in rapid social change, must be taken into account. (On this see Breton 1972.) In particular these changes led to the strengthening of nationalist and separatist sentiments and organizations, with the latter continuously reinforcing the former.

It was hypothesized earlier that sentiments of strong loyalty to Quebec should lead to the development of desires for greater autonomy for Quebec. Indeed, this is what the data in the first panel of Table 8.8 indicate. While among Francophone Quebecers who think first of the government of Canada as their government, only 22 percent would want to grant more powers to the provincial government, that proportion increases to 56 percent among those who think first of the government of Quebec. Though no doubt the causal relationship runs both ways, it is clear that loyalty to Quebec, which is presumably the result of a high degree of segmentation, does in turn promote pro-Quebec political options (see also Forbes 1976, pp. 307–08).

THE INDEPENDENCE OF QUEBEC

Currently the alternative proposed to federalism, whether "renewed" or not, is the independence of Quebec, either in its pure form or with an economic association, as in the sovereignty-association platform of the PQ. Let us first examine the strength of support for the independence option proper and

TABLE 8.5

Trend of Opinions in Quebec Relative to Independence, 1962-79 (percent)

	In Favor	Opposed	Undecided	N[a]
	All Adult Quebec Citizens			
1962	8	73	19	(998)
1965	7	79	14	(6,910)
1968A	11	71	18	(202)
1968B	10	72	18	(746)
1969	11	75	15	(367)
1970A	14	76	10	(820)
1970B	11	74	16	(1,974)
1972	10	68	22	(778)
1973	17	64	19	(1,006)
1974	15	74	11	(349)
1976B	18	58	24	(1,095)
1976C	12	66	22	(1,042)
1977A-D[b]	19	66	15	(1,750)*
1977E-F	19	71	10	(1,100)*
1977G-J	18	70	12	(3,500)*
1978B	14	79	7	(972)
1978C	12	74	14	(300)*
1978D	17	73	10	(721)
1979B	19	64	17	(300)*
1979C	19	76	5	(1,199)
1979D	16	72	12	(773)
1979E	18	70	12	(561)
1979F	19	73	8	(995)
1979H	19	72	9	(928)
	Adult French Canadians Only[c]			
1962	8	71	20	(880)
1963	13	43	23[d]	(987)
1965	8	76	16	(5,488)
1967	7	87[e]	6	(502)
1968A	13	65	22	(165)
1968B	11	68	21	(624)
1969	13	70	17	(294)
1970A	16	73	11	(696)
1970B	13	70	17	(1,513)

(continued)

TABLE 8.5 (continued)

	In Favor	Opposed	Undecided	N[a]
1972	11	65	24	(660)
1973	19	60	21	(860)
1974	17	70	13	(277)
1976A	23	63	13	(725)
1976B	20	55	25	(954)
1976C	13	65	22	(755)
1977B	19	61	20	(600)
1977F	23	68	9	(700)
1977G	18	64	18	(616)
1978D	20	70	10	(617)
1979C	22	72	5	(942)
1979D	18	70	13	(672)
1979F	23	68	9	(801)
1979H	23	68	9	(791)

[a]Whenever indicated in the sources, the "no answer" category was eliminated from the total before computing the percentages. The N's for the 1965 study are weighted. In a few cases indicated by * (some 1977 polls, 1978C, 1979B), the N's are approximate due to lack of exact information.

[b]In 1977, following the election victory of the PQ in November 1976, ten polls were made public in Quebec on independence and related issues. In the first panel of the table, rather than present all of them, or a sample of them, we present the average of the proportions for each category for the polls of January to April (four polls), May to August (two), and September to December (four). In the second part (French Canadians only) we have simply reproduced the results of the 1977 polls that gave the data for that group separately.

[c]Ethnicity is variously defined by self-identification, mother tongue, language at home, and other measures. Our experience is that any one of these measures in Quebec yields the same results.

[d]The total in this row is 79 percent; the rest (21 percent) said they were not aware of separatist activities.

[e]In that study separation was one among other less radical options; 87 percent is the total for the other options (for details see Table 8.2, 1967 column).

Sources: For the data from 1962 to 1976A, the sources and questions asked, as well as other details, see Pinard and Hamilton 1977, pp. 247-48. From 1976B to 1978, see Hamilton and Pinard 1980. Sources for other data are as follows: 1979A, CROP report (December 1978-January 1979); 1979B and E, Gallup poll by Canadian Institute of Public Opinion (January and June); 1979C, Centre de Sondage, Université de Montréal, and ADCOM, for Radio Canada/CBC (February); 1979D, Institut québécois d'opinion publique (IQOP) for *Dimanche-Matin* (March); 1979F, CROP for MAIQ (June); 1979H, CROP for Radio Canada (November).

the changes in this support since the early 1960s. Then support for sovereignty-association will be examined.

By the end of 1979 support for the independence of Quebec stood at about 18 percent, with more than 70 percent of the population opposed and about 10 percent undecided (see the last six lines of the first panel of Table 8.5). If one were to consider only the French Canadians in Quebec, one would currently obtain figures about 3 to 4 percent higher for those supporting independence and about 4 percent lower for those opposed to it. The proportion of French Canadians supporting this more radical option is therefore much smaller than that favoring greater autonomy for Quebec. Indeed, French-Canadian support for each of these options, as revealed by the data of a 1972 study presented in Table 8.3, is strikingly different. In particular, it is worth noting that whenever there is opposition to separatism, this opposition tends to be very strong. Consider, for instance, the very skewed distribution among French Canadians who would feel disappointed if Quebec were to become independent: more than half of them would feel extremely disappointed (37 percent out of 63 percent).

In this regard the situation is still the same. In the October 1977 study of the Centre de Sondage for Radio Canada, it was reported that among the 79 percent who did not favor independence, only 12 percent had answered "not particularly in favor," while 67 percent had answered "not at all in favor." (For other results along the same lines, see Pinard and Hamilton 1977, pp. 228n, 229n.) Surprisingly, as in the past, opposition to independence remains more intense than support for it. For instance, in the aforementioned poll, the group favoring independence (17 percent) split more evenly between those "very much in favor" (8 percent) and those "more or less in favor" (9 percent). The fact that opposition to independence is stronger than support for it suggests that independence is perceived as a serious threat by many; and people tend to react more strongly to threats than to new opportunities (Tilly 1978, pp. 134-35).

TRENDS IN SEPARATIST SUPPORT

At least since 1962, people in Quebec have been polled repeatedly about their opinions on separatism, often with identical questions or at least with similar ones. This is one of the rare instances in the field of public opinion in Quebec in which such a long series of comparable data is available. The data have been assembled and are presented in Table 8.5, first for all Quebecers, then for French Canadians only. Thirty-one polls that sampled all Quebecers are available—ten in 1977 alone, the year following the election that brought the PQ to power—and data for French Canadians alone are available from 23 polls (some of which did not sample all people in the province).

Contrary to what is often believed, and no doubt due to "overestimation of the dramatic," as Seymour Lipset would put it, the proportion of people favoring separatism in Quebec has never been very high, nor has it grown dramatically since the early 1960s. Even when only French Canadians are considered, the group of people favoring separation has always been a relatively small minority: they constituted slightly below 10 percent of French Canadians in the early 1960s and around 20 percent at the end of the 1970s.

In the total population the group of those opposed to separation has always constituted a clear majority; their proportion has dropped below 60 percent only once—during the 1976 provincial election, when it reached 58 percent. What has been happening since the 1976 election, however, is a reversal: the data indicate an increase of those opposed to above 70 percent in 1978 and 1979, these net gains coming from a decrease in the proportion of those undecided. For the latter the figure used to stand between 15 and 20 percent or thereabouts; since the middle of 1977, it has been around 10 percent. Even among the French, two-thirds or more have recently been opposed to total independence. (For studies between 1972 and 1974 that diverge from the data of Table 8.5, and the apparent reason for this, see Pinard and Hamilton 1977, pp. 244-46.) The PQ strategy of getting elected first, and then making the demonstration, by being "a good government," of the validity of its independence option has not worked so far.

On the whole, the data of Table 8.5 indicate that up to 1973 there was a slightly increasing trend in the proportion of people in favor of separation (a little less than 1 percent per year) and a slightly decreasing trend of about the same magnitude for those opposed to it. Since 1973, however, the independence forces seem to have made little gain. And, as we have seen, the proportion of those opposed has increased since 1976.

Also, peaks in separatist support appeared during the provincial electoral campaigns of 1970 and 1973 (see 1970A and 1973 rows in Table 8.5), when the PQ was campaigning on that issue. The party's mobilization drives during those campaigns seem, therefore, to have had some impact on people's choice of options with regard to independence. No substantial increase appeared in 1976, however (see row 1976B), when the party campaigned on the issue of good government, leaving the independence issue to be decided later in a referendum. Conversely, the data indicate that during the 1972 and 1974 federal elections (see 1972 and 1974 data in Table 8.5), support for separation declined, suggesting an opposite effect in periods of intense activity by federal parties. (For more details see Pinard and Hamilton 1978; for other data on people's expectations and assessments concerning independence, see Manzer 1974.)

SOVEREIGNTY-ASSOCIATION

The official PQ proposal is sovereignty-association. The party returned to it unambiguously after its election, following a stress on independence only

TABLE 8.6

Trend of Opinion in Quebec Relative to Sovereignty-Association, 1970 and 1977–79
(percent)

	In Favor	Undecided/ Opposed	No Answer	N[a]
		All Adult Quebec Citizens		
1970 A (opinion)[b]	35	53	12	(820)
1970 B (opinion)	28	49	23	(1,982)
1977 B (vote)	32	52	16	(742)
1977 F (opinion)	40	46	14	(823)
1977 F (vote)	38	44	18	(823)
1977 G (vote)	26	56	19	(729)
1977 H (opinion)	40	50	10	(1,458)
1977 J (vote)	42	55	4	(1,000)
1978 A (vote)	25	58	17	(714)
1978 B (opinion)	39	50	11	(972)
1978 B (vote)	33	53	14	(972)
1978 D (opinion)	35	48	17	(721)
1978 E (vote)	31	53	16	(856)
1979 A (opinion)	33	60	7	(947)
1979 A (vote)	31	54	14	(947)
1979 B (vote)	46	39	15	(300)
1979 C (opinion)	35	46	21	(1,199)
1979 C (vote)[c]	37	36	27	(1,133)
1979 D (opinion)	30	40	30	(782)
1979 F (opinion)	41	44	16	(1,004)
1979 F (vote)	41	41	18	(1,004)
1979 G (vote)	25	56	19	(574)
1979 H (opinion)	31	50	19	(928)
1979 H (vote)	23	39	38	(928)
		Adult French Canadians Only[d]		
1970 B (opinion)	33	43	24	(1,511)
1977 B (vote)	38	46	16	(600)
1977 F (opinion)	45	40	15	(700)
1977 F (vote)	44	37	19	(700)
1977 G (vote)	29	51	20	(617)
1978 A (vote)	28	55	17	(593)
1978 D (opinion)	39	45	17	(617)
1978 E (vote)	35	47	18	(740)[*]

TABLE 8.6 continued

	In Favor	Undecided/ Opposed	No Answer	N[a]
1979 C (opinion)	41	38	21	(942)
1979 C (vote)[c]	42	30	28	(890)*
1979 D (opinion)	32	36	32	(679)
1979 F (opinion)	45	39	16	(806)
1979 F (vote)	49	33	18	(806)
1979 H (opinion)	36	43	21	(791)
1979 H (vote)	28	34	38	(791)

[a]N's with * are estimated.

[b]Many polls asked both how favorable people were to sovereignty-association (opinion) and how they would vote on the issue in a referendum (vote).

[c]A group of 5 percent of the respondents, who had not heard about the referendum, were not asked this question.

[d]Limited to those polls for which the breakdown by language group was available to the author.

Sources: See Table 8.5.

between 1970 and 1976. Sovereignty-association implies that Quebec would become an independent country, but one that would establish an economic association with the rest of Canada. While support for independence is very low, support for sovereignty-association is higher, though generally short of a majority in the Quebec population as a whole (except in a few 1979 polls). It is difficult to locate the exact level of support, given the variations that can be observed in Table 8.6. These could result from the complexity of the issue, which in turn affects the formulation of the questions. In particular, while sovereignty could theoretically be decided in a Quebec referendum, the association with the rest of the country could not. Negotiations to create it would be necessary, and their course must perforce remain uncertain.

According to the results presented in Table 8.6, support for sovereignty-association in the total adult population of Quebec since 1977 stands somewhere between 25 and 40 percent and opposition stands somewhere between 40 and 60 percent. Among French Canadians only, support is from 2 to 8 percent higher, and opposition from 3 to 8 percent lower. As a result, in one 1977 poll and in two 1979 polls, there is a plurality in favor of sovereignty-association among French Canadians.

With regard to question formulation, it seems that if people are asked an opinion—how favorable they feel toward that option (see, for instance, row 1977H), support appears at times to be higher than when they are asked whether they would vote for it. Given that people are likely to favor more than one option, this is not surprising. Second, it seems that the more the association is presented as possibly separable from sovereignty, the lower the support tends to be.[10] This would be due to the fact that many of the supporters of sovereignty-association are what we call "conditional sovereignists." That is, they support that cause only if the association is a certainty; otherwise they move to the opposition. In this they are quite different from the "unconditional sovereignists," who support that option even if the association were not to be arranged. (Incidentally, as of 1979 the official PQ position was that the party would consult the population again if the association could not be arranged.) On the whole, studies from which such distinctions can be made indicate that sovereignists divide about evenly into unconditional and conditional (Hamilton and Pinard 1980). Depending on whether a question takes the association for granted, some conditional sovereignists apparently become nonsupporters.

With regard to the trend of opinions on this option, there are no long series available. The only data we know of are for 1970 and 1977–79 (see Table 8.6). With regard to opinions, what they seem to indicate is at most a modest increase in the support for sovereignty-association between 1970 and 1977, from somewhere around 32 percent to somewhere around 40 percent, and a slight decrease afterward to around 37 percent in 1978 and 34 percent in 1979, these figures being the average of the proportions for each year. With regard to a vote, and leaving the undecided within the distributions, the proportion who would vote for sovereignty-association appears to have dropped from about 35 percent in 1977 to about 30 percent in 1978, and to have increased to 34 percent in 1979. The proportion of those who would vote against it has gone from about 52 percent in 1977 to about 55 percent in 1978 and to 44 percent in 1979 (these again being averages).

We suspect that some of these variations, particularly in the vote series, do not reflect true changes in the state of mind of the population but, rather, vagaries of question formulation and polling errors (note in particular the erratic figure in 1979B, based on a very small sample, and the exclusion of part of the population in 1979C (vote), as indicated in the notes to Table 8.6). Our own estimation is that support for sovereignty-association, in terms of vote intentions, stood in 1979 somewhere between 30 and 35 percent, and opposition to it slightly above 50 percent.

Despite all of this, the 1977G results indicate that opinions on sovereignty-association are held no less strongly than those on independence by a large number of voters. In that study, among French Canadians who said they would vote either for or against sovereignty-association (that is, not considering the 20 percent undecided), 48 percent said they held their opinion "very strongly" and another 38 percent answered "somewhat strongly," for a total of 86 percent. Only 12 percent chose the categories "not too strongly" and "not strongly at all" (2 percent, not stated; N = 493). Here again, among French Canadians opponents were more intense in their views than supporters were: among the former, 54 percent held their opinion "very strongly" and 36 percent "somewhat strongly," while among the latter the figures were 38 and 41 percent, respectively.

A MANDATE TO NEGOTIATE SOVEREIGNTY-ASSOCIATION

In the middle of 1977, pollsters started using a new question to measure the support for sovereignty association. In addition to the usual question asking respondents whether they favored, or would vote for, sovereignty-association, polls often contained a question asking respondents whether they would instead be willing to give the government a "mandate to negotiate" sovereignty-association. At first sight such a question might not appear very different from previous questions, but it consistently yielded different results: with that question a larger proportion of respondents appeared to be ready to support the PQ option.

This is interesting in terms of the influence polls may have on debates of this nature. While poll results with such a "mandate" question started accumulating in 1977 (see Table 8.7), it was only around the beginning of 1979 that the PQ started using the terms "mandate" and "negotiation" (Parti Québécois 1979). Later, in November 1979, the PQ government published its White Paper on sovereignty-association (Government of Quebec 1979) that requested a "mandate" "to make this new agreement a reality through negotiation." Finally, in December, they made public the referendum question which simply asked for a "mandate to negotiate" sovereignty-association between Quebec and Canada.[11] According to some, this formulation probably originated with pollsters favoring the PQ option and searching for a question yielding stronger support. It appears that the results presented in Table 8.7 strongly influenced the PQ leaders in their formulation of the official referendum question.

The new poll question was usually put along the following lines, this one being from a 1978 poll (1978E): "If, in this referendum, the Quebec government were to ask you instead for a mandate to negotiate sovereignty-association for Quebec with the rest of Canada, would you, yes or no, give it this mandate?" The results obtained with questions such as this are presented in Table 8.7.

In eight of the nine polls reported in Table 8.7, it can be seen that a plurality of the Quebec voters—a majority of those who were decided—would be willing to give the PQ government in Quebec a mandate to negotiate sovereignty-association, despite the fact that in the same polls a plurality of the same respondents generally expressed opposition to sovereignty-association as such (compare Tables 8.6 and 8.7, for corresponding polls). Moreover, there is no

TABLE 8.7

Trend of Opinion in Quebec Relative to Willingness to Give Government a Mandate to Negotiate Sovereignty-Association, 1977-79 (percent)

	In Favor	Opposed	Undecided/ No Answer	N
	All Adult Quebec Citizens			
1977 F	50	34	16	(823)
1978 D	39	36	25	(721)
1978 E	44	39	17	(856)
1979 A	43	39	18	(947)
1979 C[a]	50	31	19	(1,133)
1979 D	44	38	18	(782)
1979 F	54	30	16	(1,004)
1979 G	36	43	21	(574)
1979 H	41	31	28	(928)
	Adult French Canadians Only			
1977 F	56	28	16	(700)
1978 D	43	31	26	(617)
1978 E	49	34	18	(740)[b]
1979 C[a]	57	24	19	(890)[b]
1979 D	47	35	18	(679)
1979 F	60	24	16	(806)
1979 H	48	24	28	(791)

[a] Excludes 5 percent who had not heard of referendum.
[b] N estimated.

Sources: 1977 F to 1978 E, Hamilton and Pinard (1980); 1979 A, CROP Report (December 1978-January 1979); 1979 C, Centre de Sondage, Université de Montréal, and ADCOM for Radio Canada/CBC (February); 1979 D, CROP for *Dimanche-matin* (March); 1979 F, CROP for MAIQ (June); 1979 G, CROP Report (September); 1979 H, CROP for Radio-Canada (November).

clearly discernible trend in Table 8.7; the results again show much variation from poll to poll. For instance, the proportion of adult Quebec citizens in favor of a mandate oscillates between 36 and 54 percent. On the average about 45 percent would vote "yes" on a mandate, while only about 36 percent would vote "no." Considering averages among the decided only, these polls suggest that a referendum would register 55 percent "yes" and 45 percent "no."

The immediate question that comes to mind is why some voters who are not in favor of sovereignty-association would nevertheless be willing to give a mandate to negotiate it. One answer probably lies in the motivation of some conditional sovereignists: they say "yes" only to a mandate question to be sure that the association will be successfully negotiated before the sovereignty is proclaimed. Another possible answer is that for many of the "inconsistent" voters, negotiations mean compromises and would lead to a result that is likely to be short of the PQ demand for sovereignty-association. This is so even if the PQ has asserted that the principle of their option is not negotiable, their position being that "a yes vote ... would ... be, in fact, a mandate ... to make this new agreement a reality through negotiation" (Government of Quebec 1979).

This last interpretation of those "inconsistent" voters' motivation is supported by the result of the 1979 Quebec government survey; it revealed that 54 percent of the respondents preferred that "sovereignty-association be a negotiation position from which the government could make compromises,"

TABLE 8.8

Pattern of Loyalty and Support for Various Constitutional Options Among French-Speaking Quebecers, 1977 (percent)

	Government Coming First to Mind When Hearing "About Your Government"		
	Gov't. of Canada	Both	Gov't. of Quebec
Greater Autonomy for Quebec			
More power to Ottawa	22	4	4
Leave as is	40	37	32
More power to Quebec	22	37	56
Don't know/other options	17	21	8
N	(143)	(94)	(346)
Independence of Quebec			
Favor independence	4	3	30
Opposed	90	74	45
Undecided/don't know	6	22	25
N	(143)	(94)	(346)
Sovereignty-Association			
Favor sovereignty-association	18	23	52
Opposed	75	52	32
Don't know/would not vote	7	24	17
N	(143)	(94)	(345)

Source: 1977 SORECOM study for the CBC.

while only 26 percent preferred that it "be a firm position from which the government should not yield." (The others rejected both of these alternatives [7 percent], accepted both [1 percent], or did not state their position [12 percent].) It is interesting to note that, in that same survey, when respondents were asked whether they would support renewed federalism or sovereignty-association if a referendum gave them that choice, 51 percent opted for the first as against 32 percent who opted for the second, the others saying neither (3 percent), don't know, or refusing to answer (12 percent) (MAIQ 1979).

At the time of this writing (December 1979), it is still uncertain whether the "inconsistent" voters will become consistent and, if so, in what direction. The referendum campaign may well be a crucial period for them.

As with the Quebec autonomy option, we find, as we have hypothesized (see also LeDuc 1977, pp. 353ff.; Elkins and Simeon 1980, ch. 1; Cuneo and Curtis 1974, pp. 19-20), that greater loyalty to Quebec is strongly related to support for both independence and sovereignty-association among French Canadians in Quebec. The data are presented in the second and third panels of Table 8.8. While almost all those whose loyalty is first to the government of Canada are opposed to independence (90 percent), that proportion drops to about half (45 percent) among those whose first loyalty is to Quebec. Among the latter about a third (30 percent) favor independence. Similarly, while three-quarters of those loyal to Ottawa first would vote against sovereignty-association in a referendum and only 18 percent would support it, among those loyal to Quebec first, about half (52 percent) intend to vote for that option and only a third (32 percent) oppose it.

A person's loyalty is clearly a strong indicator of that person's constitutional preferences. It should also be noted that those loyal to Quebec are the least uncertain concerning greater autonomy, but tend to be more uncertain about independence and sovereignty-association. Conversely, those loyal to Ottawa first have made up their minds about the last two options, but are more uncertain about autonomy.

COLLECTIVE INCENTIVES: BENEFITS AND COSTS OF VARIOUS OPTIONS

Another interesting finding in Table 8.8 is that loyalty to Quebec is about equally related to all three options considered, as indicated by percentage differences in levels of support of pro-Quebec options between those loyal to Quebec and those loyal to Ottawa. But within each loyalty group the levels of support for the various options are different. Why is this so?

In order to answer that question, it is necessary to consider the impact of another factor. While the motivational mechanisms traditionally stressed in theories of social movements have been internal states, such as deprivations and beliefs (for example, ethnic grievances and ethnic loyalty), these theories have recently been criticized for, among other things, their neglect of the role of another category of factors affecting one's motivation to participate in a movement. These are the various external incentives, either of a collective or of a selective sort, that incite one to get involved or not to get involved in a social movement. That is, people try to assess the rewards and costs resulting from support or nonsupport of a movement, or from the movement's success or failure (Oberschall 1973; McCarthy and Zald 1973, 1977; Tilly 1978).

While previous work on social movements stressed the role of internal motives, writers on the resource-mobilization approach now tend to go to the other extreme, stressing mainly the role of external incentives. It is our contention that both internal motives and perceived external incentives must be considered to account for the support of a movement.[12]

We suggest that the differential support received by the various constitutional options is due to the very different costs and benefits that people expect to flow from the implementation of each of these options. While greater autonomy is probably seen as the source of important status and cultural rewards, with no great costs of any kind, sovereignty-association and, in particular, independence are seen as the sources of the same rewards, even on a larger scale, but also the sources of important economic costs. For many this cost/reward ratio is very high, and impedes the unobstructed impact of ethnic loyalty (or ethnic grievances) on a decision for independence or sovereignty-association.

TABLE 8.9

Expected Costs and Rewards of Independence Among French-Speaking Quebecers, 1977 (percent)

	Economic Conditions	Self-Development of French-Speaking People	French Language and Culture
The situation would			
Get better	18	42	44
Remain the same	19	26	26
Get worse	46	12	12
Don't know/ no answer	17	20	18
(N)	(600)	(600)	(600)

Source: 1977 SORECOM study for the CBC.

In the 1977 SORECOM study for the CBC, respondents were asked: "According to you, if Quebec should become an independent country, would the situation in Quebec in the following areas get better, get worse or remain the same?" This was asked for economic conditions, self-development *(épanouissement)* of French-speaking people, French culture and language. The survey made it clear that among the French in Quebec, few people (12 percent) perceived independence as involving any cost to the self-development of the French or to French culture and language, while 42 and 44 percent, respectively, expected benefits in these areas. However, the situation is strongly reversed with regard to economic costs: close to half (46 percent) expected worse economic conditions under an independent Quebec, with only 18 percent expecting improvements, and barely more than a third (37 percent) expecting either improvements or no change. (For qualitative data of the same nature, see Pinard 1975, pp. 87-90.) The proportion of French Canadians in the general population who perceive an economic cost to independence is therefore quite high.

This should not be too surprising when one realizes that even the early leaders (such as Pierre Bourgault) and many early militants of the movement acknowledged that there would be at least a short-term decline in the standard of living in Quebec after independence,[13] (although only 10 percent of those favoring independence in the present study say that economic conditions in Quebec would be worse).

As hypothesized, the impact of these collective goods and bads on support for independence is very important, as can be seen in Table 8.10. The "Total" rows and columns present the zero-order relationship between each incentive separately and support for, or opposition to, independence. It can be seen that both incentives are related to views on independence, though the economic variable is more strongly related. It makes sense, since material incentives are more important for most people than more intangible ones relating to status or culture.

Further, both of these variables exert independent effect, and together their impact is quite strong, as can be seen from the body of Table 8.10. While the overall support for independence was at 19 percent among French Canadians, the proportion increased to 63 percent among those who expected both economic and status (self-development) rewards, but decreased to zero among those who expected costs on both of these counts. Conversely, opposition to independence, which was at 61 percent overall, dropped to 12 percent in the former group and increased to 97 percent in the latter. Therefore, knowing these expectations permits one to predict almost perfectly whether a person will oppose independence. To be sure, there is a high degree of proximity between these variables, but the results are nevertheless quite illuminating concerning people's motivations.

If the independence of Quebec is generally perceived as economically costly, the expectations are less pessimistic with regard to sovereignty-association.

TABLE 8.10

Support for Independence Among French-Speaking Quebecers, by Expected Economic and Status Rewards and Costs, 1977

	Self-Development of the French Would			
	Get Better	Remain Same	Get Worse	Total
	Percent Favoring Independence			
Economic conditions would				
Get better	63 (90)	47 (17)	* (1)	61 (108)
Remain same	31 (94)	7 (56)	10 (10)	18 (214)
Get worse	15 (66)	1 (148)	0 (63)	4 (277)
Total	38 (250)	7 (275)	3 (74)	
	Percent Opposed to Independence			
Economic conditions would				
Get better	12 (90)	29 (17)	* (1)	15 (108)
Remain same	36 (94)	73 (56)	50 (10)	49 (214)
Get worse	68 (66)	93 (148)	97 (63)	88 (277)
Total	36 (250)	76 (275)	89 (74)	

*Too few cases for a meaningful percentage.

Notes: Those who answered "don't know" are included with those who answered "remain same," unless they answered "don't know" on both variables; in that case they are excluded from the body of the table, though they are included with "remain same" for the "Total" rows and columns.

Percent undecided on independence in each cell can be calculated by adding percentages of those in favor and opposed, and subtracting that total from 100. The proportion of undecided tends to be lowest in the bottom row ("Worse"), particularly in the right-hand corner.

Source: 1977 SORECOM study for CBC.

French-Canadian respondents are much less likely to perceive that option as costly, according to the CROP/*Reader's Digest* study of 1977. Respondents were asked: "In your opinion, if Quebec were to obtain political sovereignty and also an economic alliance with Canada, do you believe that the following aspects of life in Quebec would be better, worse or neither better nor worse: ... the economic situation in Quebec?" Only 27 percent of the French Canadians answered that the economic situation in Quebec would be worse (compare this with our result of 46 percent for independence); 27 percent said it would be better; and 27 percent said neither better nor worse, the rest (20 percent) being undecided (N = 700). On the other hand, they perceived the "development of the French language and culture in Quebec" in no less rewarding terms than under independence, 57 percent saying it would be better, and only 4 percent worse (26 percent said there would be no change, and 13 percent were undecided).

Thus, while the status rewards are perceived in the same way as with independence, or even better, fewer perceive any economic costs under sovereignty-association.[14] Hence, we suggest, the higher level of support for the latter option. Indeed, our results indicate that it is precisely the expected economic costs of independence that lead about half the sovereignists to be only conditional ones. In the 1977 SORECOM study, we found that among the supporters of sovereignty-association who anticipated better economic conditions after independence, 57 percent were unconditional sovereignists, and only 26 percent were conditional ones; among those anticipating economic costs from independence, however, the respective proportions are strongly reversed, with 81 percent being conditional sovereignists and only 10 percent unconditional ones (N = 88 and 52).

There are no data on the presumed costs and rewards of greater autonomy for Quebec, but we may presume that very few, if any, would expect economic costs from that option, while many could expect some status and cultural rewards. Hence, we suggest, the even stronger appeal of decentralized federalism.

LOYALTY, INCENTIVES, AND INDEPENDENCE

It was suggested earlier that support for a movement required both internal motives, such as ethnic loyalty, and external, objective incentives. This suggests that support for independence would be strong only among those who are loyal to Quebec first and who expect economic conditions to be better in an independent Quebec. The data of Table 8.11 are arrayed in such a way as to test that hypothesis. They indicate that support for independence among those loyal to Quebec first is very high (65 percent) among those who expect the economic situation to be better. Opposition to independence in that group drops to a meager 9 percent. The converse is not supported. Those loyal to Canada first, or to Quebec and Canada equally, are also relatively strong supporters of independence (41 percent) under the same rewarding conditions. Thus, loyalty to Quebec would appear to be less of a necessary condition. Notice, however, that there are relatively few people falling in these combined categories, so that our percentage is based on a total of only 17 individuals. Finally, under conditions of loyalty to Canada (or to Canada or Quebec) and expectations of bad economic conditions— that is, when both favorable conditions are missing—almost everyone opposes separation (95 percent). These data, therefore, once more indicate the strong impact of positive or negative economic incentives on support for independence,

TABLE 8.11

Support for Independence Among French-Speaking Quebecers, by Pattern of Loyalty and Expected Economic Rewards and Costs, 1977

Gov't. They Think of First	Economic Conditions Would Get	Independence of Quebec			
		In Favor	Opposed	Undecided	N
Quebec	Better	65	9	26	(91)
	Same*	24	44	32	(147)
	Worse	10	77	13	(108)
Canada or both	Better	41	47	12	(17)
	Same	2	62	36	(55)
	Worse	1	95	4	(165)

*Those who answered "don't know" are included with those who answered "the same."

Source: 1977 SORECOM study for CBC.

while giving at least partial support to the hypothesis that both internal motives and external rewards are necessary to back a social movement.[15]

CONCLUSION

We have seen that among French Canadians in Quebec, the system of dual loyalties is characterized by the fact that loyalty to one's own group is often stronger than loyalty to the nation as a whole. Moreover, there are indications that loyalty to Quebec increased during the 1970s. Such loyalty affects people's constitutional preferences: whether they are of a more moderate nature, such as the desire for greater autonomy for Quebec within Confederation, or of a more radical kind, such as political independence, with or without economic association. Support for all these options has been slowly increasing since the early 1960s, though supporters for the last two options seem to remain a minority. These options also appear to have been stagnating since the PQ came to power in 1976. On the other hand, there is, as of this writing, a majority in favor of a mandate to negotiate sovereignty-association, which in general corresponds to the official PQ government question for the 1980 referendum.

One is then led to ask what the future holds for Quebec and Canada. On the one hand, some quite easily make a linear projection of recent trends into the future, and foresee the independence of Quebec as inevitable. (For a discussion of this, see Hamilton and Pinard, 1980). The more enthusiastic among them see it around the corner; the more realistic, at some greater distance in the future. For both groups the rapid growth of electoral separatist forces from around 9 percent in 1966 to 23, 30, and finally 41 percent in 1976—a growth of 32 percent in just ten years—and the accession of the PQ to power in Quebec surely presage what is to come in the constitutional field.

On the other hand, we know that easy linear projections in social life often go unfulfilled. For one thing, support for total independence has leveled off since 1973; and opposition to it has recently returned to a level as high as that of the early 1960s. For another, support for the PQ and its rapid rise to power cannot be easily equated with support for specific constitutional options; the latter, in particular, has grown much more slowly, and the growth of the PQ was in good part a response to forces other than independence (Lemieux 1970; Pinard and Hamilton 1977, 1978). Nothing so far seems to vindicate the PQ strategy built around the assumption that by coming to power and providing a good government, they would more easily make new converts to the independence cause. On the contrary, as we have seen, even support for sovereignty-association does not appear to be making major gains.

Finally, as documented here, the presence of strong negative economic incentives (in the case of independence) and, at the very least, the absence of strong positive economic ones (in the case of sovereignty-association), combined with the possibility that such negative incentives may be increasing because of difficult economic conditions currently prevailing, suggests that winning the forthcoming referendum would have represented a very serious challenge for the PQ, *if that referendum had applied directly to the approval or disapproval of the option it favors.* However, the party has now decided to ask simply for a mandate to negotiate, and the data so far suggest it could win. But, as seen before, the difference between the party's and many a voter's reading of such a mandate leaves the question of Quebec's future very indeterminate.

Such a mandate, given its imprecise character, will no doubt weaken the PQ's hand in future negotiations with the rest of the country. Given the party's commitment to an association, it finds itself in a quandary: for the best results in a referendum, it would require the assurance from the rest of Canada that the association will take place if Quebecers opt for sovereignty—an assurance it cannot get beforehand. On the other hand, it would need a Quebec vote for sovereignty if the rest of Canada is seriously to consider the negotiation of an association, a vote it cannot get beforehand either. The party has been trying unsuccessfully since 1976 to solve this dilemma. How successful the operation will be remains to be seen.

POSTSCRIPT

The Quebec referendum was finally held on May 20, 1980. The unofficial results gave to the YES side only 40.5 percent, with 59.5 percent to the NO side, despite a "soft" question on a mandate to negotiate, made even softer by the promise of a second referendum before any change could be effected as a result of negotiations (see the official question in note 11). This question thus led to a much less positive response than polls could have led one to anticipate. At first glance, the results appear to have fallen somewhere between what a hard question on sovereignty-association would, in our opinion, have yielded, and what the soft question had previously yielded in many polls; indeed the results appear to be closer to the former than to the latter.

Part of the explanation seems to be that many people gave a "hard" answer to this "soft" question when they finally came to vote. In so doing they may have been striving to resolve the inconsistency (discussed above) that resulted from the willingness of some to support a mandate to negotiate sovereignty-association while at the same time opposing sovereignty-association itself. As we had suspected all along, this inconsistency appeared to have been resolved mainly by adopting a consistent NO position, though early in the campaign tendencies to achieve consistency by moving in the other direction appear to have prevailed. There was, for instance, an increase in support for the PQ and even for independence.

It should also be noted that some YES voters—probably more than a third of them—seem to have been motivated primarily by a desire to set in motion negotiations for a renewed federalism. This is suggested by the results of a poll carried out by Richard Hamilton and the author toward the end of the campaign.

In response to this defeat, the PQ has as of this writing stated that it will participate "in good faith" in a new round of constitutional negotiations, along with the federal government and the other nine provinces, within the framework of federalism. These negotiations are expected to start in the very near future.

(Editor's Note: The above addendum was written by Professor Pinard immediately after the May 20 referendum and before either official figures or more detailed analyses of the referendum results were available.)

NOTES

1. For such a view of loyalty, see Stinchcombe 1975, esp. pp. 599ff. This view of loyalty emphasizes its rational, self-interested character. For views stressing instead its "primordial," irrational character, its "unaccountable absolute impact," see Geertz 1963; Isaacs 1975, ch. 3.

2. On patterns of dual loyalties in other multicultural Western societies, see, for example, Rose 1971, pp. 56-57, 200ff.; Rose 1975, pp. 3, 18ff.; Hechter 1975, pp. 34-39 and ch. 10; Kerr 1974, pp. 20ff.; Greenwood 1977; Lorwin 1966; Lijphart 1968. On the compatibility or exclusiveness of ethnic and national loyalties, see Greenwood 1977, esp. pp. 93ff., 98ff.; Elkins and Simeon 1980, ch. 1; Elkins 1979.

3. For our theoretical examination of the role of some of these factors, see Pinard 1973, 1976.

4. The figures presented are for all citizens of Quebec, but the sample was about 94 percent French-speaking. On the basis of the data presented, these figures would be almost exactly the same for French-speaking citizens only (Lemieux, Gilbert, and Blais 1970, tables 4.9 and 4.11, pp. 90-92).

5. Similar variations have been found in English-speaking Canada (Meisel 1977, p. 19).

6. With regard to the question on the government taking best care of people's interests, however, almost all groups, except those in Quebec who identified themselves as "English" and "Canadian," are more likely to name the provincial than the federal government. Similar findings were reported in a study of Canadian youth (Forbes 1976, pp. 304-05). But compare with Meisel 1977, p. 19.

7. Space limitations prevent us from examining another possibly important factor in these patterns of loyalties: the provincial and federal party systems prevalent among each group. (See, for instance, Jenson and Regenstreif 1970, p. 321; Lemieux et al. 1970, pp. 90-91; Schwartz 1967; Elkins and Simeon 1980, ch. 1.)

8. In particular the first category in the 1967 study (Quebec should strengthen its links with Canada) is of a different nature than in the other three studies.

9. A trend toward greater support for provincial rights also was found for all of Canada during 1945-60 (Schwartz 1967, pp. 92-95). Consistent with the data just analyzed is the finding, in John Meisel's 1968 study, that 36 percent of the French-speaking Quebecers thought that Quebec should "occupy a special position in the Canadian Confederation," while 55 percent thought Quebec should be "treated exactly like the other provinces" (Manzer 1974, pp. 158, 164; Elkins and Simeon 1980, ch. 1). For the more recent period, polls sharply disagree: Gallup reported in 1977 that the proportion of all Quebecers in favor of "special status" for Quebec was only 22 percent, as opposed to 72 percent in favor of the "same powers" for Quebec "as all the other provinces" (*Montreal Star*, August 20, 1977). But CROP also reported in 1977 that 44 percent of all Quebecers were "in favor" of "special status" for Quebec, while only 37 percent were "opposed" (*Chronique CROP*, May 5, 1977). And a more recent Quebec government survey found that 48 percent of all Quebecers would vote in a referendum for a special status for Quebec within Confederation, while only 33 percent would be opposed (MAIQ 1979). (For all Quebecers the proportion favoring a special position in Meisel's study is 35 percent—as recomputed for Manzer 1974.)

10. For instance, in 1977B the question was: "The Parti Québécois wants Quebec to become an independent country, economically associated with the rest of Canada. If the government of Quebec were able to arrange such an economic association with the rest of Canada, would you then vote in a referendum *for* or *against* Quebec becoming an independent country?" In 1977F (opinion) it read: ". . . For each of these possibilities tell me whether you personally would be very much in favor, reasonably in favor, not particularly in favor or not at all in favor: . . . sovereignty-association, in which a politically independent Quebec would have an economic alliance with Canada."

11. The actual question as finally adopted by the Quebec National Assembly on March 20, 1980, reads as follows: "The Government of Quebec has made public its proposal to negotiate a new agreement with the rest of Canada, based on the equality of nations; this agreement would enable Quebec to acquire the exclusive power to make its laws, levy its taxes and establish relations abroad—in other words sovereignty—and at the same time to maintain with Canada an economic association including a common currency; no change in political status resulting from these negotiations will be effected without approval by the people through another referendum; on these terms do you give the Government of Quebec the mandate to negotiate the proposed agreement between Quebec and Canada? Yes. No."

12. For the argument that both high grievances and low expected costs are necessary factors (the function is a multiplicative one), see Korpi 1974, p. 1573. A more elaborate model of motivation will be presented elsewhere.

13. In François-Pierre Gingras's study of the Rassemblement pour l'Indépendance Nationale in 1968, he found that 58 percent of the members expected a decrease in people's standard of living at least during the first few years, and that 33 percent expected a decrease in their own standard of living over a period of ten years. For some of these data, see Gingras 1971, pp. 204-56.

14. In the most recent study (MAIQ 1979) 51 percent of the French-speakers agreed with the statement "Sovereignty-association is an adventure which would run the risk of great costs for Quebecers," while only 34 percent disagreed and 15 percent were undecided.

15. Space limitations have prevented us from examining the bases of support for the independence options; for some data on this, see Irvine 1972; Cuneo and Curtis 1974; Hamilton and Pinard 1976, 1980; Ornstein, Stevenson and Williams 1978.

9

PUERTO RICO'S STATUS DEBATE

Sonia Marrero

In 1898, during the Spanish-American War, American troops landed on Puerto Rico. Spain subsequently ceded the island to the United States under terms of the Treaty of Paris. A civil government was established in 1900, but Washington reserved the right to appoint the governor. Puerto Ricans were granted American citizenship in 1917, and in 1947 they started electing their own governor (Cruz Monclova 1957).

The present commonwealth status is based on a U.S. Congressional act of 1950, and was approved by a plebiscite held in Puerto Rico in 1951. This political formula was designed as a third alternative to decolonization that would involve neither independence nor statehood and would permit the island to retain its beneficial economic ties to the United States without losing its national identity. The commonwealth's constitution provides for a governor and bicameral legislature, both elected by universal suffrage. Though American citizens, Puerto Ricans do not vote in U.S. elections, and are represented in the U.S. Congress by a resident commissioner who has no vote. Federal taxes do not apply in Puerto Rico unless the commonwealth government agrees, as in the case of Social Security taxes.

Despite a falling birth rate and substantial emigration to the mainland, the island's population has risen from 2,712,000 at the time of the 1970 census to well over 3 million; its population density per square mile is one of the highest in the world. It is estimated that some 2 million Puerto Ricans live in the mainland United States, most of them in eastern industrial cities (Banks 1976).

The Puerto Rican economy, traditionally based on sugar, tobacco, and rum, has been industrialized since 1948 under a self-help program known as

"Operation Bootstrap," which has stressed the use of special incentives to attract private investment. As industry grew in importance, agriculture declined; and the island now imports nearly all its food, including tropical fruits, from the United States. Per capita income more than doubled between 1964 and 1973, and as of 1977 was approximately $2,500, the highest in Latin America. Despite this apparent prosperity, the commonwealth is burdened by inflation, increasing debt, and very severe unemployment (Curet Cuevas 1976).

There are four major political parties. Puerto Rican politics was dominated from 1940 through 1968 by the Popular Democratic party (Partido Popular Democrático), led for many years by former Governor Luis Muñoz Marín. In 1968, Luis A. Ferré of the New Progressive party (Partido Nuevo Progresista) was elected governor. The present governor, Carlos Romero Barceló, also is a member of that party. The Popular Democratic party has traditionally supported commonwealth status; the New Progressive party has advocated statehood.

Two other parties favor independence for Puerto Rico. The Puerto Rican Independence party (Partido Independentista Puertorriqueño), which at one point was the second largest party, espouses a social-democratic political system and opposes political violence. The Puerto Rican Socialist party (Partido Socialista Puertorriqueño) works for the establishment of a Marxist-Leninist state. This party does not actively advocate violence, but rather considers it a necessary evil in the struggle to change political and economic structures.

DISSATISFACTION WITH AN AMBIGUOUS RELATIONSHIP

Puerto Rico's ambiguous relationship with the United States has increasingly become the subject of bitter criticism from the most disparate sectors of the island's political spectrum. The commonwealth status, which at the time of its creation in 1952 was considered one of the most ingenious political experiments, is now crumbling under the weight of political and economic stagnation. What was conceived to be a permanent solution to Puerto Rico's colonial relationship to the United States does not seem to work anymore and all political parties, including advocates of statehood, commonwealth, and independence, are voicing their discontent.

The resentment that is building up in Puerto Rico was best expressed in August 1978, when leaders of the various parties told their grievances to the U.N. Decolonization Committee for the first time. Representing the New Progressive party (statehood) was Governor Carlos Romero Barceló. The Popular Democratic party was represented by former Governor Rafael Hernández Colón and the party's president, Miguel Hernández Agosto. Ruben Berríos Martínez spoke for the Puerto Rican Independence party, and Juan Mari Bras for the Puerto Rican Socialist party. Excerpts from their testimony were published by all newspapers in Puerto Rico and in a special supplement of the New York-Spanish daily, *El diario-La prensa*.

Their appearance before the committee was an indictment of the long-established American policy that the question of Puerto Rico's political future is a domestic issue, and that the status dilemma was solved with the creation of the commonwealth in 1952 and its acceptance at the United Nations in 1953. However, both the majority in the United Nations and Puerto Rican political leaders are now saying that the status problem has not been solved, and that the word "colonialism" can still be used to describe the relationship between a powerful United States and an impoverished overseas dependency.

The debate on the status problem and the resulting tensions have been aggravated by Puerto Rico's serious economic crisis. The island has been devastated by inflation, recession, and rising energy costs. Labor unrest and demands for higher wages have chased away U.S. corporations that were initially attracted by "Operation Bootstrap," which offered tax exemptions, other industrial incentives, and, above all, cheap labor. The closing of hundreds of factories has pushed unemployment figures to a record. The official rate of unemployment is now 21.9 percent, but even Puerto Rican government officials agree that real unemployment may be twice as great.

Puerto Rico's economic situation is inevitably linked to the status question. Constrained by a vast array of federal laws and regulations, the island's government cannot come to grips with its multiple problems. The Popular Democratic party's demands for greater autonomy are one aspect of an attempt to deal with these problems more effectively. Statehooders, for their part, reason that incorporation into the Union would bring more federal funds into the island. Political stability resulting from statehood, they say, would certainly stop the mass exodus of corporations and would encourage investments from the United States and from other countries. Governor Carlos Romero Barceló has asserted:

> We are disenfranchised American citizens. Our situation is an affront to the principles of American democracy and is a source of embarrassment in the United Nations, as the U.S. is perennially accused of "colonialism." Only by statehood can a sense of political dignity be restored to the Puerto Rican people. . . . Also, as I have mentioned, in exchange for not having to pay taxes, we are disenfranchised U.S. citizens, we have no vote for President or Vice-president of the nation of which we are citizens, nor do we have the right to elect Representatives and Senators to our National Congress. We are discriminated against for geographical reasons. (Barceló 1977)

Advocates of independence claim that Puerto Rico's problems are caused by economic exploitation of its resources and its people by American corporations, as well as by political oppression. Independence and socialism, in their view, would permit true development without any external interference.

Perhaps the worst consequence of Puerto Rico's unresolved conflict is the violence and bloodshed generated over the years, on the island and in the United States. In 1950, for instance, the Puerto Rican Nationalist party led an uprising that left 30 people dead and 90 wounded. Hundreds of others were arrested. Also in 1950 members of the same party attempted to kill President Harry Truman in Washington, D.C. Four years later a group of Puerto Ricans opened fire in the chamber of the U.S. House of Representatives, wounding five congressmen.

More recently terrorist groups, including the Movimiento Independentista Revolucionario Armado (MIRA) and the Fuerzas Armadas de Liberación Nacional (FALN), have claimed responsibility for bombings of banks, hotels, stores, and government installations in the United States and in Puerto Rico. The worst incident reported occurred in New York, when a bomb went off in Fraunces Tavern in the Wall Street area on January 2, 1975. Four people were killed and 53 injured in that explosion. In 1977 police ordered the evacuation of the World Trade Center after a bomb was found in one of its twin towers. In early December 1979 pro-independence Puerto Rican gunmen, armed with M-16 rifles, ambushed a bus loaded with U.S. Navy personnel on the outskirts of San Juan, killing two sailors and wounding ten others. Three Puerto Rican clandestine nationalist groups joined in the attack and said that the raid was in retaliation for the death of a pro-independence militant in a U.S. prison.[1]

Violence also comes from the right wing. There have been numerous incidents involving bombings and shootings against pro-independence organizations and individuals by terrorist groups that favor statehood. Other attacks have been carried out by Cuban exiles. Puerto Rican independence groups and civic organizations often complain of police persecution and harassment, including instances of civil rights violations. Through the 1974 Freedom of Information Act, prominent citizens of Puerto Rico and independence leaders found out that they have been investigated by the F.B.I. Documents released under that law revealed the existence of plots organized by government agencies that were designed to harm the reputations of some Puerto Rican political leaders. Those actions by American authorities have created further tensions between the two nationalities.

DIFFICULTIES OF ASCERTAINING
PUBLIC ATTITUDES

Prominent Puerto Ricans, including political leaders, businessmen, and intellectuals and several professional groups, have clearly expressed their positions regarding the island's political status and future. Their opinions have been widely publicized by the mass media on the island. These opinions have also been recorded during special debates on Puerto Rico's status, such as the symposium sponsored by the University of Puerto Rico in November 1979 entitled "Puerto Rico: Its People and Its Future." Political parties also publicly favor one or another of the three alternatives in question. But what are the wishes of the general public regarding the status question? How do the Puerto Rican people *really* feel about the U.S. government and its treatment of the island? What is the public's preference for Puerto Rico's political future?

Public opinion on these questions is difficult to determine for various reasons. Puerto Rican electoral parties claim allegiance to different political status alternatives, but electoral campaigns in Puerto Rico are conducted mainly on administrative issues and not on the status question. Political parties carefully point out during elections that a vote for them is not a vote for a particular political status, but for a style of government. This limits the usefulness of electoral results as an instrument to measure mass attitudes regarding status. Another factor that makes it difficult to ascertain the attitudes of the public is the low level of sophistication regarding public opinion surveys. (We will deal with this issue later.)

One of the first opportunities to measure status preferences among the public, although in a partial way, was afforded by the 1967 plebiscite. On that occasion the political parties involved specifically campaigned on the status formula. The alternatives offered were statehood and commonwealth, since the pro-independence parties boycotted the plebiscite, on the ground that it was a fraud committed against the right to self-determination of the Puerto Rican people. They claimed that the plebiscite did not follow the objectives and procedures specified by the United Nations.

In the 1967 plebiscite, commonwealth status led with 425,081 votes, statehood had 274,312, and independence 4,204. Participation was relatively high, considering the boycott of the pro-independence parties and the fact that the usual political favors expected by many people after an election did not play a role. Out of 1,067,000 island residents over the age of 21 who were eligible to vote, 703,597 cast their ballots.

Even though the commonwealth forces won the plebiscite, statehood made a large leap forward by capturing 38.98 percent of the registered votes, a significant advance compared with performances in past gubernatorial elections. But the most interesting thing revealed by the plebiscite was the demographic and geographic voting pattern of each faction. While support for commonwealth came principally from rural and semirural areas, the statehood vote was concentrated in the large urban centers. In the principal cities of San Juan and Ponce it won impressive majorities. The tendency of statehood to draw support from large urban areas was reflected also in the 1968, 1972, and 1976 gubernatorial elections.

It is also noteworthy that the urban areas won by statehood are the ones

TABLE 9.1

Election Results in Puerto Rico 1952-76 (percent)

	Popular Democrats (Commonwealth)	Statehood Republican	Puerto Rican Independence	Socialist	Other
1952	64.8	12.8	19.0	3.3	—
1956	62.5	25.0	12.5	—	—
1960	58.2	32.1	3.1	—	6.6
1964	59.3	34.6	2.7	—	3.3
1968	31.3	.4	2.2	—	—
1968	32.8a	33.3b			
1972	51.3	44.1b	4.4	—	.2
1976	45.2	48.2b	5.7	.7	.1

aPeople's Party.
bNew Progressive Party.
Source: Board of Elections, Commonwealth of Puerto Rico.

that have registered the largest population increases since the mid-1960s. In 1960, San Juan's metropolitan area had 648,000 inhabitants. Recent estimates by the Puerto Rico Planning Board assigned the area 1,037,700 persons. We can conclude that sentiment favoring annexation is concentrated in the areas registering the largest population growth and having the most intensive economic activity. The pro-statehood party has consistently appealed to the new and growing urban middle class. This situation was well described by Gov. Carlos Romero Barceló when he remarked that Puerto Rico's explosive economic growth had created a middle class where none existed before, and that this new middle class eagerly embraced many values from the mainland—especially where consumer goods were concerned.

Despite the fact that gubernatorial elections in Puerto Rico are not a reliable measure of status preference among the masses, election results from 1952 to 1976 do reflect growing support for the pro-statehood party, as shown in Table 9.1.

The growing popularity of annexation (statehood) in Puerto Rico can be explained by various factors: the long years of association with the United States and the consequent social, political, and cultural contacts; the integration of Puerto Rico into the U.S. economic structure; the island's ideological domination by representatives of U.S. interests; and the increasing dependency on federal aid to support the government and the economy, and to secure the subsistence of thousands of Puerto Ricans.

Of all the factors mentioned, perhaps the most significant is the increased dependency on federal funds. In 1970 Puerto Rico received $767 millions in federal funds. By 1979 the amount reached a dramatic $3.54 billion. It is worth noting that most of the money destined for individuals is not aid, but is in payment for services rendered. This includes pensions for veterans and federal employees and social security payments. In 1979, for instance, the island paid $720 million to the United States in Social Security taxes. In the aid category are Medicare and Medicaid payments, rent subsidies and food stamps. Federal food stamp aid in 1976 came close to $600 million; by 1979 it was $821.8 million. About two-thirds of the population qualify to receive food stamps and the bulk of this money is, ironically, spent on food products imported from the United States (*Journal of Commerce*, February 1, 1980).

The decline in the vote for pro-independence parties could be attributed to several factors: political persecution of independence supporters after the 1950-54 nationalist violence; unequal access to resources during electoral campaigns; and a general defeatist attitude that an electoral victory under present political conditions is too remote.

The reasons for the failure of the pro-commonwealth party to retain its support could be explained by the sudden and drastic decline of the island's economy and by charges of government corruption and incompetence when the Popular Democratic party was in power.

Regardless of the 1952 creation of commonwealth, the 1967 plebiscite, and the conduct of elections every four years (all of which are arguments used to support the idea that Puerto Rico already has exercised its right to self-determination), the fact is that Puerto Rico continues to dwell in political limbo. Puerto Ricans continue to be divided among three status alternatives, and the island's relationship to the United States has undergone few modifications. In view of these facts the United Nations decided to reevaluate the Puerto Rican case.

On September 12, 1978, the U.N. Decolonization Committee[2] approved a resolution that "reaffirms the inalienable right of the people of Puerto Rico to self-determination and independence in accordance with General Assembly resolution 1514 (XV)," and states that

> Self-determination by the people of Puerto Rico in a democratic process should be exercised through mechanisms freely selected by the Puerto Rican people in complete and full sovereignty which establishes the complete transfer of all powers to the people of the territory, and that all determinations concerning status should have the approval of the Puerto Rican people. (United Nations 1978)

The Decolonization Committee, created in 1961, adopted the resolution on Puerto Rico by a vote of 10 in favor to none against, with 12 abstentions.

(Representatives of two member nations did not participate.) The outcome of its 1978 deliberations culminated a decade-long effort to include Puerto Rico on the list of territories that still have not obtained their independence. The resolution was approved by the General Assembly ipso facto when it accepted the committee's report.

The resolution made clear that only free association with the United States—after transfer of all powers—and independence were alternatives in the decolonization process. The Decolonization Committee discarded statehood as an alternative for the political future of Puerto Rico. In the committee's view, statehood under present circumstances would constitute the culmination of colonialism.

Perhaps for that reason Gov. Carlos Romero Barceló, though admitting that Puerto Rico's relationship to the United States has vestiges of colonialism, at the same time opposed any U.N. interference in Puerto Rico's affairs. Romero Barceló is proposing the celebration of another plebiscite on status in 1981 but thinks it unnecessary for the United States to transfer all legal, political, and economic power to the island beforehand.

There have been a number of attempts to make direct measurements of popular preferences on the status question, but most of these have produced results of doubtful reliability. Public opinion polls on controversial political issues are rarely conducted; and when they do take place, they are usually commissioned by bodies with vested interests—often a political party or its advertising agency. The results of these surveys are infrequently published (Falcón 1970). In those cases where the findings are made public, it often turns out that the polling methods used were dubious, and sometimes the publication of partial results gives an erroneous impression about the opinion distributions that were found.

An example was offered by the pro-statehood newspaper, *El nuevo día*, when it published the results of a survey on September 21, 1978. The survey attempted to measure the reaction of Puerto Ricans to the resolution approved by the U.N. Decolonization Committee regarding self-determination. According to the story, the majority of the Puerto Rican people repudiated the resolution, and the front-page headline featured this finding. A close look at the information provided showed, however, that such a conclusion could not necessarily be drawn from the poll.

The newspaper account of the survey provided little information about the methods used, but the data that were provided revealed serious deficiencies. According to the story, 1,746 Puerto Rican adults were interviewed at 17 locations in the San Juan metropolitan area. The sample did not, therefore, represent the country's entire population, although the story suggested that it did because people who were visiting the metropolitan area from other parts of the island, as well as San Juan residents, were interviewed. In addition, there seems to have been insufficient attention to controlling for such factors as sex, age, education, and economic status. For instance, 31 percent of the people interviewed said they had incomes of $1,000 or more per month. This is not a reflection of the country's economic reality! Furthermore, the survey was conducted in areas won by the pro-statehood New Progressive party in the last election. It might have a certain degree of validity if used as an indicator of how *certain sectors* of the Puerto Rican people think, but it could scarcely be considered to represent a cross section of Puerto Rican opinion.

Another example of the ease with which poll results may contribute to misunderstanding concerns a survey conducted in the United States. On October 21, 1979, *El nuevo día* carried a front-page article stating that 59 percent of the U.S. public favored statehood for Puerto Rico.[3] But the story, which was based on a Gallup Poll release, did not give prominence to the fact that the full question asked of the sample inquired whether Americans would support statehood for Puerto Rico *if the Puerto Rican people asked for it*. Even more significant was that the story buried the fact that an even larger proportion of the American public—67 percent—supported independence for Puerto Rico, again *if the Puerto Rican people asked for it*. The full Gallup release on the poll and the story about it thus gave two quite different impressions.[4]

Even if public opinion were to be measured utilizing the most scientific and strict criteria, measuring mass attitudes concerning a future political status of Puerto Rico presents problems. According to one authority on public opinion research:

> Surveys made in times of stability may give little hint of the unarticulated needs and frustrations and the potential for action of the public. Asking people who have never experienced different institutions what they would like in the way of television programs, jobs, city planning, schools, tax systems, or economic systems tends to produce answers with a status quo bias. When confronted by real alternatives, the answers and the actions may be different. This means that surveys have to be used in close relation to actual crises, social movements and campaigns for change, if the potential for change in the public is to be understood. (Barton 1971)

The results of one of the few systematic studies of public opinion done in Puerto Rico on status preference seems to confirm the status quo bias argument. The study was conducted by Professor Luis Nieves Falcón of the Center of Social Research of the University of Puerto Rico and financed by the Puerto Rican Senate. The study attempted to measure the opinion of Puerto Ricans on a variety of issues, including preference for a status formula. It was based on interviews with 1,300 persons representing the adult population of the island, and seemed rigorously designed and executed.

TABLE 9.2

Political Solutions Preferred by the Puerto Rican When Presented with the Choice of Alternatives (percent)

Alternative	Favored Political Formula					
	Improved Commonwealth	Statehood	Independence	None	Other	Total
Improved commonwealth vs. statehood	51	32	–	10	7	100
Improved commonwealth vs. independence	62	–	11	20	7	100
Statehood vs. independence	–	46	16	27	11	100
Improved commonwealth, statehood or independence	45	33	13	1	8	100

Source: Luis Nieves Falcón, *La Opinión Pública y las Aspiraciones de los Puertoriqueños*. (Rio Piedras: Editorial Universitaria, 1970), p. 162. Percentages based on 1,300 cases.

CAN SURVEYS MEASURE PREFERENCES FOR SOMETHING THAT HAS NEVER BEEN EXPERIENCED?

The results of the study by Professor Nieves Falcón show a fairly close association with those of the 1967 plebiscite and with electoral voting patterns. One should not, however, ignore the possibility that all these results are misleading. Can respondents in a survey, or voters in an election, give informed judgments and express preference about status formulas that imply profound changes in the country's future political structure and in its relationship with the United States? This question applies particularly to the independence option. One must remember that Puerto Ricans have never experienced political independence. That status formula is therefore totally hypothetical.

Another point frequently emphasized by social scientists is that the quality of public opinion depends to a large extent on the quality of public discussion of alternatives. If certain of the contending views are barred from effective presentation to the public, or suffer from discrimination in the resources available to present them, then there is interference with public discussion (Blumer 1946). Public opinion that results from a one-sided consideration of alternatives may be misleading.

This consideration is relevant to the situation in Puerto Rico, where interference with the formation of opinion takes many forms. The mere fact that a substantial part of the mass media is owned and controlled by U.S. corporations constitutes a serious problem. Also, any Puerto Rican efforts to regulate television or radio programming must deal with the Federal Communications Commission, which has jurisdiction over all electronic communications on the island (Mays 1976). In Puerto Rico, the ownership and content of the popular communications media are heavily controlled or influenced by the U.S. media giants. With significant exceptions, most of the remaining Puerto Rican-owned media are closely tied to or dependent on the island's wealthy families. The messages, values, and attitudes that are transmitted through these media in turn reinforce and justify the aims, world-view, and ideology of these controlling forces (Paláu 1979).

Free discussion indispensable for the formation of a considered opinion is hampered by the severe discrimination directed against the pro-independence minority. Independence advocates claim they are persecuted and harassed by the authorities. They are also barred from equal participation in the electronic and printed media.

The measurement of public opinion concerning status preference among Puerto Ricans under the present circumstances presents serious problems. It is hard for any Puerto Rican to visualize what an independent republic of Puerto Rico would be like, after being hammered on the head for years with the idea that independence can only bring havoc to the island. One of the most serious obstacles in the path toward self-determination is the heavy economic dependence on the United States. Regardless of whether federal funds are aid or payments, the point is that people depend on them for their subsistence. Those dependent on federal funds clearly will be reluctant to risk them, even for independence. This is perhaps one of the major points the island has to negotiate with the United States in order to create the right conditions for the exercise of self-determination. Independence will have to be accompanied by a long period of considerable U.S. assistance. Federal payments, pensions, and aid cannot be cut off overnight without severe economic and social damage.

For these reasons, any serious attempt to measure public opinion on a preferred future political status in Puerto Rico, through surveys, elections, or plebiscites will have to be preceded by a series of conditions. It is impossible for a nation to determine its future within the context of foreign domination, and it is impossible to know the opinions of a people within that same context. Any attempt to legitimize the political status quo of Puerto Rico by invoking the results of public opinion surveys is fraudulent.

Using the U.N. guidelines for a genuine decolonization process may be the best way to start to solve the problems faced by the Puerto Rican nation. For a legitimate self-determination process to occur, the United States has to take the first indispensable step, by complying with the U.N. requirement that all power be transferred to the Puerto Rican people, without conditions or reservations, before they are asked to make a decision.

This does not mean that surveys are of no value in connection with a search for a solution to Puerto Rico's status problem. But the emphasis should be on asking people questions that they can answer on the basis of their own experience. Little is to be gained by posing theoretical alternatives. Instead, the questions should focus on the actual problems and personal hopes and fears of members of the population. This is a task to which surveys are better suited (Cantril 1965). The political form should be a servant of the needs and aspirations of a people; it is not an end in itself. Just as the political institutions on the U.S. mainland have grown in accordance with the country's needs, so the political institutions of Puerto Rico should be allowed to grow in the same way.

NOTES

1. The militant in question was Angel Rodríguez, a leader of the Puerto Rican Socialist League, who was arrested during a protest demonstration on Vieques Island and was sent to a federal prison in Tallahassee, Florida. On November 4, 1979, he was found dead in his cell, with bruises and cuts on his body. Prison officials said he had committed suicide. The death, suicide or not, enraged people in Puerto Rico. Protests against the U.S. Navy's use of Vieques escalated. The navy owns or controls 27,000 acres on Vieques, leaving 6,000 acres for the 9,600 residents. Use of this area for gunfire practice and amphibious exercises has caused extensive environmental damage and has deprived hundreds of fishermen of their livelihood.

2. The Decolonization Committee, made up of 24 countries, was charged with implementing U.N. Resolution 1514, which says: "Immediate steps should be taken, in trust and non-self-governing territories that have not yet attained independence, to transfer all powers to the peoples of those territories, without any conditions or reservations . . . in order to enable them to enjoy complete independence and freedom." The creation of the committee was a reflection of the changes in the world's power balance that took place after World War II when many nations in the third world became independent.

3. The fact that both examples are taken from the same newspaper is coincidental. Other examples can be found in other newspapers.

4. The Gallup release contributed its share to the misunderstanding. It was headed "Poll Refutes Castro's Charge of Colonialism," and its content appeared to equate the opinions of the U.S. public with the position of the U.S. government. The headline obscured the fact that the charge of colonialism stemmed from the United Nations, not only from Fidel Castro. And it is obvious that the results of a poll do not necessarily reflect official Washington policy.

10

APPLICATION OF OPINION RESEARCH TO CONFLICT RESOLUTION

W. Phillips Davison

In the foregoing pages we have looked briefly at eight situations, each unique. But one characteristic is common to all of them. In all situations there is a group of people, sometimes numerous and sometimes relatively few in number, who believe that the political or economic status quo does not give their group adequate opportunities to realize its collective aspirations. In most cases, group members also feel that they are subject to personal disabilities caused by their being born in a particular community—one that speaks a different language, practices a different religion, or suffers from historically sanctioned social and economic discrimination.

Dissatisfaction with the status quo leads to protests and to demands for changes. The desired changes may be in the political structure of the state, as in the case of demands for independence or autonomy, or they may involve alterations in the rules of the game that would give the group more political power, greater economic opportunity, or additional freedom to follow and develop its own cultural patterns. Memories of the past usually strengthen current dissatisfactions: this land previously belonged to us, group members may say, why should it now be ruled by those whose language and religion are different from ours?

The desired changes and efforts to bring them about nearly always lead to counteraction from other groups within the same state. Members of these groups may see their own liberties curtailed, their own opportunities limited, by the changes that are demanded. They, too, can cite historical precedents, rational arguments, and sometimes even divine law as favoring their position.

Confrontations are likely to result. These may be violent, in some cases leading to full-scale war, or they may be confined to the political arena. We are concerned here with both types of confrontation, recognizing that avoidance of violence requires an understanding of the alternatives to it.

How can public opinion research help in the solution of nationality or communal problems of the kinds we have been discussing? If it can help at all, the way it is used and the extent of its contribution will vary from case to case. There is no single formula that can be counted upon to produce given results. Nevertheless, we believe that most nationality or communal confrontations involve certain common elements, and that awareness of these may facilitate the adaptation of opinion research to the individual situation.

These common elements can, perhaps, be more clearly distinguished if we look at two hypothetical cases.

TWO SCENARIOS FROM THE LAND OF IMAGINATION

Sometime during the fifth century the Blue people, who lived on the banks of the Spectrum River, began to migrate westward into an area inhabited mainly by Greens. The Blues were resourceful artisans and traders, and at the end of several hundred years had established a substantial number of towns and villages in what had been Green farmland. In the seventeenth century a vigorous Blue king united all the lands containing Blue settlements into the country of Spectrum.

At first all went well. The Greens cultivated the land. The Blues pursued their activities in the towns. Both peoples retained their own language and religion. There were sometimes misunderstandings when Greens and Blues met in the marketplaces, but these were regarded as matters concerning only the individuals involved.

After several generations, however, troubles became more widespread. Young Green people who had completed a higher education complained that they could not get jobs commensurate with their ability and training. The civil service was dominated by Blues. Green farmers felt that Blue bankers were discriminating against them when it came to extending credit. With the rise of the mass media, nearly all located in Blue-speaking cities, Green parents worried that their children would abandon the traditional language, and forbade them to listen to the Blue radio or attend Blue movies.

By this time the course of future development in communal relations was not difficult to predict. Several young Greens who had been abroad to study formed a nationalist Green party. They recalled that the eastern half of Spectrum had formerly been exclusively Green, and demanded that the five predominantly Green-speaking provinces secede and form an independent state. The Blues retaliated by moving troops into the five provinces. Violent incidents became more common, and civil war a possibility.

There were, however, a number of prominent Greens and Blues who were reluctant to see Spectrum plunged into violence. They agreed that it would be a disaster for both sides, although they agreed on little else, and formed an informal Conciliation Committee to see what could be done. After the existence of this committee was reported by the international press, its two cochairmen received a letter from a public opinion research institute in a distant country.

"We have heard about the difficulties in Spectrum," the letter began, "and also that your committee is exploring possibilities for peaceful solutions. While we do not claim to have detailed knowledge of conditions in Spectrum, we believe that an impartial study of popular attitudes on Blue-Green relations might be useful to you."

The letter went on to say that, thanks to a grant from the Rockaford Peace Endowment, it could send two specialists in public opinion research to Spectrum for a period of several months. If the committee could find the funds necessary to pay interviewers, the research institute would conduct the survey, analyze the results, and prepare a report. The only conditions of the offer were that the report must be made public and that the raw data from the survey would be available to anyone else who wished to make an independent analysis, although the identity of respondents would of course be protected.

After considerable hesitation the Conciliation Committee decided to accept the offer. Both Blue and Green members became convinced that any impartial investigation would favor their own side, and each group within the committee agreed to raise half of the necessary funds to cover interviewing expenses.

Other reactions in Spectrum were more acerbic. The government, dominated by Blues, issued a statement saying that any investigation conducted by an institute from another country would have no value, and that the government was already in possession of all necessary information about popular attitudes. It agreed, however, not to interpose any difficulties, since Spectrum had nothing to hide and—as was well known—all citizens of Spectrum were free to voice their opinions on any question. A number of Green spokesmen said angrily that the proposed opinion survey was undoubtedly a plot designed to discredit legitimate and internationally recognized Green grievances, and that they could take no responsibility for the safety of either the researchers or the interviewers.

After the researchers arrived, they devoted several weeks to framing a questionnaire, pretesting questions, and arranging for interviewing. Parts of the interviewing task were subcontracted to experienced market research organizations in Spectrum and to a local university research bureau. Pretesting was done by interviewers hired directly by the researchers; some of these men and women were Blues, some were Greens, and some were foreign students studying at Spectrum universities. Comparison of the results obtained by different interviewers made it possible to correct for interviewer bias when necessary, although this did not prove to be a major problem.

When the results of the poll were published, they proved mildly surprising to most citizens of Spectrum. It turned out that the Blue-Green controversy preoccupied many fewer people than had been popularly supposed. While a large majority of Greens were concerned with what they regarded as economic and cultural discrimination, they were even more worried about agricultural policies that, they felt, gave an undue advantage to foreign producers, and about the need for better roads and schools in Green-speaking areas. Approximately 15 percent said that they favored independence for the five predominantly Green provinces; nearly 60 percent wanted greater autonomy. Among the Blues, interest in the controversy was even less widespread; a substantial minority was unaware that many Greens regarded the existing situation as discriminatory, and most of those who were aware of this thought that the Greens were concerned mainly with the central government's language policy. Nearly three-quarters of the Blues said that they would be willing to see their income taxes go up by 5 percent if this would lead to peace in the country. The report issued by the research institute made no specific recommendations for action, pointing out that these could be framed more appropriately by political bodies and other groups within Spectrum.

After a lengthy discussion of the report, the Spectrum Conciliation Committee advanced a number of proposals for reform, and these were given extensive attention by the Spectrum press. The government officially disregarded the research report, as it had promised to do; but the Conciliation Committee proposals led to intense discussion within the Blue party, which split into moderate and radical wings. A similar development could be observed within the Green party; a group of moderates expressed support for the Conciliation Committee proposals, but the radicals said that they would continue their struggle for independence, using whatever means were necessary.

According to the latest reports from Spectrum, incidents of violence involving Greens and Blues continue to occur, but less frequently than before. It appears that confrontation is shifting from the streets to the political stage. Some observers in Spectrum think the outlook for peace in the country is now more promising. They are not sure, however, whether the survey conducted by the research institute had anything to do with subsequent developments; these might have occurred in any case. It will probably be many years before a consensus on the question is reached.

So much for the imaginary country of Spectrum. Now let us consider more briefly the case of another fictitious state, which can be called Nova.

The country of Nova emerged during the mass decolonization process that followed World War II. It was composed of a number of diverse peoples, speaking different languages, that had been grouped together by the colonial power for administrative convenience. There was little communication among these peoples; and the country's borders cut through the territories of several tribes, so that some citizens of Nova felt closer ties with those outside the country than with traditionally alien peoples who were formally their fellow countrymen.

Nova was rarely mentioned in the world press. No correspondents were stationed there, and such reports as filtered out were usually based on stories carried by a small European-language weekly that was published in the country's capital, or originated from accounts of occasional travelers who visited Nova's back country.

It is probable that a decade or more would have passed before Nova emerged as a major subject for attention in the world press, had it not been for the work of the Globewatch Monitoring Service. This service, a private body with headquarters in Geneva, issued a report in 1981 that pointed out that Nova was one of three countries in the world where large-scale hostilities were likely to break out at almost any time. The report, based mainly on information from travelers and Nova students who were studying abroad, described the situation there as critical. What amounted to private armies had been formed by ambitious leaders in two of the country's language areas. One of these armies was receiving significant help from abroad. People who migrated from one part of the country to another were frequently murdered or forced to flee; the central government's authority was tenuous outside the immediate area of the capital. While violent incidents based on ethnic factors had been frequent in the days of the colonial regime, the main difference now was that the two ambitious leaders had worked out political programs calling for expansion of the territories in which their peoples lived and for the establishment of independent states within these enlarged territories. Unless something was done soon, the report concluded, widespread bloodshed was a near certainty.

This report was initially sent only to the government of Nova, the secretary-general of the United Nations, and several organizations concerned with peace research. Its contents soon leaked out, however, and angry reactions were reported by the world press. The government of Nova said that it would tolerate no interference in its internal affairs; the Organization of Non-Aligned States concluded that this was a thinly disguised attempt to reintroduce colonialism and undermine Nova's territorial integrity; the representative of an Eastern European government, speaking in the U.N. General Assembly, noted that the difficulties in Nova—if indeed there were any difficulties—were caused by a class struggle and had nothing to do with ethnic differences.

Several research organizations took the monitoring service report more seriously, and one of them decided to explore the situation in Nova in greater detail. Since the government of Nova would not allow a study to be conducted within the country, and travel and security difficulties would have made this impossible in any case, the organization stationed research personnel in two neighboring countries that were frequently visited by travelers from Nova and where a number of refugees from Nova were living. The researchers interviewed as many Novitiates as possible, as well as foreigners who had visited Nova, and compiled a picture of attitudes in the country on the basis of the available information.

Their report, which was released to the press within a year, pointed out that the nationalist and expansionist ideologies of the two rebellious leaders had, in fact, little popular support. The principal problem facing all the peoples of Nova was extreme poverty. And poverty led to interethnic violence for two reasons. One was that each of the principal groups in the country thought, erroneously, that other groups were more prosperous and were being given preferred treatment by the central government. This belief explained at least some of the attacks on people from other provinces. The other reason that poverty tended to lead to interethnic violence was that the two nationalist leaders were somehow able to pay the members of their private armies. (The report speculated that both these individuals were subsidized from abroad.) Many men joined the armies mainly because they were unable to feed their families by working in their traditional occupation. The report also confirmed that the various peoples making up Nova had little sense of a common nationhood. Whether they could better attain self-realization as members of a single country or whether parts of Nova should become independent or be joined with neighboring countries was a question that would have to be decided by the peoples themselves.

When the report was released to the press, it received widespread coverage in mass media outside Nova, but little attention from media within the country. Newspapers in the capital dismissed it as idle speculation by foreigners who knew nothing about the real situation. It was ignored by the government-controlled radio, and there were no newspapers outside the capital. A substantial number of people in Nova did, however, learn about the report's contents by listening to foreign radio stations, and this information was spread further by word of mouth. Outside Nova the report aroused considerable unofficial discussion in the corridors of the United Nations, and Nova's delegation was frequently asked by representatives of other countries whether something could be done to reduce the risk of widespread fighting, which easily could spill over Nova's borders into neighboring states.

Shortly thereafter the government of Nova announced that a constitutional convention would be held to write a new basic law for the country—one that would take better account of the diverse ethnic groups than the constitution that had been composed with the advice of the former colonial power. It has not been determined whether this development was related to the report of the Globewatch Monitoring Service or to the later report on popular attitudes in Nova.

Be this as it may, preparations for the convention began to claim more and more of the attention of leaders from all the country's ethnic groups. At the same time the World Bank agreed to reconsider a loan application from Nova, which had previously been turned down, thus making it possible to start construction of irrigation canals in several sections of the country. According to some observers, the bank's decision was influenced by the research organization's conclusion that poverty in Nova was a major factor in interethnic violence.

The danger of widespread conflict in Nova now seems to have abated somewhat, although the two private armies continue to exist and a final solution to the country's problems still appears to be in the distant future.

CONSIDERATIONS AFFECTING THE USE OF OPINION RESEARCH

The hypothetical cases presented above illustrate some of the problems, difficulties, and possibilities of applying public opinion research to the resolution of conflicts based on nationality differences.

The problems and difficulties are fairly obvious. Every situation is unique; an approach used in one case will not necessarily be helpful in the next. There is likely to be resistance to the conduct of opinion research by a third party, and there is sure to be skepticism. Those who intervene in a nationality conflict within another country will certainly be exposed to the charge of meddling, even if their good intentions are acknowledged. As outsiders, they can usually do no more than provide new ways of thinking about the problem and point out a number of possible solutions. The specific measures to be taken will have to be worked out and accepted by the parties to the conflict.[1] Even if the intervention is successful, this is usually difficult to prove after the fact. Many groups and individuals are ready to accept the credit for successful solutions; responsibility for failure is much easier to pinpoint.

An immediate problem is that the use of opinion research as an aid to the resolution of nationality conflicts requires money. Not a great deal is required, if one compares the cost of a sample survey with that of a military aircraft, but those concerned with conflict resolution have to contend with the deep-seated assumption that expenditures for peacemaking should be very small, or it should cost nothing, while at the same time it is accepted that the costs of war are very large. It is difficult to put a precise figure on the investment required for the approach described here, but it is unlikely to fall below $100,000 (in 1980 dollars) for any single case, and could rise to several times that amount.

The funds for an approach along these lines should come from private sources, if possible. Government funds would be regarded with suspicion. This requirement flies in the face of the current tendency to leave more and more functions to governments. If it should prove necessary to use public monies, these should come from several different countries, or at the very least they should be mixed with private funds.

The application of public opinion research to conflict resolution presupposes the availability of one or more research organizations qualified to perform the necessary work, the existence of mass media that will disseminate the results of this work, and the presence of groups and agencies that are willing and able to apply the findings—to turn them into concrete proposals and to act on the basis of these proposals.

It would not be difficult to equip a research organization to perform the kind of task outlined here. What is necessary is to combine skills in international survey research with expertise in international relations and experience in conflict resolution. Personnel with these qualifications are available in many countries; the problem is to bring them together and to make it possible for them to prepare themselves to act in any situation where their services might prove useful. This is, in part, again a question of money.

It goes without saying that any research organization that intervenes in delicate political situations should be recognized as one that is free from political bias and that observes the highest professional standards. Not only should its financial resources come from nonpolitical sources, or from a number of diverse sources, but its personnel should, if possible, represent several different nationalities. There might be some advantages in locating the headquarters of such a research organization in a traditionally neutral country. A university affiliation might also be useful. It should have the capability to issue press releases and to make information about its research available to interested persons and groups. Most important, however, is that all its work be open to public inspection and to review by other professional bodies.

Contacts with mass media that are able and willing to give wide dissemination to research results are important because official channels for such dissemination will in many cases be blocked. A government-controlled press will be unlikely to carry information that would strengthen the case of the opposition. And in parts of the world where the mass media are poorly developed, it is possible that those to whom the survey results would be most relevant would never hear about them. This is especially likely to be the case in new countries comprising a number of diverse peoples, where modern communication systems may be rudimentary while antagonisms among the component nationalities are robust. Fortunately, short-wave broadcasting, combined with the transistor radio, has greatly reduced the number of areas that are not served by the international information network, although the quality of information provided often leaves much to be desired.

Mass media have a particularly vital role in this approach to conflict resolution, since one cannot know in advance which individuals and groups will be able to make use of the researchers' findings. Will it be an obscure politician in a remote province who sees that there are grounds for forming a coalition with leaders of another nationality group in a neighboring province? Will it be a group of shopkeepers who realize that their business would improve if they put up signs in two languages? Or will it be a scholar at a local university who discovers a political formula that will satisfy the aspirations of two diverse peoples, and presents it to his government? Only the mass media can reach large numbers of human beings dispersed over huge distances and locate those few whose work or thought can make a significant difference.

Another important function performed by the press is to focus public attention on conflicts and potential conflicts. Even powerful decision makers behave differently when they know they are being observed. If they are reluctant to engage in negotiations, or to make a generous gesture toward political opponents, the consciousness that the eyes of the world are on them may be a factor in persuading them to make conciliatory moves that they otherwise would not consider.[2]

But will there in every case be some people who are desirous of finding a peaceful solution to a nationality conflict? This approach assumes that there will be. It seems reasonable to make such an assumption, since in all the situations we have examined or heard of, there are individuals and groups who are conducting negotiations and seeking formulas that would forestall conflict. Indeed, in all of the eight cases summarized in this volume, the governments of the states in question have been involved in seeking solutions. One might object that in some instances they were not sufficiently active or were not always negotiating in good faith, but there is no indication that they would ignore ideas that might contribute to communal peace in their countries.

In addition to research organizations, mass media, and potential users of survey results, there is another institution whose existence would greatly enhance the utility of the approach to conflict resolution suggested here. This is a monitoring service, an organization that would keep close watch on ongoing and potential nationality conflicts in all parts of the world, and would maintain a running "fever chart" on these situations.

Such an organization could be of particular value in providing early warning of developing conflicts. Early warning would make it possible for groups interested in conflict resolution to prepare the way for peaceful solutions before the positions of the contending parties had become solidified, before masses of people had been inflamed by fiery rhetoric, and before bloodshed had occurred. Conflict has a momentum of its own; it is far less difficult to stop in its early stages than after it has become a way of life within a society.

It is surprising that such a monitoring service does not exist. We had assumed that something like it could be found within either the U.N. headquarters or the U.S. government, or possibly both. It probably would be called by a different name, and its functions might well include more than monitoring only nationality conflicts; but one of these huge bureaucracies would certainly be able to provide a current inventory of major ethnic or communal disturbances in progress throughout the world. We were, however, unable to find any body that maintained such an inventory, either at the United Nations or in Washington. Most of the necessary information probably was at one location or the other, but to dig it out (assuming that considerations of secrecy would not cause problems) would have been a most time-consuming task. To judge on the basis of our exploratory investigation, the attention of governments and international organizations is focused primarily on other governments and other international organizations. Except when the danger of international conflict is immediate, or when there are massive flows of refugees, attention to ethnic minorities or communal disturbances is peripheral.[3]

Establishment of a global service to monitor potential threats to the peace, as well as ongoing conflicts, would not necessarily be an expensive undertaking, although its cost would rise sharply if it were to conduct its own investigations in the field. At a minimum such a service might consist of three or four persons who kept close watch on newspaper and broadcast reports from all parts of the world, periodically consulted appropriate U.N. offices and major governments, and maintained correspondence with volunteers in various parts of the world. At a maximum the service would have the capability of sending its own personnel to look into situations that had not yet come to the attention of the press or other major information sources.[4] Whatever the cost, it would be modest in view of the savings in lives, suffering, and money that could be achieved by an institution able to provide more adequate early warning of potential conflicts than is now the case.

Like the institute to conduct research on ethnic and communal differences, the monitoring organization should preferably be financed from private sources and have an international staff. It is possible that at some point its responsibilities could be assumed by an agency of the United Nations, but this would be appropriate only if this agency could be guaranteed freedom from interference by the governments represented in the world body.

POSSIBILITIES OF CONDUCTING PUBLIC OPINION RESEARCH WORLDWIDE

The approach to the resolution of nationality conflicts that we are suggesting assumes that it is possible to conduct research on popular attitudes in all parts of the world, or almost all parts. Is such an assumption correct? There obviously are areas where conventional polling would not be permitted by governments, or where people would be afraid or reluctant to give their honest opinions to interviewers, or where poor security and nonexistent transportation facilities make it impossible for interviewers to function. Nevertheless, such areas are becoming fewer; and when conventional polling methods cannot be used, other techniques of assessing mass attitudes can be employed. Some of these will be mentioned below. There are now very few parts of the world that are beyond the reach of public opinion research.

The capabilities of international public opinion research are illustrated by a global survey conducted between 1974 and 1976 by the Gallup International Research Institutes as part of a project sponsored by the Charles F. Kettering Foundation (Gallup and Kettering 1977). It covered 59 countries on all continents. The researchers were unable to arrange for interviewing in the Soviet Union, Eastern Europe, and the People's Republic of China, although negotiations were continued in the hope of including some of these areas in future surveys (Gallup 1976-77).

Some researchers have assumed that meaningful surveys cannot be conducted in areas where illiteracy and lack of education in general are prevalent. It is sometimes thought that people with a very low educational level will be unable to verbalize their wants, needs, and concerns. This, however, turns out not to be the case, at least in areas where surveys have been conducted. The Institute for International Social Research, which undertook a study of the concerns of the Nigerian population in the 1960s, expected to find a low level of political awareness because of the extent of illiteracy among the general population. The Institute found, however, that its expectations were not borne out.

Its report on the survey stated: "The list of problems ... about which the Nigerian people expect the government to act is the most extensive we have encountered in our studies to date" (Free 1964). Among these problems were education, public health, agrarian reform, and improvement of internal trade. The more recent Gallup-Kettering global survey, which included some of the least educated populations of the world, also found that illiteracy was no bar to the expression of opinions about needs and satisfactions (Gallup and Kettering 1977).

Useful information can be obtained by surveys even when people are reluctant to state their opinions frankly. Polls conducted among the black population of the United States in the 1940s and 1950s found that white interviewers would obtain one response; black interviewers, another. During World War II, for example, a survey that asked a sample of the black population in Memphis, Tennessee, whether blacks were fairly treated in the army found that 11 percent said "no" when white interviewers asked the question; 35 percent said blacks were not treated fairly when black interviewers asked the question (Hyman 1954). While it is more difficult to gauge attitudes in such situations than in situations where less sensitive questions are involved, the use of different interviewers and various devices that help to correct for understatement or overstatement makes it possible to form fairly accurate estimates.

A more recent example of the willingness of respondents to speak out in a tense situation is afforded by the Gallup-Kettering global survey, which included Uganda as one of the countries sampled. Although the government of Idi Amin, in power at that time, was not considered to be particularly tolerant of criticism—

indeed, it was known as tyrannical and arbitrary—about a quarter of the respondents volunteered that they were dissatisfied with the regime. "Military rule has caused fear among the people and has hurt trade," said one. "The lack of a civilian government is a major problem facing the country," said another. There may, of course, have been some who feared to speak out, but the fact remains that many did.[5] Later events in Uganda attest to the fact that the poll made it possible to construct a reasonably true picture of attitudes prevailing there.

In other words, while we cannot be sure that people always will be frank in stating their opinions, previous experience makes it appear likely that they do so in enough cases to make the poll a useful tool for uncovering hardships and grievances.

Where direct interviewing within a country or region is impossible, researchers have frequently been successful in constructing accurate pictures of public opinion by interviewing residents of that country or region at other locations. This technique was developed during World War II, and has since been used most extensively by the audience research department of Radio Free Europe to measure opinion in a number of Eastern European countries. The audience research ddepartment has contracted with Western European polling organizations to interview travelers from Eastern Europe who visit Western Europe. Naturally these travelers do not represent a random sample of the home populations. The researchers therefore use a statistical technique to increase the weight in the sample of the population groups that are underrepresented and to decrease the weight of those that are overrepresented. In this way a "synthetic" random sample is constructed. Radio Free Europe has attempted to verify the accuracy of this polling method by asking some of the same questions that are asked by survey organizations within the countries concerned, then comparing the answers they receive with reports published in Warsaw, Prague, or Budapest. In general the two sets of results have agreed fairly closely (Radio Free Europe 1976). The availability of census data from the countries of Eastern Europe naturally facilitates this method of analysis.

Even when census data are not available, or are very unreliable, a fair picture of opinion distributions on many questions can be obtained by pooling information from a wide variety of sources: travelers, newspapers, refugees, older demographic studies, radio monitoring reports, and so on. This approach was used by American and British researchers during World War II, in an effort to obtain insights into public opinion in areas to which they did not have direct access (Mcad and Métrau 1953; George 1959).

Market research organizations in the United States have experimented with using "surrogates" for the people they would like to interview, when the latter are not available. This involves finding individuals who resemble the desired respondents in as many respects as possible, and then asking for their opinions or reactions. A related approach is involved in "political gaming"—that is, asking people who are highly knowledgeable about a given society to play the roles of certain individuals in that society and to respond in the way they think these individuals would respond (Goldhamer and Speier 1959; Guetzkow 1972).

These substitutes for direct opinion polling are naturally not as satisfactory as a carefully planned and executed survey on the spot. Nevertheless, they have, in many cases, produced usable results. There are still areas of the world where it is impossible to obtain a sufficiently detailed picture of the state of public opinion on major questions, but they are relatively few.

WHAT INFORMATION ABOUT POPULAR ATTITUDES IS DESIRABLE?

Since each nationality conflict is unique, the surveys used in searching for ways to resolve such conflicts must be designed individually, following a careful study of the situation. There is no standard research design that can be used, no set questionnaire that can be administered.

Nevertheless, the case studies in this volume and the literature on nationality conflicts in general suggest a number of areas that might fruitfully be explored in any given conflict. One can construct a check list of subjects that might repay investigation. Some of these subjects are likely to prove relevant to a particular case; others may not. At this point we will suggest only three categories of questions that might be considered for inclusion in any particular survey. One category has to do with the attitudes of individuals toward the conflict; the second concerns leader-follower relationships; and the third includes questions regarding alternative courses of action.

In any long-standing ethnic or communal conflict, individuals are likely to have a rich assortment of attitudes of varying strength about their own group and about other parties to the conflict. Most obviously, one would want to know to what extent people felt that they were deprived, or exploited, or discriminated against—either individually or collectively—because of their nationality or group membership. And one would also want to know whether members of some groups felt that they were superior to other groups within the same state, and therefore were entitled to greater advantages.

In regard to feelings of deprivation, the literature on nationality problems has enumerated a large number of the grievances and fears of individual groups in plural societies. One study, focusing on developing countries, identifies the following as issues that are likely to divide one national community from another: language policy, recruiting of civil servants, education policy, political organization (for example, how administrative boundaries are drawn), and the division and distribution of resources (where roads are built or factories are located) (Heinz 1973). Another study, dealing with the problems of staffing a government bureaucracy in a multinational state, says pessimistically that a politically neutral bureaucracy simply is not attainable in a plural society, and notes that the composition of the security forces in such states is a particularly thorny problem:

> Any military, paramilitary, or police force whose racial or ethnic make-up might assure the government that its orders would be carried out promptly and efficiently would, for that reason, be regarded as an instrument of oppression by the other community. (Dishman 1978, p. 282)

Other analyses explore the strains that modernization and industrialization impose on multinational states, some concluding that modernization tends to encourage conflict and others holding that it does not (Morrison and Stevenson 1972). In this connection, television has been singled out as a factor that may promote dissatisfaction on the part of cultural minorities and at the same time may encourage the nationalists among them. Television can promote dissatisfaction among minorities "because it brings the majority culture, often in its least attractive form, right into the living room" (Birch 1978, p. 336). It may be feared that children will be weaned away from the language of their parents; the possibility arises that the cherished tongue of the smaller national group may become extinct within a generation. And television encourages protest by minority nationalities because television news offers an audience of millions to any protest activity that journalists regard as newsworthy.

Numerous other examples of issues that can divide nationalities in multinational states have been mentioned in the case studies presented in this volume. In Israel ownership of the land is a major issue. In Canada and Belgium the opportunity to develop a distinctive culture has been a cause of controversy. Among the Harijans of India, deprivation of dignity, as well as curtailment of economic opportunity, have been serious grievances. In contrast, some of the Basques of Spain seem to have been spurred to revolt, in part, by a feeling of superiority. The number of issues that can lead to friction among nationalities appears to be almost infinite.

It is important to note, however, that none of these issues is a cause for conflict except when people perceive it as such. Unequal economic opportunities do not cause discontent unless the disadvantaged group is resentful of them. And this usually occurs when members of one nationality compare their situation with that of another. Indeed, some scholars believe that a feeling of *relative* deprivation is the explanation of why some groups enter the political struggle (Stokes 1978). It is not that *I* have too little, but that *he* has more. Thus, we find that Arabs in Israel frequently compare their situation not with that of the less prosperous Arabs in neighboring countries, but with that of the more prosperous Israelis. If they used another standard of comparison, the political implications might be quite different. Even when a minority is actively persecuted, when its language is forbidden, its leaders imprisoned, and its members denied attractive jobs—even then the degree of protest depends on the attitudes of the members of the minority.

One role for surveys in conflict resolution, then, is to establish how many people experience a particular grievance or deprivation; who these people are in terms of age, occupation, place of residence, and other variables; and how strong their attitudes are. If surveys were more widely utilized, we would find, I suspect, that grievances that receive the most publicity are not always the most widely held or most strongly felt. A corollary is that surveys may also be able to identify problems and grievances that are widespread but have not yet become the subject of public discussion. If a problem can be identified before it becomes a basis for public outcry, the solution presumably will be that much easier, assuming that there is a will to find a solution.

A second category of information that can be obtained through the study of popular attitudes has to do with the identification of leaders and followers, activists and passive elements, in ethnic or communal groups, and the relationships among them.

The importance of this kind of information is illustrated by an election in the late 1960s in France, when Basque nationalists ran candidates for Assembly seats in the Basque area. In no case did one of these candidates win more than 5 percent of the total vote. The isolation of the nationalist leaders from the population was dramatically demonstrated, even though they previously had been recognized, by the outside world at least, as spokesmen for the Basques of France (Jacob 1975). A less dramatic but still interesting illustration is provided by the case of Quebec, where repeated surveys found sentiment favoring independence to be less widespread than had been popularly supposed.

Since leaders of nationality groups may play a part in resolving nationality conflicts, as well as in making them worse, those involved in conflict resolution often find it useful to know who these leaders represent. This can sometimes be learned from election results. But in areas that do not have free elections, or where various ethnic or communal groups are so intermingled that interpretation of election results is difficult, the opinion poll is a useful diagnostic tool.

Surveys can, indeed, provide much more information about leader-follower relationships than can elections, since in the former it is possible to relate the expression of attitudes to other characteristics of each individual. For example, a study carried out in the early 1970s among the French-speaking population of Quebec identified two rather different groups: one was oriented more toward cultural than political matters, and was made up largely of blue-collar workers born before 1925; the other was composed mainly of younger white-collar workers and was more interested in changing Canada's political structure (Irvine 1972). A somewhat similar division is suggested by Maurice Pinard's analysis (see Ch. 8). Whereas fewer than a quarter of French-speaking Quebecers thought that the province should have complete power over such matters as defense, monetary policy, or immigration, about half favored exclusive provincial authority with respect to education and cultural affairs. Such differentiations probably would not be made clear by election returns, since members of both groups might vote for the same candidates—although for different reasons.

Several studies of national movements have noted that the personal fortunes of nationalist leaders often are dependent on the extent of national consciousness among members of the population (Wallerstein 1972). Or, to put it another way, national movements offer some people a means to get ahead, to establish claims to jobs and prestige. These people thus have a vested interest in promoting ethnic consciousness. If I campaign for political office as a spokesman for a particular ethnic group, then it is to my advantage to encourage as many people as possible to think of themselves as members of this group. As one authority has put it, "Democratic elections create new rewards for ethnic solidarity" (Enloe 1972, p. 270). Consequently, it may be in the interest of ambitious individuals to cultivate a sense of grievance among members of the group to which they belong, in order to establish themselves as the leaders who can remove that grievance.

But the reverse is also possible. The grievance may exist and may be keenly felt long before leaders emerge to give it public expression. This was certainly true with respect to the black population in the United States for many years. It probably is still the situation of minorities in many parts of the world. (Chapter 5 of this volume suggests that it has been true for the Harijans of India.)

The question is sometimes a critical one. Are leaders, spokesmen, and activists seeking to promote a national identity largely because they themselves have a keen sense of being deprived of jobs and advancement, or are they (reluctantly, perhaps) giving expression to deep feelings of discontent among broad masses of a downtrodden people? In the former case the avenue to avoiding or resolving conflict would seem to be to find jobs for the relatively small number who are vocal and unhappy.[6] In the latter case more far-reaching changes would be called for.

It goes without saying that idealism and self-interest may be mixed together in varying proportions, and usually are. Few leaders of nationality groups speak only for themselves. Even fewer may be completely selfless spokesmen for silent and suffering populations. The mixture of motives is not easy to untangle, but the survey offers one method of approximating the extent to which the statements of a leader actually represent the thoughts or desires of a larger number.

There is another reason for resorting to surveys to gain information about leader-follower relationships, and it has nothing to do with the motivations of the leaders. Numerous investigations have shown that those who speak for large groups often have incorrect information about the sentiments of the group membership. Union leaders may misjudge the political preferences of the rank and file; religious spokesmen may not be aware of the attitudes of church members; and so on. This is not to say that leaders are invariably out of touch with their followers, but it happens often enough to make it inadvisable to accept the statements of official or unofficial group spokesmen without a further check.[7] If we have to rely on statements of leaders alone, some large areas of popular discontent may remain hidden, and consequently be impossible to deal with.

One researcher, who compared the views of a sample of American Indians with those of prominent Native American spokesmen, found that the "rank and file" had rather heterogeneous attitudes on outstanding issues, and tended toward ideological moderation. There did not appear to be much support for the more radical "third world" ideas expressed by many Native American leaders (Ritt 1979). Similarly, the analysis of sentiment in the Basque country of Spain presented in Chapter 2 of this volume by Juan Linz suggests that some prominent Basque spokesmen are rather far out of step with most people of Basque descent.

This is not to suggest that leaders should necessarily hold opinions that are identical with those of the majority in their constituency. By definition a leader seeks to influence those who are led. Nevertheless, when it comes to conflict resolution, it is important to know the actual distribution of opinions within a population. If solutions are based on the views expressed by prominent spokesmen, the existing aspirations and grievances of large numbers within a nationality group may not be taken into account. Well-designed surveys can usually give a sufficiently accurate picture of these aspirations and grievances.

A third task for opinion research in the resolution of nationality conflicts would be to identify attitudes and behaviors that might facilitate peaceful solutions. This probably will be more difficult, because relatively little prior thought has been given to it. The literature on societies that have experienced nationality conflicts is voluminous; less attention has been given to attitudes in multinational societies where people have lived together in peace. Ethnic, racial, religious, and other differences obviously are not invariably a cause for hostility. Indeed, it may be that under some conditions they facilitate peaceful and constructive cooperation. If so, what are these conditions? At present there are few answers.

Nevertheless, the literature does include a number of suggestions. Some of these seem rather minor in comparison with the problems to be solved, but even small steps should not be ignored if they promise to be of any help at all. For example, several students of multinational societies mention the importance of good manners in relations between members of different communities. The Swiss are characterized in one study as a people "who are so accustomed to order and civility" that they have been deeply shocked by the occasional violence of the Jura separatists (Steiner and Obler 1977). Another study notes that the Malays and Chinese in Malaysia have a low opinion of each other, but also states that "a high degree of civility accompanies most exchanges and interactions" between them (Esman 1972). Although racial prejudice is a problem in Brazil, according to a third study, the prevailing norms call for "essentially courteous, harmonious, interracial behavior" (Saunders 1972).

The concept of civility crops up with sufficient frequency to suggest that it is important in the maintenance of working relationships, even if not cordial relations, among nationality groups. It would appear that studies of communal tensions might well give some attention to the prevalence, or the absence, of good manners in the relations between members of various nationality groups.

Another attitudinal factor that seems to help plural societies function as viable states is respect for diversity. This respect must first of all be shown by the government. In some cases it is enjoined in laws and constitutions. An analysis of ethnicity and politics in Yugoslavia notes that a guiding principle of Yugoslav leaders is "A united community of Yugoslav peoples can best be assured through governmental respect for ethnic diversity" (Bertsch 1977, p. 88).

Indeed, one formula for multinational states that seems to be favored by a number of political scientists is known as "consociationalism." This involves recognition by all concerned that the population is composed of a number of cultural elements, an agreement to disagree about some cultural issues, and willingness to allow the leaders of the constituent groups to negotiate among themselves the distribution of public benefits and the resolution of communal differences (Lijphart 1968). One variant of this formula seems to have worked reasonably well in Malaysia for over a decade, although it then broke down as more and more members of the Chinese and Malay communities became convinced that their leaders were not pressing vigorously enough for the interests of their constituents (Esman 1972).

As the example of Malaysia suggests, the consociational formula rests to a considerable degree on public attitudes. There must be at least a minimum degree of popular respect for diversity, as well as governmental recognition and respect for it. The Institute for International Social Research found a realization of this among some members of its Nigerian sample, although as recent history lamentably records, this realization was not sufficient to avert a destructive civil war. One respondent remarked that if national unity was to be maintained, "We must love our neighbors." Another said: "I would not like Nigeria to become a country where there is no love but where there is war and misunderstanding as they have in Ghana or the Congo" (Free 1964). A respondent to the Gallup-Kettering global survey in Tanzania, asked what he would do if he had all the money he wanted, replied: "I should like to travel within the country and meet all the tribes that there are."

The spirit of respect for and curiosity about those who are different is present throughout the world, even if not always in generous quantities, but it can be discovered and built upon.

The account of the Harijans in India (see Chapter 5) reminds us that surveys can do more than tap popular attitudes. They can also suggest actions that people can take themselves to overcome grievances. The Harijans, in many cases, decided to dig their own wells, thus avoiding the hardship and humiliation involved in being denied permission to draw water from the wells of higher-caste

Hindus. Family decisions with regard to education and birth control were also found to be important in relieving the disabilities of the Harijans. Civility, which was mentioned earlier, is of course another quality that is built on the behavior of individuals.

An advantage of using surveys to uncover alternative courses of action that are likely to enjoy at least some measure of support among the populations affected is not only that one or more of these alternatives might ultimately provide a formula for conflict resolution. It is also that discussion of alternatives in itself may provide a substitute for violence by diverting energies into political channels.

In the Belgian situation, for example, it would appear, at least to one standing on the outside, that the complexity of the problems faced by the Flemings and Walloons has been one reason the Belgians have dealt with those problems so successfully. There have been heated discussions, and fiery slogans, but these have mostly been directed for or against particular formulas that looked toward a solution and not toward conflict. One might speculate that many Belgians became so fascinated with the checker game of politics that there was no room for a military game. In Canada, similarly, the multiplicity of formulas through which the French-speaking population might achieve self-realization has tended to fascinate those who feel themselves involved in the controversy. There has been violence in Canada, but it proved to be a minor episode.

One of the principal purposes of the survey is thus to present a menu of possibilities that are alternatives to conflict. Each of these possibilities must be acceptable to at least some segments of the populations in question, and should be regarded as at least worthy of debate by other segments. Perhaps none of the alternatives will prove to be the one that will bring communal peace, but as long as they are being discussed, the chances of peace are improved.

THE NEXT STEPS

Whether public opinion research will, in fact, prove to be a useful tool for the resolution of ethnic or communal conflicts depends in large measure on whether those who are in a position to act are willing to make use of its results. We believe that some would be willing if these results were provided to them. Users might include governments, political parties, leaders of nationality or communal groups, diplomats, or private individuals, depending on the unique characteristics of each situation.

It seems safe to assume, further, that opinion polls or other studies of popular attitudes will be of the greatest utility in areas where they are least likely to be made, either because the facilities for making them do not exist or because no existing group or organization has an interest in doing so. The southern islands of the Philippines offer an example of an area in which the conduct of such a study would presumably be difficult; Puerto Rico appears to be a case where good facilities exist but where there has been a lack of interest.

Satisfactory tests of the utility of public opinion research as a tool for the resolution of nationality conflicts are therefore likely to occur only if research organizations select situations where they think attitude surveys could help in bringing about peaceful solutions, conduct these surveys, and offer the results to any who would be willing to make use of them. In other words, the initial surveys would probably have to be done as speculative enterprises. An organization conducting one would be somewhat like an author who writes a book without assurance that a publisher will accept it or that anyone will want to read it. Nevertheless, some of the best books have been written under these conditions.

The next steps, therefore, are for qualified persons of good will to frame proposals regarding specific areas and take them to likely funding sources. We have attempted to suggest some of the ways public opinion research might be applied to the resolution of nationality conflicts and to do some of the initial spadework for its application. We hope that others will improve on our design and will be able to put this instrument to work in the search for a more peaceful world.

NOTES

1. The principle that outsiders can often be useful in bring contending parties together and helping them to formulate solutions is widely accepted in diplomacy and labor relations. Outsiders are less likely to be successful when they formulate a compromise solution themselves, unless they are very powerful. The usefulness of social scientists as "catalysts" in helping others to find formulas to resolve international conflict has been partially demonstrated in a number of experimental situations (Burton 1969; Doob 1969).

2. It is recognized in the literature on labor disputes that the mere presence of a third party, even if this third party does nothing, will often promote agreement by moderating expressions of hostility. The same is true even in domestic disputes. A recent study of the functions of the police contains the following suggestive passage:

> One police officer I really admired, he'd come into a family beef with a husband and wife throwing and yelling at each other. Then he'd set down on the couch and take his hat off, and he didn't say a word. Sooner or later the couple felt kind of silly. He'd take 45 minutes in each of these situations, but he never had to come back. (Officer Mark Rockingham, Laconia Police Department; Muir 1979, p. 82)

The press can sometimes perform a similar role as a "third party" in domestic disputes between nationalities within a single country. Under some conditions the contending parties will act with more moderation if they know that they are the focus of public attention. An analogous role for the press in international negotiation has also been noted (Davison 1974).

3. Our observations tended to underline a remark made by Edvard Hambro, former permanent representative of Norway to the United Nations. Referring to international political disturbances in general, he said: "We do not have the intelligence machinery necessary to prepare in advance for a reasonable solution of the problems. . . . The agenda is so crowded that all but the most urgent problems are put aside." Hambro also referred to the necessity of outside pressure on political leaders: "We need . . . leaders of public opinion who can force statesmen and politicians into action before it is too late" (Stanley Foundation 1972).

4. There have been a number of proposals for the establishment of somewhat similar monitoring services, although with different specific assignments. For example, a fairly recent proposal envisages a global monitoring system to assess the degree to which governmental actions further the attainment of basic human values or the realization of human dignity (Snyder, Hermann, and Lasswell 1976). What these proposals have in common is the assumption that current surveillance systems maintained by the mass media, various governments, and the United Nations are not adequate to satisfy important human needs.

5. A somewhat similar observation is made in chapter 3, with regard to the willingness of most people in the Basque country of Spain to answer questions about terrorist activities.

6. One examination of ethnic conflict finds that ethnic activists come primarily from groups that have had access to higher education, "but that are frustrated by limited career opportunities in the peripheral regions or by a sense of ethnic discrimination." Many are underpaid teachers "suffering from both declining prestige and low incomes" (Esman 1977, p. 374).

7. Examples of subjects on which spokesmen for religious or labor groups did not reflect the majority opinion of their constituents are fairly common in the public opinion research literature. An example from the realm of urban politics is given by Elisabeth Noelle-Neumann and Hans Kepplinger (1978). They found that a sample of prominent figures in Mainz, Germany, did not know what the working people in Mainz considered the principal problems of the city to be, although the working people were well acquainted with the views of the civic leaders.

BIBLIOGRAPHY

Agirre, Julen. *Operation "Ogre." Comment et pourquoir nous avons exécuté Carrero Blanco*. Paris: Editions du Seuil, 1974.

Akenson, Donald H. *Education and Enmity*. New York: Barnes and Noble, 1973.

Apalategui, Jokin. *Los vascos. De la nación al estado. PNV, ETA, Embata. . . .* N.p.: Elkar, 1979.

Arès, Richard. *Les positions ethniques, linguistiques et religieuses des Canadiens français à la suite du recensement de 1971*. Montreal: Les Editions Bellarmin, 1975.

Azaola, José Miguel. *Vasconia y su destino*, 2 vols. Madrid: Revista de Occidente, 1976.

Banco de Bilbao. *Renta nacional de España y su distribución provincial*. Bilbao: the Bank, 1975.

Banks, Arthur S. *Political Handbook of the World*. New York: McGrawHill, 1976.

Barbancho, Alfonso G. *Las migraciones interiores españolas*. Madrid: Instituto de Desarrollo Económico, 1967.

Barton, Allen H. "Empirical Methods and Radical Sociology: A Liberal Critique." In J. David Colfax and Jack L. Roach, eds., *Radical Sociology*. New York: Basic Books, 1971.

Beltza, [Emilo López]. *Nacionalismo vasco y clases sociales*. San Sebastián: Txertoa, 1976.

Bertsch, Gary K. "Ethnicity and Politics in Socialist Yugoslavia." *Annals of the American Academy of Political and Social Science* 433 (1977).

Birch, Anthony H. "Minority Nationalist Movements and Theories of Political Integration." *World Politics* 30, no. 3 (1978).

Blishen, Bernard. "Perceptions of National Identity." *Canadian Review of Sociology and Anthropology* 15, no. 2 (1978).

Blumer, Herbert. "The Mass, the Public, and Public Opinion." In *New Outline of the Principles of Sociology,* edited by Alfred McClung Lee. New York: Barnes and Noble, 1946.

Boal, F. W. "Close Together and Far Apart." *Community Forum* 2, no. 3 (1972). Northern Ireland Community Relations Commission, Belfast.

Bordegarai, Kepa, and Robert Pastor. *Estatuto vasco.* San Sebastián: Ediciones Vascas, 1979.

Breton, Raymond. "The Socio-Political Dynamics of the October Events." *Canadian Review of Sociology and Anthropology* (Feb. 1972).

Buchheit, Lee C. *Secession. The Legitimacy of Self-Determination.* New Haven: Yale University Press, 1978.

Burton, John W. *Conflict and Communication: The Use of Controlled Communication in International Relations.* New York: Free Press, 1969.

Cantril, Hadley. *The Pattern of Human Concerns.* New Brunswick, N.J.: Rutgers University Press, 1965.

Careless, J. M. S. "'Limited Identities' in Canada." *Canadian Historical Review* (Mar. 1959).

Caro Baroja, Julio. *Los vascos.* Madrid: Minotauro, 1958.

———. *Estudios vascos.* San Sebastián: Txertoa, 1978.

Castells Arteche, José Manuel. *El estatuto vasco.* San Sebastián: Haranburu, 1976.

Charlot, J. "Le sondage d'opinion publique: instrument de recherche ou de gouvernement?" In *Actes du Colloque Sondage et opinion publique, IIème Rencontre inter-universitaire des politistes francophones, Liège, April 22, 1975.* Départment de science politique de l'Université de Liège.

Cibrián Uzal, Ramiro. "Notas sobre el sistema electoral y partidos en Euskadi." (Unpublished paper, 1978).

Connor, Walker. "A Nation Is a Nation, Is a State, Is an Ethnic Group, Is a . . . . *Ethnic and Racial Studies* 1, no. 4 (1978).

Corcuera Atienza, Javier. *Orígines, ideología y organización del nacionalismo vasco (1876-1904)*. Madrid: Siglo veintiuno, 1979.

Cornez, E. "Un péril national: La désindustrialisation de la Wallonie." *Revue du conseil économique Wallon*, no. 41 (1959).

Corpuz, Onofre D. *The Philippines*. Englewood Cliffs, N.J.: Prentice-Hall, 1965.

Cruz Monclova, Lidio. *Historia de Puerto Rico en el siglo XIX*. Rio Piedras, P.R.: Editorial Universitaria, 1957.

Cuneo, Carl J., and James E. Curtis. "Quebec Separatism: An Analysis of Determinants Within Social-Class Levels." *Canadian Review of Sociology and Anthropology* (February 1974).

Curet Cuevas, Elieser. *El desarrollo económico de Puerto Rico: 1940 a 1972*. Santurce, P.R.: Santana, 1976.

Dabin, J. *Doctrine générale de l'état. Eléments de philosophie politique*. Brussels and Paris: Bruylant-Sirey, 1939.

Dahl, Robert A. *Polyarchy, Participation and Opposition*. New Haven: Yale University Press, 1971.

Darby, John. *Conflict in Northern Ireland: The Development of a Polarized Community*. New York: Barnes and Noble, 1976.

Davison, W. Phillips. *Mass Communication and Conflict Resolution*. New York: Praeger, 1974a.

——. "News Media and International Negotiation." *Public Opinion Quarterly* 38, no. 2 (1974b).

de Miguel, Amando. *Sociología del franquismo*. Barcelona: Euros, 1973.

——. "Estructura social e inmigración en el país vasco." *Papers* 3 (1974).

Departamento de Estudios y Desarrollo Regional de Caja Laboral Popular. *Aproximación a la estructura industrial del país vasco*. Durango, Mexico: Leopoldo Zugaza, 1976.

Díez, Miguel, Francisco Morales, and Angel Sabín. *Las lenguas de España.* Madrid: Alcalá, 1976.

Dishman, Robert. "Cultural Pluralism and Bureaucratic Neutrality in the British Caribbean." *Ethnicity* 5, no. 3 (1978).

Doob, Leonard W., William J. Foltz, and Robert B. Stevens. "The Fermeda Workshop: A Different Approach to Border Conflicts in East Africa." *Journal of Psychology* 73 (1969).

Duocastella, Rogelio. "Sociometría del cambio social en España" and "El mapa religioso de España." In *Cambio social y religión en España*, edited by Paulina Americh, Jose Luis Aranguren, Rogelio Duocastella, and Lorente Ruíz Rico. Barcelona: Fontanella, 1975.

Eisenstadt, S. N., and Y. Peres. "Some Problems of Educating a National Minority (a Study of Israeli Education for Arabs)." U.S. Department of Health, Education and Welfare, Project OE 6-21-013. Washington, D.C.: the Department, 1968. (mimeo)

Elkins, David J. "Regional and National Identities in Canada: Conflict and Complementarity." Paper presented to the Shastri Indo-Canadian Institute Conference on Indian and Canadian Perspectives on Regional Development. New Delhi, 1979.

Elkins, David J. and Richard Simeon. *Small Worlds: Parties and Provinces in Canadian Political Life.* Toronto: Methuen, 1980.

Elorza, Antonio. *Ideologías del nacionalismo vasco, 1876-1978.* San Sebastián: Haranburu, 1978.

Enloe, Cynthia H. *Ethnic Conflict and Political Development.* Boston: Little, Brown, 1972.

———. "Developments Viewed from the Palace: Political Scientists Look at Southeast Asian Ethnic Politics." *Ethnicity* 6, no. 1 (Mar. 1979).

Escudero, Manu, and Javier Villanueva. *La autonomía del país vasco desde el pasado al futuro.* San Sebastián: Txertoa, 1976.

Esman, Milton J. "Malaysia: Communal Coexistence and Mutual Deterrence." In *Racial Tensions and National Identity*, edited by Ernest Q. Campbell. Nashville: Vanderbilt University Press, 1972.

———. ed. *Ethnic Conflict in the Western World*. Ithaca, N.Y.: Cornell University Press, 1977.

Falcón, Luis Nieves. *La Opinión Pública y las Aspiraciones de los Puertorriqueños*. Rio Piedras, P.R.: Editorial Universitaria, 1970.

Forbes, H. D. "Conflicting National Identities Among Canadian Youth." In *Foundations of Political Culture*, edited by Jon H. Pammett and M. S. Whittington. Toronto: Macmillan of Canada, 1976.

Free, Lloyd A. "The Attitudes, Hopes and Fears of Nigerians." Princeton: Institute for International Social Research, 1964.

Friedrich, Carl J. *Trends of Federalism in Theory and Practice*. New York: Praeger, 1968.

———. "Fédéralisme," *Encyclopaedia Universalis*, vol. VI. Paris: 1970.

Fusi Aizpurua, Juan Pablo. *El problema vasco en la II república*. Madrid: Turner, 1979.

Gallup, Goerge H. "Human Needs and Satisfactions: A Global Survey." *Public Opinion Quarterly* 40, no. 4 (1976-77).

Gallup International Research Institutes and Charles F. Kettering Foundation. *Human Needs and Satisfactions: A Global Survey* (Summary volume). Princeton: Gallup Organization, 1977.

García Venero, Maximiano. *Historia del nacionalismo vasco*. Madrid: Editora Nacional, 1968.

Gastil, Raymond D. "The Comparative Survey of Freedom—VIII." *Freedom at Issue* no. 44 (1978).

Geertz, Clifford. "The Integrative Revolution." In *Old Societies and New States*, edited by Clifford Geertz. New York: Free Press, 1963.

George, Alexander L. *Propaganda Analysis: A Study of Inferences Made from Nazi Propaganda in World War II*. Evanston, Ill.: Row, Peterson, 1959.

Gingras, François-Pierre. *Contribution à l'étude de l'engagement indépendantiste au Québec*. Ph.D. dissertation, Université René Descartes, Paris, 1971.

Goldhamer, Herbert, and Hans Speier. "Some Observations on Political Gaming." *World Politics* 11, no. 1 (October 1959).

González Blasco, Pedro. "Modern Nationalism in Old Nations as a Consequence of Earlier State-Building: The Case of the Basque–Spain." In *Ethnicity and Nation-Building: Comparative, International and Historical Perspectives*, edited by Wendell Bell and Walter E. Freeman. Beverly Hills, Calif.: Sage, 1973.

Government of Quebec. *Quebec-Canada: A New Deal. The Quebec Government Proposal for a New Partnership Between Equals: Sovereignty-Association*. Quebec: Editeur Official, 1979.

Greenwood, Davydd J. "Continuity and Change: Spanish Basque Ethnicity as a Historical Process." In *Ethnic Conflict in the Western World*, edited by Milton J. Esman. Ithaca, N.Y.: Cornell University Press, 1977.

Guerrero, León María. *Today Began Yesterday: A Historical Approach to Martial Law in the Philippines*. Manila: National Media Production Center, 1975.

Guetzkow, Harold S., Philip Kotler, and Randall L. Schulz. *Simulation in Social and Administrative Science*. Englewood Cliffs, N.J.: Prentice-Hall, 1972.

Hamilton, Richard, and Maurice Pinard. "The Bases of Parti Québécois Support in Recent Quebec Elections." *Canadian Journal of Political Science* 9, (Mar. 1976).

———. "The Quebec Independence Movement." In *National Separatism*, edited by Colin H. Williams. Cardiff: University of Wales Press, forthcoming.

Harris, Rosemary. *Prejudice and Tolerance in Ulster*. Manchester: Manchester University Press, 1972.

Hechter, Michael. *Internal Colonialism: The Celtic Fringe in British National Development, 1536–1966*. Berkeley: University of California Press, 1975.

Heinz, Walter R. "Integration und Konflikt im Prozess des sozialen Wandels: Zur Theorie des ethnischen Pluralismus in Entwicklungsgesellschaften," pt. II. *Dritte Welt* 2, no. 2 (1973).

Heisler, Martin O. "Ethnic Conflict in the World Today: An Introduction." *Annals of the American Academy of Political and Social Science* 433 (1977).

Henripin, Jacques. "Quebec and the Demographic Dilemma of French Canadian Society." In Dale C. Thomson, ed., *Quebec Society and Politics: Views from the Inside*. Toronto: McClelland and Stewart, 1973.

Hills, George. "ETA and Basque Nationalism." *Iberian Studies*, 1, no. 2 (1972).

Hyman, Herbert L. *Interviewing in Social Research*. Chicago: University of Chicago Press, 1954.

Institut belge de science politique. "Comment les Belges voient leur système socio-politique," *Res Publica*, no. 4 (special issue, 1975).

Institut national de statistique. *Annuaire de statistiques régionales*. Brussels: 1976.

Irvine, William P. "Recruitment to Nationalism." *Canadian Journal of Political Science* 5, no. 4 (1972).

Isaacs, Harold R. *Idols of the Tribe*. New York: Harper & Row, 1975.

Jacob, James E. "The Basques of France: "A Case of Peripheral Ethnonationalism in Europe." *Political Anthropology* 1, no. 1 (1975).

Jenson, Jane, and Peter Regenstreif. "Some Dimensions of Partisan Choice in Quebec, 1969." *Canadian Journal of Political Science* (June 1970).

Johnstone, John C. *Young People's Images of Canadian Society: An Opinion Survey of Canadian Youth 13 to 20 Years of Age*. Ottawa: Queen's Printer, 1969.

Joy, Richard. *Languages in Conflict: The Canadian Experience*. Toronto: McClelland and Stewart, 1972.

———. "Languages in Conflict: Canada, 1976." In James E. Curtis and W. G. Scott, eds., *Social Stratification: Canada* (second edition). Scarborough, Ontario: Prentice-Hall of Canada, 1979.

Kerr, Henry H., Jr. *Switzerland: Social Cleavages and Partisan Conflict*. Sage Professional Papers in Contemporary Political Sociology. Beverly Hills, Calif.: Sage, 1974.

Korpi, Walter. "Conflict, Power and Relative Deprivation." *American Political Science Review* (Dec. 1974).

Ladrière, J. "Le système politique belge." CRISP Report, no. 500 (1970).

Lamy, Paul G. "Political Socialization of French and English Canadian Youth: Socialization into Discord." In *Socialization and Values in Canada: I. Political Socialization*, edited by Elia Zureik and Robert M. Pike. Toronto: McClelland and Stewart, 1975.

Larronde, Jean-Claude. *El nacionalismo vasco. Su origen y su ideología en la obra de Sabino Arana-Goiri*. San Sebastián: Txertoa, 1977.

LeDuc, Lawrence. "Canadian Attitudes Toward Quebec Independence." *Public Opinion Quarterly* 41, no. 3 (1977).

Lemieux, Vincent, Marcel Gilbert, and André Blais. *Une élection de réalignement: L'élection générale du 29 avril 1970 au Québec*. Montreal: Editions du Jour, 1970.

Lertxundi, Roberto. "Perspectivas inmediatas del país vasco." *Hemen ETA Orain* no. 5-6 (1978).

Lieberson, Stanley. *Language and Ethnic Relations in Canada*. New York: Wiley, 1970.

Lightfoot, Keith. *The Philippines*. New York: Praeger, 1973.

Lijphart, Arend. *The Politics of Accommodation: Pluralism and Democracy in the Netherlands*. Berkeley: University of California Press, 1968.

Linz, Juan J. "Early State-Building and Later Peripheral Nationalisms Against the State: The Case of Spain." In *Building States and Nations*, edited by S. N. Eisenstadt and Stein Rokkan. Beverly Hills, Calif.: Sage, 1973.

———. "Politics in a Multi-Lingual Society with a Dominant World Language: The Case of Spain." In *Les états multilingues, problèmes et solutions*, edited by Jean-Guy Savard and Richard Vigneault. Quebec: Presses de l'Université Laval, 1975.

———. "Il sistema partitico spagnolo." *Rivista italiana di scienza politica* 8, no. 3 (1978).

———. "Religion and Politics in Spain: From Conflict to Consensus Above Cleavage." Paper presented at International Conference of Religious Sociology, Venice, 1979.

———. "The New Spanish Party System," In *Political Participation*, edited by Richard Rose. London: Sage, 1980.

Lorwin, Val R. "Belgium: Religion, Class, and Language in National Politics." In *Political Oppositions in Western Democracies*, edited by Robert A. Dahl. New Haven: Yale University Press, 1966a.

———. "Conflits et compromis dans la politique belge." CRISP Report, no. 323 (1966b).

Lyons, F. S. L. *Ireland Since the Famine*. Glasgow: Fontana/Collins, 1973.

MacGreil, Micheál, *Prejudice and Tolerance in Ireland*. Dublin: Research Section, College of Industrial Relations, 1977.

Manzer, Ronald. *Canada: A Socio-Political Report*. Toronto: McGraw-Hill Ryerson, 1974.

Marcos, Ferdinand E. *Notes on the New Society in the Philippines*. Manila: Marcos Foundation, 1973.

Mays, Jeb. *Puerto Rico: A People Challenging Colonialism*. Washington: Epica Task Force, 1976.

McAllister, Ian. *The Northern Ireland Social Democratic and Labour Party: Political Opposition in a Divided Society*. London: Macmillan, 1977.

McCarthy, John D., and Mayer N. Zald. *The Trend of Social Movements in America: Professionalization and Resource Mobilization*. Morristown, N.J.: General Learning Press, 1973.

———. "Resource Mobilization and Social Movements: A Partial Theory." *American Journal of Sociology* 82, no. 6 (1977).

Mead, Margaret, and Rhoda Métraux, eds. *The Study of Culture at a Distance*. Chicago: University of Chicago Press, 1953.

Medhurst, Kenneth. "The Basques." Report no. 9. London: Minority Rights Group, 1972.

Meisel, John. "Who Are We? Who Are They? Perceptions in English Canada." In *Options, Proceedings of the Conference on the Future of the Canadian Federation*. Toronto: University of Toronto, 1977.

Melson, Robert, and Howard Wolpe. "Modernization and the Politics of Communalism: A Theoretical Perspective." *American Political Science Review* 64 (December 1970).

Meynaud, J., J. Ladrière, and F. Perin. *La décision politique in Belgique: Le pouvoir et les groupes.* Paris: A. Colin, 1965.

Ministerio de Economía, Instituto Nacional de Estadística. *La renta nacional en 1976 y su distribución.* Madrid: The Institute, 1977.

Morrison, D. G., and H. H. Stevenson. "Cultural Pluralism, Modernization, and Conflict: An Empirical Analysis of Sources of Political Instability in African Nations." *Canadian Journal of Political Science* 5, no. 1 (1972).

Muir, William Ker, Jr. *Police: Streetcorner Politicians.* Chicago: University of Chicago Press, 1977.

Ninyoles, Rafael. *Cuatro idiomas para un estado.* Madrid: Cambio 16, 1977.

Noelle-Neumann, Elisabeth, and Hans Mathias Kepplinger. "Communication in the Community." Paper presented at the 1978 meetings of the American Association for Public Opinion Research, Roanoke, Virginia.

Nordlinger, Eric. "Conflict Regulation in Divided Societies." Harvard University Center for International Affairs, Occasional Paper no. 29. Cambridge, Mass.: The Center, 1972.

Núñez, Luis C. *Clases Sociales en Euskadi.* San Sebastián: Txertoa, 1977.

———. "Base social de las candidaturas en las elecciones legislativas de 1977 en Guipúzcoa." *Saioak* 2 (1978).

Oberschall, Anthony. *Social Conflict and Social Movements.* Englewood Cliffs, N.J.: Prentice-Hall, 1973.

O'Brien, Conor Cruise. "Third Ewart-Biggs Memorial Lecture: Ireland, Britain and America." Delivered at New York University, Nov. 30, 1978.

Ornstein, Michael D., H. Michael Stevenson and A. Paul M. Williams. "Public Opinion and the Canadian Political Crisis." *Canadian Review of Sociology and Anthropology*, 15, no. 2, 1978.

Ortzi, [Francisco Letamendía]. *Historia de Euskadi: El nacionalismo vasco y ETA.* Paris and Barcelona Ruedo Ibérico, 1977.

———. *El no vasco a la Reforma*. San Sebastian: Txertoa, 1979.

Osborne, R. D. "An Industrial and Occupational Profile of the Two Sections of the Population of Northern Ireland: An Analysis of the 1971 Population Census." Belfast: Fair Employment Agency for Northern Ireland, 1977.

Ouellet, Fernand. "Le nationalisme canadien-français: De ses origines à l'insurrection de 1837." *Canadian Historical Review* 45 (Dec. 1964).

Paláu, Awilda. *Prensa Comercial: Posiciones de Clase Ante la Situación Laboral*. Guaynabo, Puerto Rico: Editorial Sagita, 1979.

Parti Québécois. *D'égal à égal*. Quebec: Parti Québécois, 1979.

Pastor, Robert, ed. *Euskadi ante el futuro*. San Sebastián: Haranburu, 1977.

Payne, Stanley G. *Basque Nationalism*. Reno: University of Nevada Press, 1975.

Pérez Calvo, Alberto. *Los partidos políticos en el país vasco*. San Sebastián and Madrid: Louis Haranburu and Tucan, 1977.

The Philippines: A Nation Reborn. 3rd ed. Manila: National Media Production Center, 1977.

Pinard, Maurice. "Communal Segmentation and Communal Conflict: A New Theoretical Statement." Paper presented at meetings of the American Psychological Association, Montreal, 1973.

———. "La dualité des loyautés et les options constitutionnelles des Québécois francophones." in *Le nationalisme québécois à la croisée des chemins*. Collection Choix, Centre québécois de relations internationales. Québec: Université Laval, 1975.

———. "Pluralisme social et partis politiques: Quelques éléments d'une théorie." In *Partis politiques au Québec*, edited by Réjean Pelletier. Montreal: Hurtubise HMH, 1976a.

———. "The Moderation and Regulation of Communal Conflicts: A Critical Review of Current Theories." Paper presented to the European Consortium for Political Research and the Canadian Political Science Association Twinned Workshop, Louvain, 1976b.

Pinard, Maurice, and Richard Hamilton. "The Independence Issue and the Polarization of the Electorate: The 1973 Quebec Election." *Canadian Journal of Poltical Science* 10, no. 2 (1977).

———. "The Parti Québécois Comes to Power: An Analysis of the 1976 Quebec Election." *Canadian Journal of Political Science* 11, no. 4 (1978).

Pirenne, Henri. *Histoire de Belgique* vol. VII. Brussels: M. Lamertin, 1932.

Quinn, Herbert F. *The Union Nationale: A Study in Quebec Nationalism.* Toronto: University of Toronto Press, 1963.

Radio Free Europe. "The Method of Comparative and Continuous Sampling." Munich: RFE Audience Research Department, Jan. 1976 (mimeo.)

Rawkins, Tudor H. "The Geographical Background of Basque Nationalism." *Iberian Studies* 1, no. 2 (1972).

Riencourt, Amaury de. *The Soul of India*. New York: Harper, 1960.

Ritt, Leonard G. "Some Social and Political Views of American Indians." *Ethnicity* 6, no. 1 (1979).

Rjasanoff, N. R. "Karl Marx und Friedrich Engels über die Polenfrage." *Archiv für Geschichte des Sozialismus und der Arbeiterbewegung* 6 (1916): 175–228.

Romero Barceló, Carlos. "Address by the Honorable Carlos Romero Barceló, Governor of Puerto Rico, before the United Press International Editors and Publishers Conference." Dorado, Puerto Rico, Oct. 10, 1977.

Rose, Richard. *Governing Without Consensus: An Irish Perspective*. Boston: Beacon Press, 1971.

———. *The Future of Scottish Politics: A Dynamic Analysis*. The Fraser of Allender Institute Speculative Papers, no. 3. Edinburgh: Scottish Academic Press, 1975.

———. *Northern Ireland: Time of Choice*. Washington, D.C.: American Enterprise Institute, 1976.

———. "Opinion over Ulster at Sword's Point." *Public Opinion* 2, no. 5 (1979).

Rose, Richard, Ian McAllister, and Peter Mair. *Is There a Concurring Majority About Northern Ireland?* Studies in Public Policy, no. 22. Glasgow: University of Strathclyde, 1978.

Ruys, Manu. *Les Flamands*. Tielt-Louvain: Lannoo-Vander, 1973.

Sarrailh de Ihartza, Fernando. *Vasconia*. Buenos Aires: Norbait, 1962.

Saunders, John. "Class, Color, and Prejudice: A Brazilian Counterpoint." In *Racial Tensions and National Identity*, edited by Ernest Q. Campbell. Nashville: Vanderbilt University Press, 1972.

Schwartz, Mildred A. *Public Opinion and Canadian Identity*. Berkeley: University of California Press, 1967.

——. *Politics and Territory: The Sociology of Regional Persistence in Canada*. Montreal and London: McGill-Queen's University Press, 1974.

Smooha, Sammy. "The Status of the Arab Minority in Israel." Israel Foundations Trustees Project no. 19. N.d. (Unpublished paper.)

Smooha, Sammy, and John E. Hofman. "Some Problems of Arab-Jewish Coexistence in Israel." *Middle East Review* 9, no. 2 (1976).

Snyder, Richard C., Charles F. Hermann, and Harold D. Lasswell. "A Global Monitoring System: Appraising the Effects of Government on Human Dignity." *International Studies Quarterly* 20, no. 2 (1976).

Stanley Foundation. *Report of the Third News Media Seminar at the United Nations*. Muscatine, Iowa: The Foundation, 1972.

Steiner, Jürg, and Jeffrey Obler. "Does the Consociational Theory Really Hold for Switzerland?" In *Ethnic Conflict in the Western World*, edited by Milton J. Esman. Ithaca, N.Y.: Cornell University Press, 1977.

Stewart, A. T. Q. *The Narrow Ground: Aspects of Ulster, 1609–1969*. London: Faber and Faber, 1977.

Stinchcombe, Arthur L. "Social Structure and Politics." In *Handbook of Political Science*, edited by Fred Greenstein and Nelson W. Polsby. Reading, Mass.: Addison-Wesley, 1975.

Stokes, Gale. "The Underdeveloped Theory of Nationalism." *World Politics* 31, no. 1 (1978).

Tasker, Rodney. Dispatches to *Far East Economic Review*. Dec. 31, 1976; Jan. 28, 1977; Apr. 8, 1977; Aug. 18, 1978.

Taylor, Donald M., L. M. Simard, and F. E. Aboud. "Ethnic Identification in Canada: A Cross-Cultural Investigation." In *Social Psychology: The*

Canadian Context, edited by J. W. Berry and G. J. S. Wilde. Toronto: McClelland and Stewart, 1972.

Tessler, Mark A. "Israel's Arabs and the Palestinian Problem." *Middle East Journal* 31, no. 3 (1977).

Tilly, Charles. *From Mobilization to Revolution*. Reading, Mass.: Addison-Wesley, 1978.

Tiryakian, Edward A. "The Politics of Devolution: Comparative Aspects of Quebec, Wales and Scotland." Paper presented at a joint Duke University/Carleton University seminar, "Canada: Centralization Versus Devolution." Ottawa, June 1979.

Tovar, Antonio. *La lengua vasca*. San Sebastián: Biblioteca Vascongada de Amigos del País, 1950.

United Nations, Special Committee of 24 on Decolonization. 1133rd meeting, GA/Col/1975. New York: United Nations Office of Public Information, September 12, 1978.

van den Berghe, Pierre. *Race and Racism: A Comparative Perspective*. New York: Wiley, 1967.

Van Kalken, F. *La Belgique contemporaine, 1780–1930: Histoire d'une évolution politique*. Paris: A. Colin, 1930.

Verdoodt, A. "Les problèmes communautaires belges à la lumière des études d'opinion." CRISP Report, no. 742 (1976).

Vreeland, Nena, et al. *Area Handbook for the Philippines*. Washington, D.C.: U.S. Government Printing Office, 1976.

Wallerstein, Immanuel. "Social Conflict in Post-Independence Black Africa: The Concepts of Race and Status-Group Reconsidered." In *Racial Tensions and National Identity*, edited by Ernest Q. Campbell. Nashville: Vanderbilt University Press, 1972.

White, Jack. *Minority Report: The Protestant Community in the Irish Republic*. Dublin: Gill and Macmillan, 1975.

Whyte, J. H., et al. "Governing Without Consensus: A Critique." Belfast: Northern Ireland Community Relations Commission, 1972.

Wigny, P. "Une Belgique en mutation." *Bulletin de l'Académie Royale de Belgique* (Classe des Lettres et des Sciences morales et politiques) 55, no. 5 (1969).

Young, Crawford. *The Politics of Cultural Pluralism*. Madison: University of Wisconsin Press, 1976.

Zavala, José María. *PC. Partido carlista*. Barcelona: Avanca, 1976.

Books written under the auspices of
THE CENTER OF INTERNATIONAL STUDIES, PRINCETON UNIVERSITY
1952–1980

Gabriel A. Almond, *The Appeals of Communism* (Princeton University Press, 1954).

William W. Kaufmann, ed., *Military Policy and National Security* (Princeton University Press, 1956).

Klaus Knorr, *The War Potential of Nations* (Princeton University Press, 1956).

Lucian W. Pye, *Guerrilla Communism in Malaya* (Princeton University Press, 1956).

Charles De Visscher, *Theory and Reality in Public International Law*, trans. by P. E. Corbett (Princeton University Press, 1957; rev. ed., 1968).

Bernard C. Cohen, *The Political Process and Foreign Policy: The Making of the Japanese Peace Settlement* (Princeton University Press, 1957).

Myron Weiner, *Party Politics in India: The Development of a Multi-Party System* (Princeton University Press, 1957).

Percy E. Corbett, *Law in Diplomacy* (Princeton University Press, 1959).

Rolf Sannwald and Jacques Stohler, *Economic Integration: Theoretical Assumptions and Consequences of European Unification*, trans. by Herman Karreman (Princeton University Press, 1959)

Klaus Knorr, ed., *NATO and American Security* (Princeton University Press, 1959).

Gabriel A. Almond and James S. Coleman, eds., *The Politics of the Developing Areas* (Princeton University Press, 1969).

Herman Kahn, *On Thermonuclear War* (Princeton University Press, 1960).

Sidney Verba, *Small Groups and Political Behavior: A Study of Leadership* (Princeton University Press, 1961).

Robert J. C. Butow, *Tojo and the Coming of the War* (Princeton University Press, 1961).

Glenn H. Snyder, *Deterrence and Defense: Toward a Theory of National Security* (Princeton University Press, 1961).

Klaus Knorr and Sidney Verba, eds., *The International System: Theoretical Essays* (Princeton University Press, 1961).

Peter Paret and John W. Shy, *Guerrillas in the 1960's* (Praeger, 1962).

George Modelski, *A Theory of Foreign Policy* (Praeger, 1962).

Klaus Knorr and Thornton Read, eds., *Limited Strategic War* (Praeger, 1963).

Frederick S. Dunn, *Peace-Making and the Settlement with Japan* (Princeton University Press, 1963).

Arthur L. Burns and Nina Heathcote, *Peace-Keeping by United Nations Forces* (Praeger, 1963).

Richard A. Falk, *Law, Morality, and War in the Contemporary World* (Praeger, 1963).

James N. Rosenau, *National Leadership and Foreign Policy: A Case Study in the Mobilization of Public Support* (Princeton University Press, 1963).

Gabriel A. Almond and Sidney Verba, *The Civic Culture: Political Attitudes and Democracy in Five Nations* (Princeton University Press, 1963.

Bernard C. Cohen, *The Press and Foreign Policy* (Princeton University Press, 1963).

Richard L. Sklar, *Nigerian Political Parties: Power in an Emergent African Nation* (Princeton University Press, 1963).

Peter Paret, *French Revolutionary Warfare from Indochina to Algeria: The Analysis of a Political and Military Doctrine* (Praeger, 1964).

Harry Eckstein, ed., *Internal War: Problems and Approaches* (Free Press, 1964).

Cyril E. Black and Thomas P. Thornton, eds., *Communism and Revolution: The Strategic Uses of Political Violence* (Princeton University Press, 1964).

Miriam Camps, *Britain and the European Community 1955-1963* (Princeton University Press, 1964).

Thomas P. Thornton, ed., *The Third World in Soviet Perspective: Studies by Soviet Writers on the Developing Areas* (Princeton University Press, 1964).

James N. Rosenau, ed., *International Aspects of Civil Strife* (Princeton University Press, 1964).

Sidney I. Ploss, *Conflict and Decision-Making in Soviet Russia: A Case Study of Agricultural Policy, 1953-1963* (Princeton University Press, 1965).

Richard A. Falk and Richard J. Barnet, eds., *Security in Disarmament* (Princeton University Press, 1965).

Karl von Vorys, *Political Development in Pakistan* (Princeton University Press, 1965).

Harold and Margaret Sprout, *The Ecological Perspective on Human Affairs, With Special Reference to International Politics* (Princeton University Press, 1965).

Klaus Knorr, *On the Uses of Military Power in the Nuclear Age* (Princeton University Press, 1966).

Harry Eckstein, *Division and Cohesion in Democracy: A Study of Norway* (Princeton University Press, 1966).

Cyril E. Black, *The Dynamics of Modernization: A Study in Comparative History* (Harper and Row, 1966).

Peter Kunstadter, ed., *Southeast Asian Tribes, Minorities, and Nations* (Princeton University Press, 1967).

E. Victor Wolfenstein, *The Revolutionary Personality: Lenin, Trotsky, Gandhi* (Princeton University Press, 1967).

Leon Gordenker, *The UN Secretary-General and the Maintenance of Peace* (Columbia University Press, 1967).

Oran R. Young, *The Intermediaries: Third Parties in International Crises* (Princeton University Press, 1967).

James N. Rosenau, ed., *Domestic Sources of Foreign Policy* (Free Press, 1967).

Richard F. Hamilton, *Affluence and the French Worker in the Fourth Republic* (Princeton University Press, 1967).

Linda B. Miller, *World Order and Local Disorder: The United Nations and Internal Conflicts* (Princeton University Press, 1967).

Henry Bienen, *Tanzania: Party Transformation and Economic Development* (Princeton University Press, 1967).

Wolfram F. Hanrieder, *West German Foreign Policy, 1949-1963: International Pressures and Domestic Response* (Stanford University Press, 1967).

Richard H. Ullman, *Britain and the Russian Civil War: November 1918-February 1920* (Princeton University Press, 1968).

Robert Gilpin, *France in the Age of the Scientific State* (Princeton University Press, 1968).

William B. Bader, *The United States and the Spread of Nuclear Weapons* (Pegasus, 1968).

Richard A. Falk, *Legal Order in a Violent World* (Princeton University Press, 1968).

Cyril E. Black, Richard A. Falk, Klaus Knorr and Oran R. Young, *Neutralization and World Politics* (Princeton University Press, 1968).

Oran R. Young, *The Politics of Force: Bargaining During International Crises* (Princeton University Press, 1969).

Klaus Knorr and James N. Rosenau, eds., *Contending Approaches to International Politics* (Princeton University Press, 1969).

James N. Rosenau, ed., *Linkage Politics: Essays on the Convergence of National and International Systems* (Free Press, 1969).

John T. McAlister, Jr., *Viet Nam: The Origins of Revolution* (Knopf, 1969).

Jean Edward Smith, *Germany Beyond the Wall: People, Politics and Prosperity* (Little, Brown, 1969).

James Barros, *Betrayal from Within: Joseph Avenol, Secretary-General of the League of Nations, 1933-1940* (Yale University Press, 1969).

Charles Hermann, *Crises in Foreign Policy: A Simulation Analysis* (Bobbs-Merrill, 1969).

Robert C. Tucker, *The Marxian Revolutionary Idea: Essays on Marxist Thought and Its Impact on Radical Movements* (W. W. Norton, 1969).

Harvey Waterman, *Political Change in Contemporary France: The Politics of an Industrial Democracy* (Charles E. Merrill, 1969).

Cyril E. Black and Richard A. Falk, eds., *The Future of the International Legal Order.* Vol. I: *Trends and Patterns* (Princeton University Press, 1969).

Ted Robert Gurr, *Why Men Rebel* (Princeton University Press, 1969).

C. Sylvester Whitaker, *The Politics of Tradition: Continuity and Change in Northern Nigeria 1946–1966* (Princeton University Press, 1970).

Richard A. Falk, *The Status of Law in International Society* (Princeton University Press, 1970).

John T. McAlister, Jr. and Paul Mus, *The Vietnamese and Their Revolution* (Harper & Row, 1970).

Klaus Knorr, *Military Power and Potential* (D. C. Heath, 1970).

Cyril E. Black and Richard A. Falk, eds., *The Future of the International Legal Order.* Vol. II: *Wealth and Resources* (Princeton University Press, 1970).

Leon Gordenker, ed., *The United Nations in International Politics* (Princeton University Press, 1971).

Cyril E. Black and Richard A. Falk, eds., *The Future of the International Legal Order.* Vol. III: *Conflict Management* (Princeton University Press, 1971.)

Francine R. Frankel, *India's Green Revolution: Economic Gains and Political Costs* (Princeton University Press, 1971).

Harold and Margaret Sprout, *Toward a Politics of the Planet Earth* (Van Nostrand Reinhold, 1971).

Cyril E. Black and Richard A. Falk, eds., *The Future of the International Legal Order.* Vol. IV: *The Structure of the International Environment* (Princeton University Press, 1972).

Gerald Garvey, *Energy, Ecology, Economy* (W. W. Norton, 1972).

Richard H. Ullman, *The Anglo-Soviet Accord* (Princeton University Press, 1973).

Klaus Knorr, *Power and Wealth: The Political Economy of International Power*

Anton Bebler, *Military Rule in Africa: Dahomey, Ghana, Sierra Leone, and Mali* (Praeger, 1973).

Robert C. Tucker, *Stalin as Revolutionary 1879-1929: A Study in History and Personality* (W. W. Norton, 1973).

Edward L. Morse, *Foreign Policy and Interdependence in Gaullist France* (Princeton University Press, 1973).

Henry Bienen, *Kenya: The Politics of Participation and Control* (Princeton University Press, 1974).

Gregory J. Massell, *The Surrogate Proletariat: Moslem Women and Revolutionary Strategies in Soviet Central Asia, 1919-1929* (Princeton University Press, 1974).

James N. Rosenau, *Citizenship Between Elections: An Inquiry Into the Mobilizable American* (Free Press, 1974).

Ervin Laszlo, *A Strategy for the Future: The Systems Approach to World Order* (George Braziller, 1974).

R. J. Vincent, *Nonintervention and International Order* (Princeton University Press, 1974).

Jan H. Kalicki, *The Pattern of Sino-American Crises: Political-Military Interactions in the 1950s* (Cambridge University Press, 1975).

Klaus Knorr, *The Power of Nations: The Political Economy of International Relations* (Basic Books, 1975).

James P. Sewell, *UNESCO and World Politics: Engaging in International Relations* (Princeton University Press, 1975).

Richard A. Falk, *A Global Approach to National Policy* (Harvard University Press, 1975).

Harry Eckstein and Ted Robert Gurr, *Patterns of Authority: A Structural Basis for Political Inquiry* (John Wiley & Sons, 1975).

Cyril E. Black, Marius B. Jansen, Herbert S. Levine, Marion J. Levy, Jr., Henry Rosovsky, Gilbert Rozman, Henry D. Smith, II, and S. Frederick Starr, *The Modernization of Japan and Russia* (Free Press, 1975).

Leon Gordenker, *International Aid and National Decisions: Development Programs in Malawi, Tanzania, and Zambia* (Princeton University Press, 1976).

Carl von Clausewitz, *On War*, edited and translated by Michael Howard and Peter Paret (Princeton University Press, 1976).

Gerald Garvey and Lou Ann Garvey, *International Resource Flows* (D. C. Heath, 1977).

Walter F. Murphy and Joseph Tanenhaus, *Comparative Constitutional Law: Cases and Commentaries* (St. Martin's Press, 1977).

Gerald Garvey, *Nuclear Power and Social Planning: The City of the Second Sun* (D. C. Heath, 1977).

Richard E. Bissell, *Apartheid and International Organizations* (Westview Press, 1977).

David P. Forsythe, *Humanitarian Politics: The International Committee of the Red Cross* (Johns Hopkins University Press, 1977).

Paul E. Sigmund, *The Overthrow of Allende and the Politics of Chile, 1964–1976* (University of Pittsburgh Press, 1977).

Henry S. Bienen, *Armies and Parties in Africa* (Holmes and Meier, 1978).

Harold and Margaret Sprout, *The Context of Environmental Politics: Unfinished Business for America's Third Century* (University Press of Kentucky, 1978).

Samuel S. Kim, *China, the United Nations, and World Order* (Princeton University Press, 1979).

S. Basheer Ahmed, *Nuclear Fuel and Energy Policy* (Lexington Books, D.C. Heath, 1979).

Robert Johansen, *The National Interest and the Human Interest: An Analysis of U.S. Foreign Policy* (Princeton University Press, 1979).

Richard A. Falk and Samuel S. Kim, eds., *The War System: An Interdisciplinary Approach* (Westview Press, 1980).

James H. Billington, *Fire in the Minds of Men: Origins of the Revolutionary Faith* (Basic Books, Inc., 1980).

Samuel S. Kim and James C. Hsiung, eds., *China in the Global Community* (Praeger Publishers, 1980).

INDEX

Acción Nacionalista Vasca/ Eusko Abertzale Ekintza (ANV), 21, 23, 25
Achenese, 4
Agirre, Julen, 52n
Aizpún, Jesús, 26
Akenson, Donald H., 75
al-'Ard, 114, 117 (*see also* Sons of the Village)
Alava, 12, 15–16, 22–24, 27–29, 34, 35
Albania, 2
al-Bayadir, 115
Alianza Foral Navarra (AFN), 24, 26
Alianza Popular (AP), 23, 24, 25, 26, 29, 34, 39, 41, 48, 49 (*see also* Unión Foral del País Vasco)
Alliance Party, 79
American Indians, xiii, 4, 206
Amin, Idi, 200
Amul prototype, 85–86
Andra Pradesh, 98
Apalategui, Jokin, 51n
Arabs, 3, 6, 7, 8, 9, 203; in Israel, 107–31
Arafat, Yassir, 117, 119
Aragón, 12
Arana, Sabino, 30, 52n
Ashoka, 84
Assam, 84, 87
atrocities, 139 (*see also* terrorism)
Austro-Hungarian Empire, 1
Autonomy Statute: Basque, 17, 30, 49, 52n
Azaola, José Miguel, 51n

Bangsa Moro Islamic Region, 138
Banks, Arthur S., 177
Barbancho, Alfonso G., 34
Barcelona, 30
Baroja, Pío, 31
Barton, Allen, 185
Basques, 6, 8, 11–51, 53, 203, 204, 206; economy, 13, 17, 44–45, 49; elections, 15, 20–30, 49–51; party system, 20–30; population, 12–13; social classes, 30, 35, 37–39
Baxter, Richard, iii
Belgium, xvi, 6, 7, 8, 19, 20, 35, 53–73, 203, 208; constitutional revision, 67, 68; economy, 57, 59, 60, 62, 63–64, 65, 68, 70, 71; elections, 60; political parties, 60, 68; population, 62; social classes, 53, 54–55, 66; unity of, 70
Benelux, 60
Berríos, Martinez, Ruben, 178
Bertsch, Gary K., 207
Biafra, 4
Bihar, 86, 104
Bilbao, 13, 17, 24, 29, 37
bilingualism: Belgium, 56, 62, 65–66, 71–73
Bilingualism and Biculturalism Commission (Canada), 145, 148
Birch, Anthony H., 203
Black, Cyril E., vi, xiii
Black Panthers (India), 91
Blishen, Bernard, 149, 151

Blumer, Herbert, 187
Boal, F. W., 78
Borbón-Parma, Carlos Huge de, 27
Bourgault, Pierre, 169
Bowen, William G., vii
Brás, Juan Mari, 178
Brazil, 206
Breton, Raymond, 150
Bretons, xiii
British Columbia, 149
British Empire, 3
Brussels, 56, 58, 60, 62-63, 64-65, 68-69, 71
Buchanan, Percy H., vi
Buchheit, Lee C., 16
Buddhism, 84
Bulgaria, 2
Burma, 2
Burton, John W., 209

Canada, 6, 7, 8, 19, 69, 82, 140-74, 203, 204, 208
Canadian Broadcasting Corporation (CBC), 151, 166, 169
Cantril, Hadley, 103, 188
Careless, J. M. S., 148
caste system, 83, 84, 103, 104
Castile, 11-12
Castro, Fidel, 188
Catalonia, 19, 30, 31, 34
Centre de Recherches Contemporaines, 151
Centre de Recherche et d'Information Socio-Politiques (CRISP), xiii
Centre de Recherches sur l'Opinion Publique (CROP), 146, 153, 170
Centre de Sondage, University of Montreal, 150, 159
Centre Harmel, 56
Chamberlin, Ward B., Jr., iii
Charlot, J., 70
China, 200
Christian Democrats: Belgium, 66; Spain, 29, 41, 43

Christian People's Party (CVP), 66
Cibrián Uzal, Ramiro, 22
civility, 206, 208
civil service, 191, 202, 203
Class of 1917, Princeton University, v-vii, xi, xiv
colonialism, 1-2, 3, 13, 194; and Puerto Rico, 179, 183, 184, 188
Colwell, Kent G., vi
Comisiones Obreras (CC.OO), 25
Committee for the Defense of Arab Lands, 120
communism: in Israel, 109, 112, 113, 117; (see also Maki Party, Rakah Party); in the Philippines, 139; in Spain (see Partido Communista de Euskadi)
Congress Party, 104
Connor, Walker, 5
consociationalism, 19, 20, 22, 44, 207
constitutional referendum: Spain, 15, 17
Cornez, E., 59
Corpuz, Onofre D., 133
Council of Ireland, 79
Croatia, 1
Crossley, Archibald M., iii, xiv
Crossley, Helen M., iii
Crossley poll, xivn
Cruz Monclova, Lidio, 177
Cuneo, Carl J., 167, 176
Curet Cuevas, Elieser, 178
Curtis, James E., 167, 176
Cyprus, 4
Czechoslovakia, 1, 2

Dabin, J., 68
Dahl, Robert A., 14
Darby, John, 76
DATA (Madrid), 11
Davison, W. Phillips, vi, xiv, 209
Dayan, Moshe, 120
decolonization, 2, 177, 193; Puerto Rico, 178, 181, 183, 184, 188

Deia, 26, 31
Delanoy, Eleanor, vi
Delruelle, Nicole, 71
De Miguel, Amando, 34
Democratas Independentes Vascos-Unión Foral para la Autonomía, 23
Democratic Front for Peace and Equality, 114
Democracia Cristiana Vasca (DCV), 23, 26, 27
Destrée, Jules, 58
de Terán, Rocío, 11
Dishman, Robert, 203
dual loyalties, 7; in Belgium, 60, 69; in Canada, 140, 141, 143-49, 172; in Israel, 117-18; in Northern Ireland, 77
Duocastella, Rogelio, 41
Duplessis, Maurice, 151

early warning of conflict, vi, 209
East Bengal, xiii
economic discrimination, 5, 193, 203; India, 85, 86, 87, 104; Northern Ireland, 76, 81; Philippines, 133
education, 191, 200, 202; Basque country, 30-31; Belgium, 56, 59, 60; India, 91, 94-98, 103, 104, 105; Northern Ireland, 77, 80; Philippines, 134, 135, 136; Quebec, 155
Egin, 26, 31
Egmont Pact, 57, 67
Eisenstadt, S. N., 118
Elkins, David J., 149, 167
el-Tahdi, 120
Engels, Friedrich, 11
Enloe, Cynthia H., xi, 205
Enrile, Juan Ponce, 136
Esman, Milton J., 206, 207, 209
Esthonia, 1
Ethiopia, 4
ethnicity, xi, xii, xiii, 4, 5, 6-7, 9, 194, 198
European Common Market, 63
Euskadi, 12, 16, 17, 22, 23, 26, 27, 29, 47, 49
Euskadi Nord, 12, 19
Euskadi Sur, 12, 22, 23, 29, 35
Euskadiko Ezquerra (EE), 23, 25, 26, 29, 34, 39, 43, 45, 47, 49
Euskadi ta Askatasuna (ETA), 18, 21, 25, 39, 41, 43, 44, 46, 48
Euskal-Herria, 11 (*see also* Basques)
Euskal Langileen Alkartasuna-Solidaridad de Trabajadores Vascos (ELA-STV), 21
Euskal Sozialista Biltzarrea (ESB), 21, 23, 25
Euskera, 31 (*see also* language, Basque)
Eyskens, Gaston, 67

Falange Española de las Juntas de Ofensiva Nacional-Sindicalista (auténtica) (FE de las JONS(a)), 21
Falcon, Luis Nieves, 184, 185, 187
family planning: India, 94-98
Fatah, 117, 119
Faulkner, Brian, 79
federalism: Canada, 153, 156, 167; Belgium, 56, 59, 61, 63, 64, 66, 67-69; Philippines, 136
Ferré, Luis A., 178
Finland, 1
Fitzgerald, Gerald A., 6
Flemings, 7, 9, 53-73
Flemish Movement, 54-57, 61, 62
Forbes, H. D., 149, 156
France, 11, 12, 13, 20, 25, 58, 204
Franco, Francisco, 30, 41, 43, 44, 47, 48, 49
Free, Lloyd A., 200, 207
French Canadians, 7, 8 (*see also* Canada, Quebec)
Friedrich, Carl J., 69
Front Démocratique des Franco-

phones bruxellois (FDF), 73
Fuerzas Armadas de Liberación National FALN), 180

Galilee, 111, 112, 120, 121, 128, 130
Gallup, Alec, xiii
Gallup, George H., iii, 200
Gallup, George H., III, iii
Gallup International Research Institutes, 300, 207
Gallup/Kettering World Survey, xiii
Gallup Poll, xiv n, 185, 188
Garaikoetxea, Carlos, 37
Garson, June, xiii
Gascony, 11
Gastil, Raymond D., 4
Gaza Strip, 115, 116, 118, 120
George, Alexander L., 201
Gérard-Libois, J., xiii
German Federal Republic, 46, 69
German Marshall Fund, 11
Germany, 1, 2
Goldhamer, Herbert, 202
Gómez Reino, Manuel, 11
Gordenker, Leon, vi, xiv
GRAPO, 46
Great Britain, 74, 75, 81, 82, 107, 149
Greece, 1
Guerrero, León María, 133
guerrillas, 117, 119, 137
Guetzkow, Harold S., 202
Guipúzcoa, 12, 16, 17, 22, 23, 24, 29, 45
Guipúzcoa Unida, 24, 27
Gujarat, 85, 86

Habbash, George, 117
Hamilton, Richard, 160, 172, 173
Hammer, Zevulun, 119
Harijans, 6, 83–105, 203, 205, 207–08; attitudes of, 98–104; discrimination against, 87–94
Harris, Rosemary, 98

Haryana, 85
Hebrew University Student Union, 120
Heinz, Walter R., 202
Heisler, Martin O., 4–5
Henripin, Jacques, 142
Herman, Fernand, xiii
Hernández Agosto, Miguel, 178
Hernández Colon, Rafael, 178
Herri Batasuna (HB), 25, 26, 27, 29, 39, 45, 47, 49
Himachal Pradesh, 84
Hinduism, 84, 85, 86, 104
Hofman, John E., 124, 126
housing: India, 94
Houwaert, Jan, 69
Hungary, 1, 2
Hyman, Herbert H., 200

Ibo, 3
identity: Basque country, 12, 30–34, 35; Belgium, 59, 61; Canada, 150–51; Israel, 114, 115–18, 122, 124, 127; Northern Ireland, 76, 77
illiteracy, 91, 95
immigration: Basque country, 13, 24, 25, 31, 34, 35, 47, 48; Philippines, 133; Quebec, 141–42, 155
India, 2, 3, 4, 6, 8, 9, 83–105, 203, 205, 207; economy, 91
Indian Institute of Public Opinion, 85, 87, 95, 104
Indochina, 2
Indonesia, 2, 4
Institut belge de Science Politique, 71
Institute for International Social Research, 200, 207
Interuniversity Institute for Public Opinion Polls, Brussels, 71
interviewing in public opinion research, 192–93, 194–95, 199, 200, 202

Ireland, Republic of, 8, 31, 74, 75
Irish Republican Army (IRA), 8, 46, 75, 76, 79, 80
Irvine, William P., 204
Israel, 2, 3, 6, 7, 8, 9, 30, 107-31, 203; agriculture, 109, 110, 111, 112; economy, 108, 110, 111, 113, 130; elections, 113, 114, 116; population, 107, 111
Israel Black Panthers, 114
Israel Council for Israel-Palestine Peace, 122
Islam, 11, 120 (*see also* Muslims)
Islamic Conference, 136, 137, 138-39
Italy, 25, 46
Izquierda Republicana (IR), 21

Jacob, James E., 204
Jackson, Norris D., vi
Janata Party, 104
Jennings, Sir Ivor, 16
Jewish Agency, 112
Jews, 8, 107-8, 121
Johnson, Joseph E., iii
Johnstone, John C., 149
Jordan, 2, 107, 111
Joy, Richard, 142

Kaira District, 85
Karnataka, 86
Kashmir, 87
Kettering Foundation, 200

Labor Party, Israel, 116, 117
labor unions: Basque country, 24-25; Belgium, 58, 59, 63
Ladrière, J., 66, 68
Lamy, Paul C., 149
Land Day, Israel, 112, 114, 119
Landon, S. Whitney, vi
land ownership: Israel, 110, 111-13, 129, 130; Philippines, 133, 134
Langile Abertzale Batzordea (LAB), 21, 25

language, 202, 203; Basque country, 12, 13, 30, 31-34; Belgium, 53, 54, 55-57, 59, 60, 61, 62, 65, 67, 71-73; Canada, 141-42; Philippines, 134
Latvia, 1
leader-follower relationships, 204-06
League of Nations, 1, 2
Lebanon, 2, 19, 120
LeDuc, Lawrence, 148, 150, 167
legitimacy: Israel, 122, 123, 124, 126
Leibowitz, Yeshayuha, 122
Lemieux, Vincent, 148, 173
Leopold I, 68
Libya, 136
Lieberson, Stanley, 141, 142
Liga Communista Revolucionaria (LCR), 21
Lightfoot, Keith, 133
Lijphart, Arend, 19, 44, 207
Linz, Juan J., 20, 41, 206
Lipset, Seymour M., 160
Literary Digest, xi
Lithuania, 1
Logan, Jan, xiii
Lorwin, Val R., 68
Louvain, 62
Luxembourg Province, 60
Lyons, F. S. L., 74

Macedonians, 2
MacGreil, Micheál, 81
Madhya Pradesh, 84
Maharashtra, 91, 104
MAIQ, 146, 167
Maki Party, 113
Makrotest, 73
Malaysia, 139, 206, 207
Manipur, 84
Manu, 84
Manzer, Ronald, 149
maratabat, 135
Marcos, Ferdinand E., 132, 133-34, 136, 137, 138, 139

Marcos, Imelda, 136, 138
Maritime Provinces, 148
mass media, v, 191, 193, 194, 195, 197, 198, 199, 203, 209; Basque country, 17, 26, 31; Belgium, 68; Israel, 129; Philippines, 132, 136, 137; Puerto Rico, 184, 185, 187
McAllister, Ian, 77
McCarthy, John D., 168
McDowall, Jane, xiii
Mead, Margaret, xi, 201
Meghalaya, 84
Meisel, John, 149
Melson, Robert, 150
Métraux, Rhoda, 201
Meynaud, J., 66
Mindanao, 132, 139
Mindanao Independence Movement, 133
Mindanao State University, 135
Misuari, Nur, 136, 137, 138
modernization, 203
monitoring mass attitudes, vi, 105, 194, 198-99, 209
Monnet, Jean, 69
Montreal, 142
Moro National Liberation Front, 133, 136, 137, 138
Moro Reform Liberation Front, 137, 138
Moros, 132-39
Morrison, D. G., 203
Movimiento Comunista de Euskadi-Organización de Izquierda Comunista (EMK-OIC), 21
Movimiento Independentista Revolucionario Armada (MIRA), 180
Muñoz Marin, Luis, 178
Museum of Philippine Traditional Cultures, 135
Muslims, 6, 132-39 (see also Islam)

Nagaland, 84
Nahada, 115

Nasser, Gamal Abdel, 114, 116
National Committee for Defense of Arab Lands, 113, 114, 115, 128
National Committee to Continue the Defense of the Lands, 112
nationalism: Basque, 11, 13, 20, 29, 30-31, 34, 41; Flemish, 54, 66; Israeli Arab, 115, 117; Palestinian, 115, 121; Spanish, 17, 18
nationality, xii, xiii
Nationalist Party, Northern Ireland, 77
Navarra, 11, 12, 15, 16, 18, 22, 23, 24, 25, 26, 27, 29, 34, 35, 45
Nazareth, 112, 115, 121
Negev, 112
Netherlands, 2
Netherlands Empire, 3
New Brunswick, 148
New Delhi, 86
Newfoundland, 82, 148
New Progressive Party, 178, 179
Nigeria, 3, 4, 200, 207
Nordlinger, Eric, 18
Northern Ireland, 4, 6, 7, 8, 74-82; economy, 77, 78
Nuñez, Luis C., 22, 41

Oberschall, Anthony, 168
Obler, Jeffrey, 206
O'Brien, Conor Cruise, 80
Occupations: Arabs in Israel, 110, 118, 130, 131
O'Neill, Terence, 78
Ontario, 148, 149
Operation Bootstrap, 179
Organización Revolucionaria de Trabajadores (ORT), 21, 24, 25
Organization for African Unity, 3
Orissa, 98
Orizo, Francisco Andrés, 11
Ortzi [Francisco Letamendia], 49
Osborne, R. D., 81
Ottoman Empire, 1
Ouellet, Fernand, 143

Pakistan, xv, 2, 3
Palawan, 132, 138
Palestine, 2, 7, 107, 108, 114
Palestine Liberation Organization (PLO), 8, 114, 115, 116, 118, 119, 120, 121, 122, 127, 129
Palestinians, 109, 114, 116–17, 118, 122, 123, 124, 127, 128, 131
Parti Québécois, 147, 159, 160, 164–66, 172, 173
Partido Carlista, 27
Partido Comunista de Euskadi (PCE), 24, 25, 26, 30, 34, 39, 41, 43, 48, 49
Partido Nacionalista Vasco-Euzko Alderdi Jeltzalea (PNV-EAJ), 22, 23, 24, 25, 26, 27, 29, 34, 39, 41, 43, 44, 45, 47, 49
Partido Socialista Andaluz, 31
Partido Socialista Obrero Español (PSOE), 22, 23, 24, 25, 27, 29, 34, 35, 39, 43, 45, 47, 48, 49
Partido Socialista Obrero Español-sector histórico (PSOE-h), 22
Partido del Trabajo de España (PTE), 22
peace movement, Northern Ireland, 8, 80
Peres, Shimon, 118
Perin, F., 66
personal relationships: Arab-Jewish, 123, 124, 126–27 (see also civility)
Philippines, 132–39; economy, 136, 138, 139
Pinard, Maurice, 159, 160, 169, 172, 173, 204
Pirenne, Henri, 55
plebiscites: Philippines, 138; Puerto Rico, 177, 181, 188; Spain, 16, 17, 18, 25–26 (see also referendums)
Poland, 1, 2
police: Basque country, 30, 44, 46, 48

polyarchy, 14
Popular Democratic Party, 178, 179, 183
Popular Front for the Liberation of Palestine, 117
Portugal, 25
power sharing, Northern Ireland, 79–80, 81
Progressive National Movement, 115, 116, 119, 121, 127
Protestants: Northern Ireland, 6, 8, 74–82
Protestant Unionist Party, 75
Provisionals (see Irish Republican Army)
public opinion research, v, vi, xi, xii, 5, 6, 9, 10, 190–209; Belgium, 70–73; Canada, 143–49, 150–72; India, 104–05; Philippines, 139; Puerto Rico, 181, 184, 185, 187–88; Spain, 44–49
Puerto Rico, 6, 8, 177–88; commonwealth status, 177, 178, 179, 181, 185; economy, 177, 178, 179, 182, 183, 188; elections, 182, 188; independence, 177, 178, 179, 181–82, 183, 184, 185, 187; population, 177, 182; social classes, 182; statehood, 178, 179, 181, 182, 183, 184, 185
Puerto Rican Independence Party, 178
Puerto Rican Nationalist Party, 180
Puerto Rican Socialist League, 180
Puerto Rican Socialist Party, 178
Punjab, 84, 85, 86, 98

Qaddafi, Muammar, 136, 137, 138
Quebec, 7, 8, 9, 140–74, 204; autonomy, 141, 151–56, 159, 163, 171; culture, 155, 169, 170–71; economic costs of independence, 168–71, 173; economy, 169, 171, 172, 173;

identification with, 140, 141, 149-51, 156, 166, 167, 168; independence, 141, 151, 156-60, 168, 169, 170, 171, 172, 173 (*see also* Quebec-sovereignty-association); population, 141, 142; sovereignty-association, 141, 152, 153, 156, 160-67, 169, 170, 172, 173
Quinn, Herbert F., 151

Radio Cairo, 118
Radio Canada, 146, 150, 152, 153, 159 (*see also* Canadian Broadcasting Corporation);
Radio Free Europe, 201
Rakah Party, 112, 113, 114, 115, 116, 120, 128
Reader's Digest, 146, 153, 170
referendums: Philippines, 138; Quebec, 160, 164, 167, 173, 174; Spain, 26, 49
refugees, 194, 199; Arab, 117, 127, 128
Riencourt, Amaury de, 84
relative deprivation, 203
religion, xiii, 5, 7; Belgium, 53, 55; Philippines, 134; Spain, 41-43 (*see also* Buddhism, Hinduism, Islam, Jews, Muslims, Protestants, Roman Catholics, Sikhism
Renard, André, 58, 59
revolution of rising expectations, 105
Ritt, Leonard G., 206
Rjasanoff, N. R., 11
Rodriguez, Angel, 188
Roman Catholics: Basque, 41-43; Belgium, 56; Northern Ireland, 7, 8, 74-82
Romero Barceló, Carlos, 178, 179, 182, 184
Roper Poll, xivn
Rose, Richard, 74, 75, 76, 77, 79, 80, 82
Rugh, William, iii

Russian revolution, 1
Ruys, Manu, 57, 61

Sabah, 139
Saunders, John, 206
scheduled tribes (*see* caste system)
Schwartz, Mildred A., 148, 149
Scotland, 53
self-anchoring striving scale, 103
self-determination, v, xi, xii, 1-10; Basques, 16-20, 29; Puerto Rico, 181, 183, 184, 188 (*see also* Arabs, Basques, Flemings, French Canadians, Moros, Roman Catholics, Puerto Rico)
Serbs, 1
Sikhism, 85
Simeon, Richard, 149, 167
Sindicato Unitario, 25
Sinn Fein, 74
Six-Day War, 116, 118
Slovaks, 1
Slovenia, 1
Smooha, Sammy, 124, 126, 127, 128, 129
social change, 8; Israel, 109-11, 129, 131; Quebec, 141
Social Democratic and Labor Party, 79
Social Research Group, 148
Société de recherche en science du comportement, 146
Somalis, 4
Sons of the Village, 114, 115, 116
SORECOM, 147, 151, 166, 168, 169, 171
sovereignty-association (*see* Quebec)
Soviet Union, 4, 113, 200
Spain, 8, 11-51, 53, 203, 206 (*see also* Basques); and Philippines, 133, 134; and Puerto Rico, 177
Speier, Hans, 202
Starkey, Gladys, xiii
Steiner, Jürg, 206
stereotypes, 124, 135, 139

Stevenson, H. H., 203
Stewart, A. T. Q., 74
Stinchcombe, Arthur L., 149
Stokes, Donald E., vii
Stormont Parliament, 75, 76, 79
students: Arab in Israel, 115, 116, 117, 118, 119, 120, 121, 122, 127
Suárez, Adolfo, 22, 44
Sulu Archipelago, 132
Sweden, 70
Switzerland, 19, 35, 206
Syria, 2, 111

Tanzania, 207
Tasker, Rodney, 136, 137
Taylor, Donald M., 149
terrorism, 6, 8; Basque country, 14, 18, 19, 30, 39, 44, 46–47, 48, 49; Israel, 115, 117; Northern Ireland, 81; Philippines, 180 (*see also* violence)
Tessler, Mark A., 122
Tilly, Charles, 168
Treaty of Paris, 177
Truman, Harry S, 180
Turkey, 1
Turrentine, Lowell, vi

Uganda, 200
Ulster, 14, 46, 74–82 (*see also* Northern Ireland)
Ulster Special Constabulary, 76
Ulster Volunteer Force, 74
Ulster Workers Council, 79
Unamuno, Miguel de, 31
Unión Autonomista de Navarra (UAN), 22, 23
Unión de Centro Democrático (UCD), 22–24, 25, 26–27, 29, 34, 39, 41, 43, 47, 48, 49
Unión del Pueblo Navarro (UPN), 26
Unión Foral del País Vasco (UFV), 24, 26, 29 (*see also* Alianza Popular)

Unión General de Trabajadores (UGT), 25
Unionist Party, 74, 76, 78, 79
Unión Nacional (UN), 27, 29, 49
Unión Navarra de Izquierda (UNAI), 14, 23
United Kingdom (*see* Great Britain)
United Nations, 2, 3, 119, 194, 195, 198, 199; and Puerto Rico, 178, 179, 181, 183, 184, 188
United States, 69, 73, 131, 198; and the Philippines, 133, 135; and Puerto Rico, 177–80, 182, 183, 184, 185, 187, 188
University of the Philippines, 136
University of Puerto Rico, 185
untouchables (*see* harijans)
untouchability, 91 (*see also* caste system)
Urquhart, Brian, vii
Uttar Pradesh, 103

van den Berghe, Pierre, 149
Van Kalken, F., 68
Vascongadas, 12, 23, 25
Verdoodt, A., 73
Versailles Conference, v, 1
Vila Carro, Dario, 11
violence, 9, 191, 192, 193, 194, 195; India, 91, 103; Israel, 108, 120, 128; Northern Ireland, 80, 81, 82; Philippines, 132, 133, 134, 136, 137, 139; Puerto Rico, 178, 180; Spain, 16, 39, 44, 45, 47–48
Vizcaya, 12, 15, 16, 22, 23, 24, 29
Vreeland, Nena, 133, 134

Wallerstein, Immanuel, 205
Walloon movement, 58–60, 62–65
Walloons, 7, 8 (*see also* Belgium)
Weathersby, William H., iii
Welsh, xiii
Wenrich, Gail, xiii
west bank, Jordan River, 112, 115,

116, 118, 120
West Germany (*see* German Federal Republic)
White, Ralph K., iii
Whyte, J. H., 78
Wigny, P., 68
Wilson, Donald M., iii
Wilson, Woodrow, v, 1, 16
Wolpe, Howard, 150

Yoruba, 4
Young, Crawford, 4
Yugoslavia, 2, 207

Zald, Mayer, 168
Zayad, Tewfik, 112
Zionism, 108, 111, 112, 114, 117, 119, 120, 127, 130

ABOUT THE EDITORS AND CONTRIBUTORS

W. PHILLIPS DAVISON is professor of journalism and sociology at Columbia University. He is a former editor of the *Public Opinion Quarterly* and served as president of the American Association for Public Opinion Research in 1972-73. His work on this volume occurred while he was a visiting scholar at the Center of International Studies, Princeton University, 1979-80.

LEON GORDENKER, professor of politics and faculty associate of the Center of International Studies at Princeton University, is the author of numerous books and articles on the United Nations and problems of international organization and cooperation. He has served as director of African Studies at Princeton, and was visiting fellow at the Netherlands Institute for Advanced Study in the Social Sciences and Humanities in 1972-73.

PAUL DABIN is principal private secretary of the President (Speaker) of the Belgian House of Representatives, and is also advisor to the Minister for Institutional Reforms. He was formerly editor-in-chief of the *Encyclopaedia Universalis* (20 volumes, Paris, 1968-73). Since 1961, he has been scientific advisor to the Centre de Recherche et d'Information sociopolitiques (CRISP) in Brussels.

ERIC P. W. da COSTA is one of India's leading economists and currently managing director of the Indian Institute of Public Opinion, which is affiliated with the International Gallup Research Institutes. He is a former George Webb Medley University Scholar in economics at Oxford University, and served as the elected President of the World Association for Public Opinion Research (WAPOR) from 1966 to 1968.

GERALD A. FITZGERALD is news editor at Religious News Service in New York City and was a general desk editor for the Associated Press from 1972 to 1978. He is also former New York State manager of News Election Service. His chapter is based on a four-month visit to Northern Ireland in 1978, while research associate at the Columbia University Graduate School of Journalism, under a grant from the Spencer Foundation.

LINDA S. LICHTER is a member of the research staff of the Institute on International Change of Columbia University, and was previously affiliated with the Center for Policy Research in New York. She is currently completing her doctoral dissertation in sociology at Columbia.

JUAN J. LINZ is Pelatiah Perit Professor of Political and Social Science, Yale University, and has been chairman of the Committees on Political Sociology of the International Sociological and Political Science Associations, and President of the World Association for Public Opinion Research. He has written extensively on Spanish politics and society and on totalitarian and authoritarian regimes, and is Chairman of DATA, a private survey organization in Madrid.

SONIA MARRERO is a free-lance journalist who has reported on Puerto Rican affairs in both New York and the Caribbean. She formerly covered the United Nations and the New York Puerto Rican community for *Unidad* (San Juan), and is currently completing a doctorate in sociology at Columbia University.

DON PERETZ, professor of political science at the State University of New York, Binghamton, is the author of *Israel and the Palestine Arabs, Government and Politics of Israel, The Middle East Today,* and several dozen articles dealing with Israel-Arab relations. A number of his studies of Israel-Arab relations have been supported by grants from the Ford and Rockefeller Foundations. In 1979 he taught at Haifa University while on a Fulbright Foundation fellowship.

MAURICE PINARD is professor of sociology at McGill University, and was elected Fellow of the Royal Society of Canada in 1973. He is the author of numerous books and articles on Canadian society and politics in both French and English, including *The Rise of a Third Party: A Study in Crisis Politics*, of which an enlarged edition was recently published by McGill-Queens University Press, and is currently working on a theory of communal conflict in multicultural societies and on an extended study of the independence movement in Quebec.